Quantum Paradoxes and Physical Reality

Fundamental Theories of Physics

An International Book Series on The Fundamental Theories of Physics: Their Clarification, Development and Application

VOLUME 35

Quantum Paradoxes and Physical Reality

by

Franco Selleri

Dipartimento di Fisica,
Università di Bari, Italy

edited by

Alwyn van der Merwe

Department of Physics,
University of Denver, U.S.A.

KLUWER ACADEMIC PUBLISHERS

DORDRECHT / BOSTON / LONDON

Library of Congress Cataloging-in-Publication Data CIP

Selleri, Franco.
Quantum paradoxes and physical reality / by Franco Selleri ; edited by Alwyn van der Merwe.
p. cm. -- (Fundamental theories of physics)
Includes index.
ISBN 0-7923-0253-2
1. Quantum theory. 2. Physics--Philosophy. I. Van der Merwe, Alwyn. II. Title. III. Series.
QC174.12.S46 1989
530.1'2--dc20 89-2630

Published by Kluwer Academic Publishers,
P.O. Box 17, 3300 AA Dordrecht, The Netherlands.

Kluwer Academic Publishers incorporates
the publishing programmes of
D. Reidel, Martinus Nijhoff, Dr W. Junk and MTP Press.

Sold and distributed in the U.S.A. and Canada
by Kluwer Academic Publishers,
101 Philip Drive, Norwell, MA 02061, U.S.A.

In all other countries, sold and distributed
by Kluwer Academic Publishers Group,
P.O. Box 322, 3300 AH Dordrecht, The Netherlands.

Printed on acid-free paper

Printed in The Netherlands

Table of Contents

Preface vii

Chapter 1 / Quantum Theorists and the Physical World 1
1.1. Three Central Questions about Physics 1
1.2. The Older Generation 4
1.3. The Middle Generation 8
1.4. The Younger Generation 17
1.5. Conclusions 24

Chapter 2 / Is Quantum Mechanics a Complete Theory? 33
2.1. The Problem of Completeness and of Hidden Variables 33
2.2. De Broglie's Paradox 36
2.3. The Spin-1/2 System in Quantum Mechanics 40
2.4. A Simple Proof of von Neumann's Theorem 44
2.5. The Theorem is not General Enough 48
2.6. Von Neumann's Theorem: Assumptions, Definitions, and Results 51
2.7. General Proof of von Neumann's Theorem 54
2.8. Jauch and Piron's Theorem 61
2.9. The Debate on Impossibility Proofs 69

Chapter 3 / The Wave—Particle Duality 73
3.1. Duality for Photons 73
3.2. Duality for Neutrons 77
3.3. Einstein's Discovery of Duality 81
3.4. De Broglie's Duality 86
3.5. Schrödinger's Waves 92
3.6. Bohr's Complementarity 97
3.7. Fock's Relativity with Respect to the Means of Observation 103
3.8. Heisenberg Beyond Complementarity 107
3.9. The Consciousness Interpretation 111
3.10. Delayed Choices 114
3.11. How to do what Complementarity Forbids 118

Chapter 4 / Properties of Quantum Waves 123
4.1. Quantum Waves and Quantum Potential 123
4.2. Experiments on the Nature of Duality 130
4.3. Stimulated Emission 137
4.4. Quantitative Empty Wave Amplification 144
4.5. Two Further Experimental Proposals 156
4.6. Triple-Slit Experiments 160
4.7. The Bohm—Aharonov Effect 167
4.8. Further Ideas about Wave—Particle Duality 171

Chapter 5 / The Einstein—Podolsky—Rosen Paradox 181
5.1. The Original Formulation 181
5.2. Bohr's Answer 187
5.3. Two Types of State Vectors 191
5.4. Spin States for Two Particles 197
5.5. Reality and Separability 200
5.6. The EPR Paradox: Quantum Mechanics Complete 204
5.7. The EPR Paradox: Quantum Mechanics not Complete 210
5.8. From Theory to Practice 216
5.9. The Experimental Information 226
5.10. Solution 1: Modifying the Past 235
5.11. Solution 2: Superluminal Connections 240
5.12. Solution 3: New Definitions of Probability 247
5.13. Solution 4: Modifications of Quantum Theory 254

Chapter 6 / The EPR Paradox in the Real World 264
6.1. Criticisms of Einstein Locality 264
6.2. Probabilistic Einstein Locality 269
6.3. New Proof of Bell's Inequality 276
6.4. Probabilities for Pairs of Correlated Systems 279
6.5. A New Factorizability Condition 285
6.6. All the Inequalities of Einstein Locality 290
6.7. Tests of the EPR Paradox in Particle Physics 302
6.8. On the Possibility of New Experiments 307
6.9. Variable Probabilities 310

Chapter 7 / Perspectives of Physical Realism 321
7.1. Objectivity of Scientific Knowledge 321
7.2. Mathematics and Reality 325
7.3. The Role of History of Physics 334
7.4. Fragmentation of Modern Physics 338
7.5. Niels Bohr and Philosophy 345
7.6. Quantum Physics and Biological Sciences 352
7.7. Forms of Physical Realism 356

Index 367

Preface

Although the debate about the true nature of the quantum behavior of atomic systems has never ceased, there are two periods during which it has been particularly intense: the years that saw the founding of quantum mechanics and, increasingly, these modern times.

In 1954 Max Born, on accepting the Nobel Prize for his 'fundamental researches in quantum mechanics', recalled the depth of the disagreements that divided celebrated quantum theorists of those days into two camps:

> . . . when I say that physicists had accepted the way of thinking developed by us at that time, I am not quite correct: there are a few most noteworthy exceptions — namely, among those very workers who have contributed most to the building up of quantum theory. Planck himself belonged to the sceptics until his death. Einstein, de Broglie, and Schrödinger have not ceased to emphasize the unsatisfactory features of quantum mechanics . . . .

This dramatic disagreement centered around some of the most fundamental questions in all of science: Do atomic objects exist independently of human observations and, if so, is it possible for man to understand correctly their behavior? By and large, it can be said that the Copenhagen and Göttingen schools — led by Bohr, Heisenberg, and Born, in particular — gave more or less openly pessimistic answers to these questions. Niels Bohr, for example, went so far as to advocate that the word *phenomenon* be applied *exclusively* to a measurement whose definition necessarily include a full description of the experimental apparatus that is being employed.

Such negative attitudes aroused a life-long opposition in Einstein, Planck, Schrödinger, Ehrenfest, and de Broglie to the Copenhagen—Göttingen approach. Referring to these dissenters, Heisenberg wrote:

> . . . all the opponents of the Copenhagen interpretation do agree on one point. It would, in their view, be desirable to return to the reality concept of classical physics or, to use a more general philosophical term, to the ontology of materialism. They would prefer to come back to the idea of an objective real world whose smallest parts exist objectively in the same sense as stones or trees exist, independently of whether or not we observe them.

The existence of such influential critical voices could not stop the general acceptance of the Copenhagen formulation of quantum theory: no one was ready to accept that the theory had to be modified at times when its triumphal successes were near, the areas to which its formalism could be applied were numerous, and the inherent difficulties and ambiguities plaguing it were still to be uncovered. In fact, even today it can be said that the successes of quantum mechanics are so numerous and its predictions so accurate that, from the computational point of view, no comparable scientific theory has ever existed.

The critical researches of Einstein, Schrödinger, and de Broglie were therefore directed much more against the *interpretation* of the theory than against its practical validity. What they repudiated was, specifically, the 'philosophical taste' of the Copenhagen—Göttingen paradigm, even though they were willing to accept the theory as correct in its quantitative predictions. The situation in regard to the latter has, however, been changing somewhat in recent times, due to experimental and theoretical developments. It remains true, now as then, that no experimental result in disagreement with quantum mechanics has yet been found. Nevertheless, researches involving neutron interferometry, ultracold neutrons, laser physics, superfluorescence, electron interferometry, low-intensity beams, and so on, have sharpened our concrete understanding of the atomic world and rendered more acute a growing awareness about the existence of numerous quantum paradoxes.

Talking about the advent of quantum theory, Heisenberg wrote:

> The strangest experience of those years was that the paradoxes of quantum theory did not disappear during this process of clarification; on the contrary, they became even more marked and more exciting.

Nor has any of the quantum paradoxes in question disappeared in the intervening time. Nevertheless, armed with an improved physical understanding, we can now distinguish three types of quantum paradoxes:

(1) those that depend only on the completeness of quantum theory and which may thus be solved simply by declaring quantum mechanics incomplete (de Broglie's box, Schrödinger's cat, Wigner's friend, Wheeler's delayed choices, etc.);
(2) the paradox of wave—particle dualism, which probably can be solved through a creative development of existing quantum theory without any change in its mathematical structure;
(3) the Einstein—Podolsky—Rosen paradox, whose resolution requires that we either give up, once and forever, some basic concepts of rational thinking (spacetime, locality, arrow of time, realism, etc.) or declare the existing quantum theory to be wrong.

The first two types of paradoxes were already known at the birth of modern quantum theory, but the paradox of Einstein, Podolsky, and Rosen has become far more problematic only in recent times (Bell's inequality was put forth in 1965), even though it was discovered (partly) as early as 1935.

The present monograph is devoted essentially to a critical discussion of all the aforementioned paradoxes from a point of view that generally sympathizes with the realistic and rationalistic outlook of Einstein, Planck, de Broglie, Schrödinger, and Ehrenfest. One might therefore view our presentation as slanted in one direction. None the less, great care has been taken to reconstruct faithfully the true opinions and insights of many other physicists who, from different perspectives than ours, have made essential contributions to the shaping of the new theory. The present author feels a deep sense of admiration for the profound achievements of these scholars, whose general viewpoint he cannot share, but whose names will certainly remain immortal in the history of physics, whatever further developments the future may bring.

Some physicists have been complaining about what they see as a slow progress in the foundations of quantum theory. One should, however, remember that important new results have been established in overcoming the von Neumann-type theorems and in the parallel surmounting of the negative content of Bohr's complementarity and of a particular interpretation of Heisenberg's uncertainty relations which purported to show that a causal representation of atomic phenomena in space and time were impossible. Also, a powerful sharpening of the paradox of dualism and that of locality has been achieved. All the same, it is perhaps true that faster progress would have been desirable. The lessons that we are learning in this regard is that the pace of scientific progress is very much a function of the number of scholars that are actively involved in the particular area under consideration: If, for example, there were only one worker in the foundations of quantum physics for, say, every fifty that are active in particle physics, then our progress would no doubt have been phenomenal. But perhaps we should not complain too much, since an increasing number of international conferences are being devoted to foundational issues — I have kept a list of 28 such conferences that were held since the first one took place in Varenna in 1970 — and the number of people interested in this area is clearly growing.

Finally, I must confess that very little space is devoted in the present book to the quantum theory of measurement. The reason for this neglect is basically that the problem in measurement theory is to get rid of a non-factorizable state vector describing both the object and the apparatus. Now, exactly the same problem, albeit in a different physical situation, plagues the Einstein—Podolsky—Rosen paradox, with however the great advantage in the latter case that there the choice will be made experimentally. My own strategy is, accordingly, to wait and see what happens with locality; if this notion survives future experimental tests, then it is possible that our difficulties with measurements will automatically be solved. But, in such an event, much deeper problems will of course make their appearance!

I wish to thank Alwyn van der Merwe for his excellent editing of a difficult typescript, Biagio Bia for the drawings, and Derek Boothman for the English translation of the final chapter.

Bari, Italy FRANCO SELLERI

Chapter 1

Quantum Theorists and the Physical World

Consider the following questions. First: Are the microscopic entities (molecules, atoms, and elementary particles) just figments of our imagination, or do they exist independently in objective reality? Second: Is matter understandable to man and, in particular, does it make any sense to describe it in space and time? Third: Do physical phenomena happen by pure chance, almost miraculously, or are they caused?

These questions rank among the fundamental problems of physics. Should the reader be inclined to believe that the great physicists who shaped twentieth-century science basically agreed on the answer to the foregoing questions, then this chapter will attempt to change his mind.

1.1. Three Central Questions about Physics

History is the most fundamental of all sciences, for there is no human knowledge that does not lose its scientific character when men forget the conditions in which it originated, the questions it answered and the functions it was created to serve. (Schrödinger, 1956, p. 132.)

These lines bespeak a cultural viewpoint totally at odds with the one still prevailing today and according to which modern scientists have little to learn from the history of science.

In fact, quantum physicists customarily present quantum mechanics from the beginning as a perfectly logical structure, without mentioning the difficulties and the contradictions the subject had to face during the years it came into being. But it should be recognized that without benefit of this historical background it is simply impossible to gain a *critical* appreciation of today's physics. The essential role of history for a deep understanding of science has been stressed especially by T. S. Kuhn (1970) in his book *The Structure of Scientific Revolutions.* Textbooks on physics are here described (correctly, I think) as merely persuasive and pedagogical, and it is stressed that any concept of science taken from such sources is not likely to be correct, any more than a national culture can be extracted correctly from a tourist brochure or a language text. As a consequence most physicists merely learn how to *use* their tools rather than really understand them.

These difficulties in learning quantum physics have been noticed by several authors. Dirac (1958, p. 10) wrote:

The reader may possibly feel dissatisfied with the attempt . . . to fit in the existence of photons with the classical theory of light. He may argue that a very strange idea has been introduced — the possibility of a photon being partly in each of two states of polarization, or partly in each of two separate beams — but even with the help of this strange idea no satisfying picture of the fundamental single-photon processes has been given . . . In answer . . . it may be remarked that the main object of physical science is not the provision of pictures, but is the formulation of laws governing phenomena and the application of these laws to the discovery of new phenomena.

In a similar vein, Feynman (1965, p. 1—6) presents a detailed description of a double-slit experiment with electrons and concludes:

It is all quite mysterious. And the more you look at it the more mysterious it seems. Many ideas have been concocted to try to explain the curve for P_{12} in terms of individual electrons going around in complicated ways through the holes. None of them has succeeded. None of them can get the right curve for P_{12} in terms of P_1 and P_2.

It is then not surprising to discover that the 'learning' of quantum mechanics takes place basically by evading the main problems, as is shown, for instance, by Dyson (quoted in Landé, 1965, p. 148), who wrote:

The student begins by learning the tricks of the trade . . . Then he begins to worry because he does not understand what he is doing. This stage often lasts six months or longer, and it is strenuous and unpleasant. Then quite unexpectedly . . . the student says to himself: 'I understand quantum mechanics,' or rather he says: 'I understand now that there isn't anything to be understood . . .' For each new generation of students there is less resistance to be broken down before they feel at home with quantum ideas.

In the present chapter we propose to show that the perennial difficulties in understanding quantum mechanics are deeply rooted in a realistic philosophical attitude even if not at the level of conscious awareness — which was at the basis of a long and still not solved dispute between the followers of the Copenhagen—Göttingen paradigm and the followers of the philosophically quite different ideas of Einstein, de Broglie, Schrödinger, Planck, and Ehrenfest.

Science is fashioned by man in his effort to comprehend nature. It is thus from the personalities themselves that one has to start for a full understanding of their creations. For this reason, the first chapter of the present monograph is devoted to portraits of those physicists who stood at the center of the most important developments in quantum physics. Only twelve quantum theorists will be discussed: Planck, Sommerfeld, Ehrenfest, Einstein, Bohr, Born, Schrödinger, Pauli, Dirac, de Broglie, Heisenberg, and Jordan. Inevitably, important names are missing from this list; those of Lorentz, Nernst, Debye, Franck, Langevin, van Vleck, Kramers, Bose, Ladenburg, Landé, von Neumann, Compton, Davidson, O. Klein, Fermi — and many other physicists easily come to mind. But even though the role of the latter was essential for quantum physics it is probably correct to say that historically the contributions of paramount importance are attributable to scholars included in the former list. Accordingly, we furnish below short biographical sketches that are designed to put these scientists in their correct human and historical perspective. Additionally, their attitudes toward physics will briefly be reviewed. Together this should provide a human background against which quantum physics might more meaningfully be understood.

We shall examine particularly the attitude of the physicists under discussion to three general scientific problems. These problems, which are so general in scope that they could probably better be called philosophical issues, revolve around the following three questions:

1. Are the basic entities of atomic physics, such as electrons, photons, and the atoms themselves actually existing independently of the human beings and of the observations they are able to perform?
2. Supposing the answer to the foregoing question is positive, is it then possible to comprehend the structure and evolution of atomic objects and processes in terms of mental images formed in correspondence with their reality?
3. Finally, should one formulate physical laws in such a way that at least one cause can be given for any observed effect?

In future sections these points will be referred to as the *reality, comprehensibility*, and *causality* questions, respectively. We shall see that an intense debate raged around these questions at the time that quantum mechanics was being formalized and that the answers given to them by the winners (the physicists of the so-called Copenhagen and Göttingen schools) were by no means optimistic.

The year 1927 saw the establishment of the *quantum mechanical paradigm*: Heisenberg's uncertainty principle and Bohr's complementarity were taken to provide the final foundations for those mathematical structures constituting what already was called *quantum mechanics* and which even then had been extensively used for the prediction of observable quantities. The future developments of quantum theory were all to be natural outgrowths of that theoretical core which had been fashioned in the first twenty-seven years of our century.

The successes of quantum theory are so numerous and so accurate that it is probably fair to say that, from this point of view, no scientific theory can even remotely compare with it. Indeed, quantum mechanics, which originated in atomic physics, can explain the properties of atomic aggregates, as well as of single atoms. Therefore, the properties of matter generally fall within the scope of quantum theory, whose successful predictions range from the minutest details of line spectra to the varied thermal and electromagnetic phenomena (including superconductivity) characteristic of the solid state.

Several properties of nuclei and of elementary particles, too, have been predicted with astonishing accuracy by quantum field theory.

But quantum theory also has its difficulties, which are connected with the problem of divergencies (treated, where possible, with the technique of renormalization), with the very large number of 'elementary' particles now known,[1] with the difficulties of 'quark' models, and so on. Elementary particle physics is, in fact, the area in which quantum theory faces its sternest test. True, there have been important successes here, from the relativistic formulation of quantum field theory and its applications (quantum electrodynamics) to numerous individual properties of particles (e.g., the K^0—$\overline{K}^0$ doublet) and of nuclei [shell structure, optical models, etc.]. Generally speaking, however, quantum theory cannot be considered as fully satisfactory in particle physics, since some fundamental problems have remained unsolved, even though an unprecedented number of investigators, backed by liberal funding and using the most sophisticated technological and mathematical techniques, have worked for about a quarter of a century now toward their resolution. Thus, rethinking about the foundations of quantum mechanics can perhaps be useful to particle physics, as well.

1.2. The Older Generation

Planck and Sommerfeld were the oldest among the physicists who made major contributions to quantum theory: Planck was 42 when he proposed his celebrated formula for the spectrum of black-body radiation and Sommerfeld had attained 47 when he discovered, and applied to the hydrogen atom his famous quantum conditions.

Their scientific world views were very different. Planck was guided in research by a belief that there is a physical world, external to man, and that physics has the ability to furnish mental pictures thereof; Sommerfeld was much more mathematically inclined and enjoyed applying equations to physical problems. Planck had a well-defined philosophy, often stated and defended, that viewed as unacceptable any lack of causality in a physical theory or any failure to provide pictures of reality; Sommerfeld was not at all interested in philosophical ideas and could accept whatever was amenable to mathematical treatment. Planck influenced, helped, and sympathized with physicists (notably Einstein, Schrödinger, de Broglie) whose views were similar to his; Sommerfeld trained a new generation of theoreticians who contributed strongly to the success of ideas opposite to those of Planck.

Max Planck was born in Kiel, Germany (now F.R.G.) in 1858.[2] His father occupied the chair of constitutional law at the Universities of Kiel and, later, Göttingen. In his *Scientific Autobiography*, Planck (1949) recalls the excellent instruction he received in the Maximilian Gymnasium in Munich, particularly from his mathematics teacher, "a past master at the art of making his pupils visualize and understand the meaning of the laws of physics". After

Gymnasium, Planck attended university, first in Munich for three years, then in Berlin for another year. He studied experimental physics and mathematics. (In those days there were not yet professorships or classes in theoretical physics as such.)

In Munich Planck learned a great deal from the physicist P. von Jolly and from the mathematicians L. Seidel and G. Bauer. In Berlin he widened considerably his scientific horizon under Helmholtz and Kirchhoff, even though he was not very satisfied with their lectures: Helmholtz was never prepared and made a lot of mistakes, so that his classes became almost deserted, while Kirchhoff was carefully prepared but always dry and monotonous in lecturing. The lectures about the treatise by R. Clausius on thermodynamics made an enormous impression on the young Planck who completed his doctoral dissertation in Munich in 1879: In it he discovered the essential meaning of the second law of thermodynamics as residing in the principle of entropy increase. Although Planck was the first to state the second law in this analytic form, he did not gain recognition for it, probably because his own teachers reacted improperly: Helmholtz probably did not even bother to read the thesis, while Kirchhoff expressly disapproved of it. Eventually the law of entropy increase was universally accepted, but only because of Boltzmann's work, so that Planck's contribution to its acceptance can be said to have been nil. Commenting on this, Planck wrote:

> It is one of the most painful experiences of my entire scientific life that I have but seldom — in fact, I might say, never — succeeded in gaining universal recognition for a new result, the truth of which I could demonstrate by a conclusive, albeit only theoretical proof. (*Ibid.*, p. 30.)

Politically Planck was a man who believed in the ideas of State and honor and found it possible in 1914 to sign a *Manifesto to the Civilized World*, which contained a defense of German militarism. Later he opposed Hitler's regime. One of his sons was executed by the Nazis for having taken part in an unsuccessful attempt to assassinate Hitler (Klein, 1970, p. 300).

Planck believed in the existence of an objective physical world:

> . . . the world of sensation is not the only world which may conceivably exist, but . . . there is still another world. To be sure, this other world is not directly accessible to us, but its existence is indicated, time and again, with compelling clarity, not only by practical life, but also by the labors of science. (Planck, 1949, p. 102.)

Planck was also convinced that it is possible for man to comprehend the real world. He wrote, for instance:

> . . . the laws of human reasoning coincide with the laws governing the sequences of the impressions we receive from the world about us . . . therefore pure reasoning can enable man to gain insight into the mechanism of the latter. (*Ibid.*, p. 13.)

He also stressed that

> . . . the plain fact that we are able, at least to a certain degree, to subject future natural

occurrences to our thought processes and to bend them to our will, would be a totally incomprehensible mystery, did it not permit us to surmise at least a certain harmony between the external world and the human intellect. (*Ibid.*, p. 149.)

Planck condemned all attempts to abandon causality. He wrote, for instance, that the validity of the uncertainty principle

has even caused some indeterminists to regard the law of causality in physics as decisively refuted. However, upon closer scrutiny, this conclusion . . . proves a rash one, to say the very least. (*Ibid.*, p. 134.)

Planck's position about comprehensibility and causality was rather elaborate and he wrote about it extensively and with great clarity.[3] Sometimes his position was close to that of Kant, particularly when he stressed the impossibility of understanding *completely* the physical world.

Arnold Sommerfeld was born in 1868 in Königsberg, Prussia (now Kaliningrad, U.S.S.R.),[4] also Kant's birthplace and permanent home. His father was a medical practitioner, devoted to science and a passionate collector of natural objects like minerals, amber, shells, beetles, etc.

Sommerfeld studied at Königsberg University, which was one of the first institutions where theoretical physics became an acknowledged branch of study. Here he worked under the direction of several brilliant mathematicians (Lindemann, Hurwitz, and Hilbert), obtaining his doctor's degree in 1891 with a thesis entitled *The Arbitrary Functions in Mathematical Physics*, which he conceived and wrote down in the course of a few weeks.

In 1893, after a year of military service, he went to Göttingen, at that time the center of mathematics in Germany. He became assistant at the Mineralogical Institute, but his real interest remained directed towards mathematics and mathematical physics. He considered Felix Klein, then professor of mathematics at Göttingen, as his real teacher, not only in pure mathematics, but also in mechanics and mathematical physics. In 1894 Sommerfeld became Klein's assistant, charged with the management of the Mathematical Reading Room. One of his duties was to work out Klein's lectures and to produce a copy for use by students in the reading room. Klein directed his attention toward the problems of mathematical physics, and so it came about that Sommerfeld in 1896 wrote a *Habilitationsschrift* entitled *The Mathematical Theory of Diffraction*, which allowed him to become *Privatdozent* in mathematics. There was a moment when he had the opportunity to go into experimental physics, namely when Voigt offered him an assistantship in his laboratory, but Sommerfeld did not accept.

In 1897 Sommerfeld became professor of mathematics at the Mining Academy of Clausthal, and in 1900 he accepted the chair of technical mechanics at the *Technische Hochschule* in Aachen. Thus he had to concentrate for a few years on the solution of technological problems, a task which he typically approached through the application of mathematical techniques. In this period he seems to have particularly enjoyed an application of mathematical analysis to the hydrodynamical theory of lubrication, a field which was previously regarded as inaccessible to exact reasoning.

In 1906 Sommerfeld was offered the chair of theoretical physics in Munich, previously held by Boltzmann. Here he remained until his death — 1951. In 1935, on reaching the age limit of sixty-seven, he had formally to retire; but no successor was appointed, and he was asked to continue his teaching. In 1940 he was succeeded by a man from the Nazi ranks, whom he called his "worst possible successor" (Kuhn *et al.*, 1967, p. 146).

Although Sommerfeld had had several opportunities to leave Munich, he refused all offers, even a call to Vienna (1916) and one to Berlin (1927) as Planck's successor.

His scientific *oeuvre* ranged over several different areas, from the theory of the spinning top, to electrodynamics, relativity, and quantum theory. In relativity he liked particularly the formal approach of Minkowski, while in quantum theory he enthusiastically embraced Schrödinger's approach, centering around the wave equation and eigenvalue problems, since it fitted nicely with his own investigations on partial differential equations which had occupied him since his early mathematical period.

In fact, according to Born, Sommerfeld's

> . . . whole scientific development was in the direction from pure towards applied mathematics and empirical . . .
>
> If the distinction between mathematical and theoretical physics has any significance its application to Sommerfeld ranges him decidedly in the mathematical section. His gift was not so much the divination of new fundamental principles from apparently insignificant indications or the daring combination of two different fields of phenomena into a higher unit, but the logical and mathematical penetration of established or problematic theories and the derivation of consequences which might lead to their confirmation or rejection. Yet it is true that in his later, spectroscopic period, he developed a gift for the divining or guessing of mathematical relations from experimental data. (Born, 1952, pp. 278 and 282.)

Very famous and influential was Sommerfeld's book *Atombau und Spektrallinien* (1924), whose successive six editions reflected the development of atomic physics between 1916 and 1946. The later editions contained brilliant presentations of the physical facts and their mathematical interpretation, but very little about the fundamental epistemological and metaphysical questions connected with quantum mechanics. This aspect of the matter was not in Sommerfeld's line (*ibid.*, p. 285). Therefore it would be very difficult to find out what Sommerfeld thought about physical reality and its comprehensibility and about causality. These matters simply did not interest him, as he devoted all of his scientific energy to the applications of mathematical techniques to physical problems. His scientific attitude was therefore akin to that later advocated by neo-positivism.

In any event, Sommerfeld's influence on modern physics has been enormous, not only through his books and his scientific papers, but also through his skillful teaching. Not only were his pupils among the most important authors of quantum physics (Debye, Landé, Heisenberg, and Pauli all were introduced to research by him), but many other famous physicists, such as H. A. Bethe, P. S. Epstein, P. P. Ewald, W. Lenz, G. Wentzel, and W. Heitler, studied under him. About his pupils Born wrote:

> It would be easier to compile a list of prominent theoretical physicists who were not Sommerfeld's pupils than of those who were. He himself has published in America reminiscences of his teaching career . . . which contains the names of a large number of well-known physicists, most of whom now occupy chairs in the U.S.A. A long list of Germans and others could be easily added. (*Ibid.*, p. 286.)

Sommerfeld thus can be said to have made an extremely important, even if indirect, contribution to the success of the Copenhagen—Göttingen formulation of quantum theory. It is interesting to note that none of Sommerfeld's pupils, except for Landé in his later years, became an opponent of this quantum theory.

1.3. The Middle Generation

The key physicists of the middle generation (Ehrenfest, Einstein, Born, Schrödinger, and Bohr) were born in the years 1879 to 1887 and obtained their university degrees between 1904 and 1911. Two of them, Einstein and Ehrenfest, came from the lower classes of society; they were in this respect alone among the main authors of quantum theory.

Here, as for Planck and Sommerfeld, one finds a sharp divergence in basic epistemological attitudes: Einstein, Ehrenfest, and Schrödinger, believing in an understandable objective reality and in causal laws, opposed the Copenhagen—Göttingen formulation of quantum mechanics; Born and Bohr, by contrast, raised the most serious arguments against causality and comprehensibility and limited in a significant way the relevance of the very concept of physical reality. The final synthesis of the complex development of quantum theory was something that Born and Bohr actively searched for and always defended; Einstein, Ehrenfest, and Schrödinger disavowed such a goal, even though they made extremely important contributions to quantum theory. In Ehrenfest's case, the refusal to accept quantum mechanics was so vehement that it probably contributed to the tragic end of his life. All five physicists of the middle generation opposed Nazism, but had otherwise rather different political opinions, ranging from progressive (Einstein and Ehrenfest) to conservative.

Paul Ehrenfest was born in Vienna in 1880.[5] His father, who came from a very poor family, was at first a laborer in a weaving mill in Moravia, but, after marrying, his situation improved and he could open a grocery business in a workers' district of Vienna.

Paul Ehrenfest was of Jewish descent and directly experienced anti-Semitism during his childhood. By the age of twelve, however, his religious faith was entirely gone, "and he took great pleasure in sharply arguing the absurdity and falseness of all organized religions" (Klein, 1970, p. 27). His readings were wide, ranging from Schopenauer, Nietzsche, and Bergson to Marx. There is little doubt that his political views were rather 'progressive', given the enthusiasm with which he received the political news from Russia in 1917 (*ibid.*, p. 309).

Ehrenfest studied at the universities of Vienna and Göttingen. Boltzmann, more than anyone else, by his teaching and his example, pushed him into becoming a theoretical physicist and left a permanent mark on his scientific personality. By contrast, it seems that he had little or no contact with Mach, who was then teaching philosophy of science in Vienna. In 1903 Ehrenfest went to Holland for a few months to attend Lorentz's lectures on theoretical physics in Leyden. In 1904 Ehrenfest received his doctorate, after having prepared his thesis under Boltzmann. He then married a Russian mathematician and in 1907 moved to St Petersburg (Russia), where he remained until 1912.

He was about to become *Privatdozent* in Munich, despite difficulties raised particularly by Sommerfeld, when he received — and accepted — an invitation from Lorentz to occupy his vacated chair at Leyden. Shortly before this happened, Ehrenfest had the opportunity to become Einstein's successor at the University of Prague. There was, however, one difficulty: no one who lacked a formal religious affiliation could become a teacher in the Austro-Hungarian Empire. This was only a formal prerequisite, with which Einstein himself had previously complied, even though he was not committed to any religion for over a decade. But Ehrenfest refused to conform, feeling that it would be hypocritical to declare his adherence to Judaism just so he might become eligible for a professorship. This incident demonstrates a deep-seated characteristic of Ehrenfest's personality, his complete honesty, which guided him both in his life and in his research. Because of this trait, Ehrenfest was viewed by his colleagues as 'the conscience of physics', and more than anybody else he found himself "at the very heart of the drama of contemporary physics" (*ibid.*, p. xv), and Langevin wrote about him in 1933.

Ehrenfest was a very close friend of Albert Einstein, whom he met in 1912. Einstein wrote about him after his death:

> He was not merely the best teacher in our profession whom I have ever known, he was also passionately preoccupied with the development and destiny of men, especially his students. To understand others, to gain their friendship and trust, to aid anyone embroiled in outer or inner struggles, to encourage youthful talent — all this was his real element, almost more than immersion in scientific problems. (*Ibid.*, p. xv.)

And, earlier, Einstein wrote Ehrenfest: "You are one of the few theoreticians who have not been deprived of their native intelligence by the mathematical epidemic" (*ibid.*, p. 190).

Ehrenfest committed suicide in 1933. About this event Einstein wrote that Ehrenfest had often felt inadequate to the tasks of research and that this feeling used to depress him. "In the last few years," Einstein added,

this situation was aggravated by the strangely turbulent development which theoretical physics has recently undergone. To learn and teach things that one cannot fully accept in one's heart is always a difficult matter, doubly difficult for a mind of fanatical honesty, a mind to which clarity means everything. . . . I do not know how many readers of these lines will be capable of fully grasping that tragedy. Yet it was this that primarily occasioned his escape from life. (Einstein, 1950.)

Albert Einstein was born in Ulm, Germany (now F.R.G.), in 1879.[6] His father was an artisan who owned a small electrochemical laboratory, which he transferred with his family to Munich in 1880. Einstein's life, marked by difficulties from the beginning to the end, was quite different from that of the other main authors of quantum theory. His studies in Munich met with little success: "You will never amount to anything, Einstein," one of his teachers used to tell him. But in reality it was the authoritarian type of teaching that Einstein could not stand:

This coercion had such a deterring effect upon me that, after I passed the final examinations, I found the consideration of any scientific problems distasteful to me for an entire year . . . It is, in fact, nothing short of a miracle that the modern methods of instruction have not yet entirely strangled the holy curiosity of inquiry. (Schilpp, 1970, p. 17.)

Einstein's personality is also well typified by his position about religion:

Through the reading of popular scientific books I soon reached the conviction that much in the stories of the Bible could not be true. The consequence was a positively fanatic orgy of freethinking coupled with the impression that youth is intentionally being deceived by the state through lies . . . Suspicion against every kind of authority grew out of this experience . . . an attitude which has never again left me . . . (*Ibid.*, p. 5.)

Einstein tried to enter the ETH of Zurich, but he did not pass the examinations on zoology, botany, and languages. Only after attending a preparatory school in Aarau was he admitted to the ETH, where he studied from 1897 to 1900. Having received a diploma, he wished to enter the teaching profession, perhaps as an assistant of a professor at ETH, but he was not accepted. After spending a year in Winterthur (near Zürich) he became a Swiss citizen and, through the intercession of a friend, secured a position at the patent office of Bern. Here he worked until 1909, finding time for research and publishing about thirty articles, among which the three papers of 1905 dealing with Brownian motion, the photoelectric effect and special relativity, became immortal. In 1905 he also obtained his doctorate under Kleiner in Zürich. Four years later he finally got an assistantship at the University of Zürich, and in 1912 he was appointed to a chair in Prague. After spending one more year in Zürich, Einstein accepted a chair in Berlin, where he remained until 1933, when the Nazis forced him to leave Germany. He settled in the United States of America, where he remained to the end of his life (1955).

One can say that success brought Einstein as many troubles as honors. In 1907 his application to become an unpaid *Privatdozent* at the University of Bern was rejected by the faculty. But, even worse, his theory of relativity met with bitter opposition among certain German intellectuals. Max Born tried to shield Einstein from some of these pseudo-scientific attacks, even by writing newspaper articles in his defance. After one of these, Einstein wrote Born in 1919:

Your article in the *Frankfurter Zeitung* pleased me very much. But now you, too, are being haunted by the press and other rabble, though to a lesser degree than I am. With me it is so bad that I can hardly breathe, let alone get down to doing sensible work. (Born, 1969, p. 158.)

But it was to become still worse: In the summer of 1920 an organization for opposing Einstein's theories was set up. In a public meeting of this group (the Anti-Einstein League), which he attended, Einstein was denounced as "a publicity hound, a plagiarist, a charlatan, and a scientific Dadaist" (Klein, 1970, p. 320), all this in a poisonous atmosphere, with swastikas and anti-Semitic pamphlets.

Nor did Einstein have an easy life in the U.S.A. His unconventional personality and his socialist ideas, defended with articles and other activities, were not calculated to please some American intellectuals, and he soon found himself almost isolated and with very few friends. He commented bitterly on this: "Here in Princeton they regard me as an old fool" (Born, 1969, p. 161).

Turning now to Einstein's epistemological position, one can say that he was basically a realist. He wrote of his curiosity about ". . . this huge world, which exists independently of us human beings and which stands before us like a great, eternal riddle, at least partially accessible to our inspection and thinking" (Schilpp, 1970, p. 5).

Einstein never accepted the acausal formulation of quantum mechanics. He wrote for instance:

This double nature of radiation (and of material corpuscles) is a major property of reality, which has been interpreted by quantum mechanics in an ingenious and amazingly successful fashion. This interpretation, which is looked upon as essentially final by almost all contemporary physicists, appears to me as only a temporary way out. (*Ibid.*, p. 51.)

And he stated several times his belief "that this theory offers no useful point of departure for future development" (*ibid.*, p. 87).

Einstein did not accept Bohr's notion of complementarity, about which he humorously remarked:

From these meager remarks one will see that to me it must seem a mistake to permit theoretical description to be directly dependent upon acts of empirical assertions, as it seems to me to be intended for example in Bohr's principle of complementarity, the sharp formulation of which, moreover, I have been unable to achieve despite much effort which I have expended on it. (*Ibid.*, p. 674.)

In fact Einstein, like Planck, believed that the task of physics is to provide a picture of physical reality, and he objected to the renunciation of quantum mechanics to provide it: "I cannot seriously believe in it because the theory is incompatible with the principle that physics is to represent a reality in time and space, without spookish long-distance effects" (Born, 1969, p. 162).

Einstein furthermore believed that

science is not just a collection of laws, a catalogue of unrelated facts. It is a creation of the human mind, with its freely invented ideas and concepts. Physical theories try to form a picture of reality and to establish its connection with the wide world of sense impressions. Thus the only justification for our mental structures is whether and in what way our theories form such a link. (Einstein and Infeld, 1938, p. 310.)

Max Born was born in Breslau, Silesia (now Wroclaw, Poland) in 1882.[7] His father was an anatomist and embriologist and his mother came from a rich family of Silesian industrialists. From 1901 to 1904 Born studied at the University of Breslau, except for a couple semesters spent in Heidelberg and Zürich. In 1904 he continued his studies in Göttingen under the guidance of the famous mathematicians Felix Klein, Hilbert, and Minkowski, receiving his doctorate in 1906 with a mathematics thesis directed by Hilbert. In the fifteen years that followed Born worked in several universities: Cambridge (1906—1907), Breslau (1907—1908), Göttingen (1908—1914), Berlin (1914—1919), and Frankfurt (1919—1921). This period, during which Born honed to a sharp edge his capabilities in mathematics and theoretical physics, was capped 1921 by his appointment to the chair of theoretical physics in Göttingen. With Born as theoretician and James Franck (author of the famous Franck—Hertz experiment) as experimentalist physics flourished and Göttingen rapidly became one of the world's foremost research centers in this discipline. Many famous physicists worked in Göttingen for some time, among them the experimentalists K. T. Compton, Condon, Blackett, von Hippel, Houtermans, and Rabinowitch as well as the theoreticians Pauli, Heisenberg, Fermi, von Neumann, Wigner, and Dirac. The roster of Born's many pupils includes the well-known names of Delbrück, Elsasser, Jordan, Maria Goeppert-Mayer, Nordheim, Oppenheimer, and Weisskopf.

In 1933 Born was dismissed from his university for racial reasons by the Nazis and forced to leave Germany. He moved to Great Britain, finding refuge first in Cambridge (1933—1936) and later, as professor of natural philosophy in Edinburgh (1936—1952). In 1954 he retired to Bad Pyrmont (Germany), where he spent the last years of his life, which ended in 1970.

Along with Bohr and Heisenberg, Born was one of the few physicists who built the *philosophical* structure of quantum mechanics. His main contribution was the probabilistic interpretation of Schrödinger's waves, an interpretation which elevated probability to the *primary* concept of quantum mechanics, leaving the behavior of a single system largely undetermined and thus open to an acausal description. It is, of course, more than a coincidence that Born started to entertain the idea of acausality in physics as early as 1920, even before his interest was aroused by quantum theory (Forman, 1971, p. 71).

Born's views about the nature of physics were opposite to Einstein's. He admitted this frankly by writing in the following way about his scientific disputes with Einstein: "It was in fact a matter of a fundamental difference in our view of nature" (Born, 1969, p. 163). Einstein himself also recognized this difference in a letter to Born:

In our scientific expectations we have become antipodes. You believe in the dice-playing God, and I in complete lawfulness of objectively existing world, which I try to catch in a speculative way. (*Ibid.*, p. 161.)

Born accepted without reserve Bohr's principle of complementarity. The following statement expresses his view with clarity: ". . . the development of modern physics has enriched our thinking by a new principle of fundamental importance, the idea of complementarity" (*ibid.*, p. 171). In view of this, it is natural to understand Born's pessimistic opinions about our prospects for comprehending the physical world:

We have reached the end of our journey into the depth of matter. We have sought for firm ground and found none. The deeper we penetrate, the more restless becomes the universe, and the vaguer and cloudier. (*Ibid.*, p. 166.)

Erwin Schrödinger was born in Vienna in 1887.[8] His father was a successful businessman with several cultural interests. Although Schrödinger entered the university of Vienna in 1906, the year that Boltzmann died, it can nevertheless be said that he was strongly influenced by Boltzmann's great personality through contact with Boltzmann's pupils Hasenhörl and Exner. Schrödinger himself declared that Boltzmann's way of thinking was his 'first love in science' and that nobody else, either before or after, could fill him so much with enthusiasm (Schrödinger, 1957, p. xiv). He received his doctor's degree in 1910 and became Exner's assistant from 1911 to 1914. In the 1919/1920 academic year he stayed in Jena and Stuttgart; thence he moved to Breslau, for the first time to fill a chair. In 1921 he was called to the University of Zürich where he remained until 1927, throughout the crucial years for the development of quantum mechanics.

In Zürich Schrödinger carried out the fundamental investigations that culminated in his famous wave equation. These investigations, which were based on the hypothesis of objectively existing waves with comprehensible and causal behavior, aroused a great deal of enthusiasm in Berlin, where Planck and Einstein were teaching at the time. Not surprisingly then Schrödinger was called upon to succeed Planck when the latter retired from his chair in 1928. Schrödinger remained in Berlin until 1933 when Hitler seized power.

As is well known, physicists of Jewish extraction lost their jobs in Germany because of the Nuremberg laws. Given his Catholic background, Schrödinger could have retained his position at the University of Berlin if he wanted, but he chose to leave Germany, thereby concretely testifying his aversion to Nazism. He moved to Oxford, where he stayed until, in 1936, he was offered a chair in Graz (Austria). But two years later, after the *Anschluss*, he felt compelled to leave again; this time he went to Rome for a short period and then to Dublin, where he became professor at the School for Advanced Studies. In 1956 he returned to Austria, there to live out the last five years of his life.

Schrödinger was a man with exceptionally wide cultural interests, which ranged from philosophy to physics, from history to politics. He wrote books on biology and on the scientific culture of the ancient Greeks. He always held firmly the opinion that nature is comprehensible. He defined comprehensibility as "the hypothesis that *the display of nature can be understood* . . . It is the non-spiritistic, the non-superstitious, the non-magical outlook" (Schrödinger, 1956, p. 103). This comprehensibility, on which he often insisted, was very much akin to the notions of Planck and Einstein, as it implied the possibility of forming mental pictures (or images) of reality:

> There is a widespread hypothesis that an objective image of reality in any previously believed interpretation cannot exist. Only the optimists among us (and I consider myself one of them) consider this a philosophical eccentricity, a desperate measure in the face of a great crisis. (Heisenberg *et al.*, 1961, p. 38.)

On reality, too, he was definitely an optimist. His atomic waves were real and his writings contain appreciative statements about reality like the following:

> Even today there are those who are haunted by the idea that chemistry is the sole and original domain of 'atom' and 'molecule'. From the very hypothetical, somewhat anemic role they played there — the school of Ostwald rejected them flatly — they were for the first time raised to physical reality in the gas theory of Maxwell and Boltzmann. (*Ibid.*, p. 40.)

Only with regard to causality can Schrodinger's position be characterized as mixed. On the problem of whether or not the behavior of a single atom is determined by rigid causality, he wrote:

> There is acarcely any possibility of deciding this issue by experiment. For the methods of pure reasoning evidently allow us either to derive chance from law or law from chance, whichever we prefer. (Schrödinger, 1957, p. 50.)

Nevertheless, the waves introduced in his scientific papers completely obeyed causal laws, and his objective historical role was therefore undoubtedly favorable to causality.

Even though Schrödinger from 1928 to 1933 accepted quantum mechanics as it was, he opposed it strongly both before and after this period. He wrote, for instance:

. . . de Broglie, so I believe, disliked the probability interpretation of wave mechanics as much as I did. But very soon and for a long period one had to give up opposing it and to accept it as an expedient interim solution. (George, 1953, p. 20.)

Niels Bohr was born in Copenhagen in 1885.[9] His father was a professor of physiology who, reacting to the materialistic trend in science at the beginning of this century, vigorously advocated the teleological point of view in the study of his own speciality. It is most likely that Bohr's father had a profound influence on the formation of his son's scientific personality, considering, for example, that Bohr later invoked his principle of complementarity in an attempt to justify his father's point of view in biology.

Bohr spent his youth in Copenhagen. He became very excited when reading a book by the existentialist philosopher Kierkegaard. Later he joined an intellectual club, called 'Ekliptica', in which philosophical and epistemological problems were discussed and whose leading member was Høffding, a pupil of Kierkegaard (Rozental, 1967, p. 24). Very probably these discussions deeply influenced Bohr's attitude to physics; for, as we shall see later, there is a close resemblance between the complementarity principle and some fundamental ideas of Kierkegaard. Also William James, the American pragmatist, had an important influence on Bohr (Jammer, 1966, p. 176).

In 1911 Bohr graduated from the University of Copenhagen with a doctor's thesis on the electron theory of metals. Soon afterwards, supported by a Carlsberg Foundation stipendium, he attended Cambridge University, where he followed lectures by Larmor, J. J. Thomson, and Jeans. In March of 1912 he left Cambridge for the University of Manchester, where the big attraction was the presence of Rutherford. He started to analyze the scattering of alpha particles, on the basis of Rutherford's atomic model, before returning to Denmark. In 1914 Bohr toured Germany, visiting the universities of Würtzburg, Göttingen, and Munich, and meeting famous physicists like Debye, Born, Wien, and Sommerfeld. He was forced, however, by the outbreak of the First World War to interrupt this tour and to hurry back to Denmark. After two more years spent in Manchester, Bohr in 1916 settled down in Copenhagen where a chair of theoretical physics was established for him. On a trip to Berlin in 1920 Bohr met Einstein, Planck, and Franck. The year 1921 saw the establishment of the Institute of Theoretical Physics, at the University of Copenhagen, with Bohr as its first Director. Due to this institute, Copenhagen and Göttingen soon became the main centers of the new quantum theory. Numerous physicists from all over the world came to work in Bohr's institute: Kramers did so for several years, while Pauli, Heisenberg, Dirac, Landau, and Oskar Klein all spent at least some time at the Institute. Bohr remained in Copenhagen until the end of his life (1962), with the exception of a period spent in England and the U.S.A. during the Second World War.

Bohr is rightly recognized as the father of quantum mechanics. His principle of complementarity was essential in providing a logical foundation for the various theoretical contributions advanced from 1924 to 1927. Bohr's definition of physics was very different from that of Planck, Einstein, or Schrödinger. He wrote in this connection: "... in our description of nature the purpose is not to disclose the real essence of the phenomena but only to track down, as far as it is possible, relations between the manifold aspects of our experience." (Bohr, 1961, p. 18).

This limitation on the meaning of physical reality accords with Bohr's pessimism about the comprehensibility of nature. The principle of complementarity was basically a postulate that declared as impossible any straightforward resolution of the contradictions (such as the particle-wave duality) that plagued atomic physics:

> Evidence obtained under different experimental conditions cannot be comprehended within a single picture, but must be regarded as *complementary* in the sense that only the totality of the phenomena exhausts the possible information about the objects. (Schilpp, 1970, p. 210.)

On the problem of causality, too, Bohr took a negative position. He wrote for instance:

> How radical a change in our attitude towards the description of nature the development of atomic physics has brought about is perhaps most clearly illustrated by the fact that even the principle of causality ... has proved too narrow a frame to embrace the peculiar regularities, governing individual atomic processes. (Bohr, 1958, p. 25.)

In his contribution to the volume *Albert Einstein: Philosopher-Scientist*, Bohr stresses that in 1920 he was already prepared to abandon the validity of the causality law.

Sometimes Bohr's epistemological position was not very different from that advocated by the positivists, as when he wrote:

> As a more appropriate way of expression I advocated the application of the word phenomenon exclusively to refer to the observations obtained under specified circumstances, including an account of the whole experimental arrangement. (Schilpp, 1970, p. 237.)

But this apparent adherence to positivism was not neutral with respect to the fundamental philosophical problems of reality, comprehensibility, and causality. Partly this has already been stressed above, but it is useful to give one more citation:

> Indeed the *finite interaction between object and measuring agencies* conditioned by the very existence of the quantum of action entails — because of the impossibility of controlling the reaction of the object on the measuring instruments, if these are to serve their purpose — the necessity of a final renunciation of the classical ideal of causality and a radical revision of our attitude towards the problem of physical reality. (*Ibid.*, pp. 232 and 233.)

1.4. The Younger Generation

The physicists of the younger generation (de Broglie, Pauli, Heisenberg, Jordan, and Dirac), with the exception of de Broglie, were born in our century and received their doctorates from 1921 to 1926 — in times, therefore, when quantum physics was already twenty years old and its main problems were becoming acute. Dirac earned his Ph.D. in mathematics, and the physicists of the German—Austrian group studied under theoreticians with strong mathematical tendencies (Jordan with Born, and Heisenberg and Pauli with Sommerfeld). Epistemologically these four physicists had views that ranged from a lack of interest in the question of physical reality to a strong urge in favor of disowning causality and even reality itself.

Very different was the case for de Broglie whose views were similar to those of Einstein. His epistemological position can be considered as an optimistic one because of his defense of causality, his belief in an objectively existing real world that is comprehensible to man, and his rejection of complementarity. Politically, Dirac and de Broglie are reputed to have been completely uninterested, while Heisenberg and Jordan were rightists; Pauli's political views are apparently undocumented. It is very probable that politics played a far greater role in culturally shaping physicists of the younger generation than it did for their older colleagues: Physicists born at the beginning of this century grew up and searched for their own identities in years that were marked by the First World War, the Russian Revolution, the establishment of parliamentary democracies in central Europe, and by strong class struggles in Germany and elsewhere. It is then not surprising to find that people like Heisenberg and Jordan held strong political convictions. Nor is it unexpected that for them this could have important philosophical implications even in physics.

Prince *Louis de Broglie* was born in Dieppe (France) in 1892, the son of Victor, duc de Broglie, and Pauline d'Armaille.[10] Brought up in a highly intellectual family with strong interests in history, politics, and literature, de Broglie's youthful curiosity had an almost universal range, and he was doubtful for some time on whether his future involvement should be in science or in literature. But suddenly, he was overcome by the idea that his true destiny in life was science (de Broglie, 1976, p. 8).

De Broglie described physics as resulting from the presence of two basic tendencies (*ibid.*, p. 10). The first one, stemming from attempts to understand physical phenomena causally in space and time, included Fresnel's theory of light, Maxwell's electromagnetism and even Fourier's theory of heat. The second tendency belonged to thinkers with abstract inclinations and centered around classical thermodynamics and its rigorous laws of energy conservation and entropy increase. Ostwald and Pierre Duhem showed the latter tendency and thus scorned intuitive hypotheses (atomism and causality). In de Broglie's formative years the first trend appeared to be winning (*ibid.*, p. 10) due to the works of Boltzmann and Gibbs on statistical thermodynamics, of Lorentz on electromagnetism, and those of Planck and Einstein on the new ideas of quantum physics and relativity. The young French physicist adhered enthusiastically to the first school of thought and throughout his life never changed this attitude. In 1968, for instance, he wrote:

> Contrary to certain extreme statements of certain idealist philosophers, I have the deep conviction that a physical reality external to us exists which is independent of our thoughts and of our imperfect means of knowing it. (*Ibid.*, p. 49.)

The problems of quantum physics were at first little known in France. But then attention was focused on them by the Solvay Congress of 1911. The young Louis de Broglie studied the proceedings of this meeting (which were being edited for publication by his older brother Maurice) with intense and ever-rising interest. The outcome of this experience was his decision to direct his future efforts entirely toward investigations in quantum physics.

During the years 1914 to 1919 de Broglie served his military duty at the center for radio telegraphy on the Tour Eiffel. He was thus compelled to learn the theory of electromagnetic waves and electronics always in close association with experience and technology. This was a very important period for de Broglie's scientific growth, since it saved him from becoming only theoretically and mathematically minded. As de Broglie stated himself, this contact with practical problems much influenced his future work.

The continuity between the scientific researches of Einstein and de Broglie was recognized and stressed by the latter:

> . . . in three papers published on *Comptes Rendus* in September and October 1923, I stated the principles of wave mechanics . . . and in the following year I made them the object of my thesis . . . The fundamental idea of this work was the following one: After Einstein's introduction of photons in the luminous wave, one knew that light contains particles which are concentrations of energy embodied in the wave; this suggests that all particles, like the electron, must be transported by a wave in which they are embodied. (*Ibid.*, pp. 11 and 12.)

One is then not surprised to find great similarities between Einstein's and de Broglie's epistemological attitudes. Both opposed a final formulation of quantum mechanics, in particular where the complementarity principle was concerned, and both favored physical reality, comprehensibility, and causality. From 1928 to 1952, it is true, de Broglie accepted quantum mechanics, but he opposed it before and after.

Concerning his early views about an objectively-existing wave and his initial acquiescence in the Copenhagen formulation, de Broglie wrote:

> However, I was soon forced to abandon this view by the great difficulties which it presented, and finally accepted the interpretation which is now orthodox . . . (de Broglie, 1964, p. vii.)

Later, however, de Broglie changed his opinion and returned to his original views. He spoke then of his

. . . conviction that the arguments on which the currently accepted interpretation of physics is based are not as decisive as they appear to be, but on the contrary, contain many significant loopholes. (*Ibid*., p. viii.)

Summarizing, de Broglie's non-acceptance of quantum mechanics can be illustrated by the following quotation:

A general criticism of the usual interpretation of wave mechanics is that it constitutes a verbose refusal to provide a real explanation and this is, it seems to me, quite contrary to the principles of sound scientific method. (*Ibid*., p. 24.)

The waves which de Broglie introduced for all microscopic entities were for him objectively real, as is clear from a statement like the following:

Faced with the phenomena of interference and diffraction, a physicist with no theoretically preconceived ideas is compelled to believe that he is dealing with the propagation of real waves and not with a simple representation of a probability which exists only in his mind. (*Ibid*., p.22.)

Talking of his early work on wave mechanics, de Broglie wrote:

I adopted the view that it was essential to retain the concept of a physical reality independent of the observer, as in classical physics, and tried to obtain a clear representation of physical processes within the framework of space and time. (*Ibid*., p. 37.)

The previous statement makes it clear that de Broglie also favored comprehensibility. This went together with his rejection of complementarity, about which he wrote:

The idea of complementarity has been very successful, and attempts have even been made to extrapolate it in a most dangerous manner out of the realm of physics into biology, sociology, psychology, and so on. I have, for a long time, adopted the idea of complementarity in the realm of quantum physics, whilst at the same time realizing that it was inadequate. In recent years, I have been led to regard the concept of complementarity with increasing suspicion. (*Ibid*., p. 7.)

Wolfgang Pauli was born in Vienna in 1900.[11] His father was professor of chemistry at the University of Vienna and his godfather Ernst Mach, the famous positivist. Pauli obtained his doctorate in 1921 in Munich (where he had gone in 1918) under Sommerfeld. About Sommerfeld's influence on his scientific personality Pauli once wrote:

My teacher in theoretical physics was Professor A. Sommerfeld, and the stimulating suggestions I received from him and his circle of pupils . . . decisively influenced my scientific outlook. (Pauli, 1964a, p. x.)

In 1921—1922 Pauli was Born's assistant in Göttingen and in 1922—

1923 Bohr's in Copenhagen. Bohr's influence on Pauli must have been very considerable if Rosenfeld, comparing Bohr's collaborators Kramers, O. Klein, Heisenberg, Pauli, and Dirac, could write: "Of this whole group the one who stood nearest to Bohr was undoubtedly Pauli . . ." (Rozental, 1967, p. 118).

Rosenfeld also stressed that: ". . . Bohr's disclosure of the complementarity features of quantum theory found full resonance in Pauli's mind" (*ibid.*, p. 119).

From 1923 to 1928 Pauli was *Privatdozent* at the University of Hamburg, but at the same time frequently traveling to Copenhagen for work and discussions with Bohr. Subsequently, from 1928 to 1958 he held a chair at Zurich's ETH.

Pauli's personality was once described in the following terms:

> Wolfgang Pauli was . . . unathletic, hedonistic, indifferent to nature, addicted to urban night life, sarcastic, cynical, incisively critical, and Jewish to boot. (Forman, 1971, p. 106.)

Pauli's epistemological attitude was close to that of Bohr: He renounced comprehensibility and causality, put limitations on the concept of physical reality, and accepted without reservation the Copenhagen—Göttingen formulation of quantum mechanics and complementarity. The following statements from his own pen, will briefly document Pauli's position:

> . . . the conception of a material physical object, as an object whose nature and constitution is independent of the way it is observed. We have seen that modern physics, forced by events, had to give up such an abstraction. (Pauli, 1964b, p. 30.)

> It was quantum mechanics that first assumed the existence of *primary* probabilities in the laws of nature, which could not be reduced, by means of auxilliary hypotheses, to deterministic laws, as is possible, for instance, with the thermodynamical probabilities of classical physics. This revolutionary development is considered as final by the large majority of modern physicists, first of all by Born, Heisenberg, and Bohr, with whom I myself agree. (*Ibid.*, p. 35.)

> For a given state of a system (object) one can in general make only statistical predictions (primary probabilities) about the results of future observations. The result of a single observation is not determined by law; it is in the final analysis, without cause. (*Ibid.*, p. 115.)

Werner Heisenberg was born in Würtzburg, Germany (now F.R.G.) in 1901,[12] the son of a professor of Greek language at the University of Munich. In 1923 he took his doctor's degree with Sommerfeld in Munich, and during the 1923/24 academic year he was Born's assistant in Göttingen. The following year Heisenberg worked in Copenhagen in regular association with Bohr and Kramers. After spending the summer of 1925 in Cambridge, where he held a series of lectures, Heisenberg went back to Göttingen as *Privatdozent* (1925/26). The next year saw him again in Copenhagen, whence he accepted the chair of theoretical physics at the University of Leipzig (1927—1941). Subsequently (1942—1945) he also taught at the University of Berlin. In the post-war years Heisenberg worked at the Max Planck Institute for Physics in Munich, becoming its Director in 1958.

Heisenberg traveled back and forth between Göttingen and Copenhagen in the crucial years when quantum mechanics was being developed. This allowed him to keep track simultaneously of both Bohr's and Born's research advances. He contributed ideas that were very essential to the creative enlargement of the new theory: To his credit stand the discoveries of matrix mechanics and the celebrated uncertainty relations. Together with Bohr and Born, Heisenberg is recognized as one of the creators of the conceptual basis of quantum mechanics.

Heisenberg's epistemological attitude was characterized by his rejection of physical reality, comprehensibility, and causality. Since he formulated his views repeatedly and carefully in several books, Heisenberg's position is one which is very well known. Thus he states that "one may say that atomic physics has turned science away from the materialistic trend it had during the nineteenth century". Furthermore, "the elementary particle of modern physics is still far more abstract than the atom of the Greeks . . ." (Heisenberg, 1959, p. 70), a statement in full agreement with the view that the elementary particle "is not a material particle in space and time but, in a way, a symbol on whose introduction the laws of nature assume an especially simple form" (Heisenberg, 1952, p. 62).

Another statement with the same philosophical flavor is the following:

> For modern natural science there is no longer in the beginning the material object but form, mathematical symmetry. And since mathematical structure is, in the last analysis, an intellectual content, we could say in the words of Goethe's Faust: "In the beginning was the word — the Logos." (Quoted by Landé, 1971.)

Also about the possibility for understanding the world, Heisenberg's ideas have been expressed rather clearly:

> Almost every progress in science has been paid for by sacrifice, for almost every new intellectual achievement previous positions and conceptions had to be given up. Thus, in a way, the increase of knowledge and insight diminishes continually the scientist's claim on 'understanding' nature. (Quoted in Jammer, 1966, p. 206.)

Heisenberg's opinion on causality was the following:

> The chain of cause and effect could be quantitatively verified only if the whole universe were considered as a single system — but then physics has vanished, and only mathematical scheme remains. The partition of the world into observing and observed system prevents a sharp formulation of the law of cause and effect. (Heisenberg, 1930, p. 58.)

Heisenberg thought that quantum mechanics reintroduced old Aristotelian concepts into science. Thus he wrote:

> The probability wave of Bohr, Kramers, Slater . . . meant a tendency for something. It was a quantitative version of the old concept of 'potentia' in Aristotelian philosophy. It introduced something standing in the middle between possibility and reality. (*Ibid.*, p. 41.)

Heisenberg did not fully accept Bohr's complementarity principle. His opposition was, however, of a very different nature from that of Einstein, Schrödinger, and de Broglie. The latter physicists fought complementarity because they were not willing to accept any severe limitations on our comprehension of nature, while Heisenberg thought that theoretical physics was essentially a human creation whose *sole* aim was to predict experimental results. He proposed therefore to avoid that use of language which makes direct reference to the reality of the atomic objects.

Pascual Jordan was born in Hanover, Germany (now F.R.G.), in 1902.[13] Enrolled at first, from 1921 to 1922, at the Technische Hochschule of Hanover he continued his studies at the University of Göttingen, where he worked initially as a mathematician, helping Courant and Hilbert in the preparation of their famous book *Methods of Mathematical Physics* (Courant and Hilbert, 1962) In 1924 Jordan received his doctorate under Born with a thesis on the mathematics of quantum theory and then stayed on as Born's assistant until 1926. He became *Privatdozent* at the same university after having spent the summer of 1926 with Bohr in Copenhagen.

Jordan accepted the main quantum mechanical ideas of Bohr, Born, and Heisenberg and made essentially mathematical contributions to the development of the new theory. His role was nevertheless very important because his papers were essential in imparting logical and mathematical rigor to the emerging structure.

Jordan was a rare physicist in that he declared his sympathy for a philosophical current. He accepted and defended positivism and saw in quantum mechanics a theory that was fully consistent with the positivistic point of view. Thus, for instance, Jordan's position on the existence of an external reality is the standard positivistic one:

> A common misunderstanding is that, according to the positivistic point of view, *the existence of a real external world should be denied.* The negation of a meaningless statement gives another meaningless statement; the opinion of the nonexistence of a 'real external world' is not more meaningful than the opinion of its existence. *The one and the other are neither right nor false*, but *meaningless* . . . (Jordan, 1936, p. 303.)

His opinion was that

> The dogmatic materialistic conception of the world cannot agree with positivism, because the former represents a special form of metaphysical and non-scientific teaching. (*Ibid.*, p. viii.)

On the other hand he was convinced that positivism could naturally accommodate idealistic conceptions:

> The positivistic conception offers new possibilities of granting living space to religion without contradiction from scientific thought. (Jordan, 1944, p. 160.)

Jordan also stressed that quantum physics could not allow an intuitive understanding of reality:

The situation born from the paradoxes of quantum physics can be considered as a breakdown of the hopes cultivated by past generations of physicists: One can say that a *renunciation* of the classical representation of reality has been introduced. (Jordan, 1936, p. 309.)

Like Bohr, Born, and Heisenberg, Jordan recognized that the new conceptions could be applied outside the physical domain; he expressed

the conviction that the complete development of the idea of complementarity opens a new era not for physics alone, but for our entire conception of natural science. (*Ibid.*, p. ix.)

On causality, Jordan stated as conceivable

that perhaps a *completion* of this [quantum] theory could be possible, such that the completed theory was again deterministic. This is however *mathematically impossible.* J. von Neumann has shown with an exact analysis that the present theory does *not* allow such a completion. The present theory should be factually wrong in part, if it were incomplete in such a sense. (*Ibid.*, p. 283.)

Paul Adrienne Maurice Dirac was born in Bristol in 1902.[14] His father was Swiss by birth, his mother English. Dirac's education started in Bristol in the same school where his father was a teacher of French. The latter appreciated the importance of a good education and encouraged his son's interest in mathematics. From 1918 to 1923 Dirac studied first electrotechnical engineering in Bristol and then mathematics in Cambridge, where he obtained his Ph.D. in 1926 with R. H. Fowler. In those years he read the works of Boltzmann, whose statistical mechanics he disliked, along with the works of Gibbs, for whose approach to thermodynamics he had a far better appreciation. Dirac's working habits in physics are best told in his own words:

I never went to theatres. I spent most of my time by myself, sitting working things out or going for walks. I used to spend every Sunday going for a long walk, a whole day walk, taking my lunch with me . . . I found these occasions (walks) most profitable for new ideas coming. It was on one of these occasions that the possibility of (*ab-ba*) corresponding to a Poisson bracket occurred — on one of these Sunday walks.[15]

In 1926—1927 Dirac did research in Copenhagen and Göttingen, supported by a fellowship. From 1927 on he remained in Cambridge, becoming full professor in 1932.

Dirac's epistemological position prior to 1970, at least, resembles that of Sommerfeld. In his papers and his books he displays little interest in philosophical questions and concentrates instead on mathematical aspects. It is nevertheless possible to find statements which reveal a substantial agreement with the Copenhagen—Göttingen ideas. The following, for instance, is directed against comprehensibility:

Nature's fundamental laws do not govern the world as it appears in our mental picture in any very direct way but instead they control a substratum of which we cannot form a mental picture without introducing irrelevancies. (Dirac, 1958, p. vii.)

Similarly, Dirac wrote against causality:

... we must revise our ideas of causality. Causality applies only to a system which is left undisturbed. (*Ibid.*, p. 4.)

Soon afterwards Dirac stresses, however, that only experimental results must interest a physicist:

Only questions about the results of experiments have a real significance, and it is only such questions that theoretical physics has to consider. (*Ibid.*, p. 5.)

It follows, implicitly, that causality is never applicable in physically meaningful situations.

The previous statements, all taken from the opening part of Dirac's famous classic, occupy only a marginal position in his scientific output. For this reason it is perhaps fair to say that Dirac's position calls to mind Sommerfeld's. It is interesting to note that Dirac's views seem to have undergone an important evolution in the last part of his life, similar to the changes of opinion that Schrödinger, de Broglie and, to some extent, Bohr went through during their lives. It is known (Hiley, 1977) that in 1972 Dirac had already expressed his dissatisfaction with the nonlocal nature of quantum mechanics and that he considered the reconciliation of quantum theory and relativity as an unsolved problem. Furthermore, in a 1974 seminar held in Rome he stated:

It seems to me to be evident that we do not yet have the fundamental laws of quantum mechanics. The laws that we are now using will need to have some important modification made in them before we shall have a realtivistic theory. It is very likely that this modification from the present quantum mechanics to the relativistic quantum mechanics of the future will be just as drastic as the modification from the Bohr orbit theory to the present quantum mechanics. When we make such a drastic alteration, of course, our ideas of the physical interpretation of the theory with its statistical calculations may very well be modified. (Dirac, 1974.)

During a 1975 conference in Australia Dirac even expressed the view that it might turn out that ultimately Einstein would prove to be right. He stated:

There are great difficulties ... in connection with the present quantum mechanics. It is the best that one can do up till now. But, one should not suppose that it will survive indefinitely into the future. And I think that it is quite likely that at some future time we may get an improved quantum mechanics in which there will be a return to determinism and which will, therefore, justify the Einstein point of view. (Dirac, 1976, p. 10.)

1.5. Conclusions

Our discussions in the previous sections concerning the scientific personalities of the main authors of quantum theory brought out a sharp division in the way these authors answered the following three questions:

1. Do the basic entities of atomic physics, such as electrons, photons, and so on, actually exist independently of the observations performed by physicists?
2. If the answer to the previous question is positive, is it possible to comprehend the structure of atomic objects and the evolution of atomic processes, in the sense of forming spacetime images somehow in correspondence with their reality?
3. Should one formulate physical laws in such a way that one or several causes are given for all observed effects?

By and large, the evidence presented in this chapter allows one to conclude that the opponents of quantum mechanics (Ehrenfest, Planck, Einstein, Schrödinger, and de Broglie) replied positively to the above questions, while the creators and defenders of quantum mechanics (Sommerfeld, Born, Bohr, Pauli, Heisenberg, Jordan, and Dirac) replied negatively.

However, not all of the foregoing opponents maintained the same stance throughout their lives. Schrödinger and de Broglie, for example, accepted quantum mechanics without any major reservations for several years after 1927, before returning to their original critical positions in later years. It should also be remembered that the answers to the first of the above questions are rarely as sharp as they are for the other two: Only Heisenberg, among the defenders of quantum mechanics, is very clearly opposed to reality; Bohr and Born are more evasive, while certainly not obliging with a straightforward 'yes'.

Notwithstanding these considerations, the evidence presented is strong enough to suggest rather directly that the main contradictions in the evolution of quantum physics centered around the issues of physical reality, its comprehensibility, and causality. It should be noted that these problems are not unrelated: It is no use, for example, to insist on the reality of atomic objects if these are in principle not comprehensible. Without comprehensibility physical reality must remain a 'ghost' in the background, mysterious and forever isolated from us. Likewise, without causality it is impossible to visualize the evolution of a physical process in a way that is complete. It should therefore not surprise us that there exists a strong correlation of positive answers, as well as of negative answers, to the three fundamental questions posed above.

These fundamental disagreements raise the problem of the nature of physics itself: If science were a *pure* reflection of objective reality, there would scarcely be any room for disagreement (except that due to error) between intelligent persons. But if science should comprise logically arbitrary elements (like philosophical prejudices) together with irreversible cognitive achievements, then disagreement becomes possible.

Einstein stated on several occasions that physics is a creation of the human intellect in its attempt to understand the physical world (cf. Einstein and Infeld, 1938, p. 310, and Forman and von Meyenn, 1984). This is essentially a double-sided definition of physics: As a *creation* of the human intellect, a physical theory is without doubt partly arbitrary (from the *logical* standpoint); as an understanding of the physical world, it must on the other hand, insofar as it is successful, contain objective elements, i.e., irrevocable achievements that can and will be enriched by future research but never invalidated.

This point can be illustrated with a specific example. Consider the Ptolemaic cosmological system: It viewed the earth as the center of the universe and the planets as rotating around it in epicycles. As is well known, this cosmology was successful for 1300 years in predicting the positions of planets, the times of eclipses, and so on.

Only during the Renaissance did Copernicus, Kepler, and Galilei formulate a completely new cosmological hypothesis which radically undermined the conceptual basis of the Ptolemaic system and led to the abandonment of epicycles and geocentrism.

Recalling today the case of Ptolemaic cosmology, it is not difficult to distinguish the logically arbitrary component of this system from its objective content. Logically arbitrary was the idea that the earth stood at the center of the universe. Centuries before Ptolemy, Aristarchus had invented a heliocentric system, thus demonstrating that it was not impossible for the human intellect to conceive of a different picture of the universe even in those early times. Logically arbitrary was also the concept of epicycles, which made out of the circle the privileged geometrical form of planetary motion. As objective can, on the other hand, be considered the notion that planets are objects moving in space — and not, for instance, human fantasies or direct manifestations of a god — and that it is possible to chart these motions with some degree of accuracy.

Characteristically, the arbitrary elements have been changed in subsequent cosmological models, while the objective ones have been enriched, but not abandoned. Thus, while nobody today will dream of giving the earth a privileged position in the universe, we still maintain that planets are objects moving in space. In fact, the latter notion has only been rendered more concrete by what we since have learnt about a host of planetary properties: mass, radius, atmospheric density and composition, surface features, and so forth.

The idea of epicycles has been abandoned, but we can still calculate to our complete satisfaction the positions of planets in space (Kuhn, 1957).

Similar discussions are possible for other scientific theories. To give just one more example, consider the atomic models of the nineteenth century: the existence of atoms is now firmly established, but the *shapes* then attributed to them (rings of ether, tetrahedrons, etc.) have by now been completely abandoned.

Random examples of the objective (i.e., irrevocable) contents of modern science are the concepts of galaxy, star, molecule, atom, atomic nucleus, and most elementary particles: it is hard to imagine that future scientists would deny the existence of any of these physical systems, although it is very likely that we still do not know many of their features.

When the *logically arbitrary* contents of physical theories are considered more closely, one sees that from a different point of view they are after all not so arbitrary. Perhaps Ptolemy did not really put the earth at the center of the universe of his own free will, but because pressure from the society in which he lived practically left him no other choice. Religious prejudices, cultural traditions, and power structures favored the geocentric view as the only *acceptable* one and strongly opposed the development of any rival cosmological system. History abounds with examples of the grave risks anyone took especially in defying the power structures, which were prepared to forcefully defend themselves, with violence if necessary, against 'revolutionary' ideas. Democritus, Anaxagoras, Galileo, Bruno, Diderot, and Voltaire were all punished in some way or other because their scientific views clashed with those of the prevailing culture.

Briefly, my thesis is that the seemingly arbitrary contents of physical theories are more often than not a reflection of pressures emanating from society and of its prevailing cultural values. Sometimes this influence is very clear, the choice made by physicists engaged in building a new theory being a conscious one; at other times the choice will be an unconscious one, arrived at through a process of induction that is still largely not understood.

The effect of society on the birth of new theories has been acknowledged, either implicitly or explicitly, by both founders and opponents of quantum mechanics. Heisenberg, for example, wrote:

> . . . modern physics is just one, but a very characteristic, part of a general historical process that tends toward a unification and a widening of our present world. (Heisenberg, 1959, p. 205.)

Obviously, if modern physics has to be a part of a more general historical process in a meaningful (rather than accidental) way, then there must be some general causes of that process which also influence physics. Or perhaps the historical process itself, once started, affected the new scientific ideas. In either case, the action of society on physics has to be recognized. Similar remarks apply to the following lines, written by Jordan in 1936:

> To me, modern physics and its accompanying revolution in century-old conceptions of physical science is an integral component of the unfolding of the new world of the twentieth century. (Jordan, 1944, p. x.)

Einstein, too, sometimes wrote things which implied a correlation between the general social attitude of a scientist and his scientific creations. He stressed that concern for man himself and his fate must always form the chief interest of all technical endeavors in order that the creations of our mind shall be a blessing and not a curse to mankind. (Schilpp, 1970, pp. 649 and 650.)

This quotation comes close to an admission that mental creations relative to a given problem may differ depending on the scientist's concern, or lack of it, for man and his future.

Much more explicit in his pronouncements was Schrödinger, who devoted an essay with the title 'Is Science a Fashion of the Times?' to the question of the interaction between society and science. About scientists he wrote:

They are men and they are children of their age. The scientist cannot shuffle off his mundane coil when he enters his laboratory or ascends the rostrum in his lecture hall. (Schrödinger, 1957, p. 99.)

Schrödinger also stressed that there are common characteristics in different human activities:

. . . there are numerous lines of activity which are attractive because they are the fashion of the age, whether in politics or in art or in science. These also make themselves felt in the 'exact' science of physics. (*Ibid.*, p. 100.)

Concluding his essay, he further wrote:

All this shows how dependent science is on the fashionable frame of mind of the epoch of which it forms a part. (*Ibid.*, p. 104.)

This kind of dependence, he stressed, is not immediately evident to everybody because

. . . when we live in the midst of a cultural epoch it is difficult to perceive the characteristics that are common to various branches of human activity within that epoch. (*Ibid.*, p. 105.)

In his very important paper on the history of quantum theory, Paul Forman found

overwhelming evidence that at the end of the First World War but before the development of an acausal quantum mechanics, under the influence of 'currents of thought', large numbers of German physicists, for reasons only incidentally related to the developments in their own discipline, distanced themselves from, or explicitly repudiated, causality in physics. (Forman, 1971, p. 3.)

The pressure resulting from the general intellectual climate was so strong that many physicists were led to "ardently hope for, actively search for, and willingly embrace an acausal quantum mechanics" (*ibid.*, p. 3).

In the aftermath of Germany's defeat in the First World War, the dominant intellectual tendency in the Weimar Republic was an existentialist *Lebensphilosophie* which strongly opposed rationalism in general but particularly in the exact sciences.

This 'philosophy of life' was not a regular philosophy developed by a particular school, but rather a general predilection of German culture at large that opposed every rational *Weltanschauung*, like dialectics, logical systematizations, causal explanation, mathematics, and so on. Perhaps the most influential single man in the dissemination of this attitude, which was promoted by intellectuals, politicians, and even scientists, was Oswald Spengler, whom *The British Encyclopaedia* calls one of the most important cultural forerunners of Nazism. His most celebrated work, *Der Untergang des Abendlandes*, known in its English translation as *The Decline of the West* (Spengler, 1926—1928), went through sixty printings in the eight years following 1918, adding up to a sales figure of about 100,000 copies "in a country with scarcely three times that number of college graduates" (Forman, 1971, p. 30).

Spengler's view of physics was that its content is *totally* conditioned by history:

There simply are no conceptions other than anthropomorphic conceptions . . . so is it certainly with every physical theory, no matter how well founded it is supposed to be. (Quoted *ibid.*, p. 32.)

One of the chief targets of Spengler's attacks was the idea of causality. He wrote, for instance:

I mean the opposition of the destiny-idea and the causality-principle, an opposition which, in its deep world-shaping necessity, has never hitherto been recognized as such . . . Destiny is the word for an indescribable inner certainty. One makes the essence of the causal clear by means of a physical or epistemological system, by means of numbers, by means of conceptual analyses . . . The one requires us to dismember, the other to create, and therein lies the relation of destiny to life and causality to death. (Quoted *ibid.*, p. 33.)

Scientists of the Weimar era frequently found themselves under heavy attack in newspapers, public meetings, and conversations. Sometimes they were even strongly criticized by members of their own family. Forman makes reference to the "accusations . . . which poor Max Born had to listen to daily from his wife, a would-be poet and playwright" (*ibid.*, p. 11). And she was certainly not satisfied with a letter that Einstein wrote her and in which it was explained that what she called 'Max's materialism' was simply the causal mode of viewing things. To this strong and pervasive pressure the scientists did not react with a united front. A few, notably Planck and Einstein, consistently defended their world views, but some converted quickly to an acausal philosophy (well before the advent of quantum mechanics), while many responded ambivalently, defending science in some instances and accepting the 'new' ideas in others. Still other people entered the field of scientific research armed with anticausal ideas that they acquired in the course of social and political activities carried out even *before* they became physicists. This might well have been the case for Jordan and Heisenberg. The latter, we know, was active in a rightist group (the *Weisse Ritter*) of the *Jugendbewegung* and here was forced to defend his decision to become a physicist. This he did on the grounds that theoretical physics had "thrown up problems that challenge the whole philosophical basis of science, the structure of space and time, and even the validity of causal laws" (quoted *ibid.*, p. 106).

Most remarkable were also the 'conversions' of people who had earlier believed in causality. Once more we let Forman tell the story:

> The quasi-religious conversions to acausality . . . became a common phenomenon in the German physical community during the summer and fall of 1921. As if swept up in a great awakening, one physicist after the other strode before a general academic audience to renounce the satanic doctrine of causality and to proclaim the glad tidings that the physicists are about to release the world from bondage to it. (*Ibid.*, p. 80.)

In 1921 Richard von Mises had already invoked quantum theory in support of his repudiation of causality and even tried to show that in his own field of classical mechanics causality could not possibly hold.[16]

Walter Schottky, also in 1921, published an acausal manifesto entitled 'The Problem of Causality in Quantum Theory as a Basic Question for Modern Natural Science as a Whole'. In this paper Schottky stresses again and again his opinion that any solution to the problem concerning the interaction of atoms and radiation must abandon causality.

As early as January 1920, Born confessed in a private letter to Einstein that he was willing to entertain the idea of acausality. In 1919 Franz Exner declared causality dead, presenting his 'evidence' from the study of a falling body. In 1920 Willy Wien, one of the most influential German physicists, also renounced causality.

Erwin Schrödinger, too, was swept along, for some time, by the anticausal wave. In his inaugural lecture as professor of theoretical physics at the University of Zürich, he stressed the importance of Exner's acausal views and essentially accepted them. In 1923, however, he reverted to causality.

Considering all these facts, and cognizant of the double-sided nature of all physical theories, quantum mechanics in particular, it seems possible to admit that the acausal views which influenced the birth of modern physics were connected with the general cultural and social climate of central Europe in the twenties.

Notes

1. For an elementary introduction to particle physics, see K. W. Ford, 1963, *The World of Elementary Particles*, Blaisdell, Waltham, Mass.
2. My main source of information on Max Planck is: Planck, 1949.

3. Besides Planck, 1949, one can read M. Planck, 1960, *A Survey of Physical Theory*, Dover, New York.

Main source(s) of information:

4. On A. Sommerfeld: Born, 1952; Kuhn *et al.*, 1967.
5. On P. Ehrenfest: Klein, 1970.
6. On A. Einstein: Born, 1969; Klein, 1970; Schilpp, 1970.
7. On M. Born: Born, 1978.
8. On E. Schrödinger: Scott, 1967.
9. On N. Bohr: Moore, 1966; Rozental, 1967.
10. On L. de Broglie: de Broglie, 1976.
11. On W. Pauli: Forman, 1971; Kuhn *et al.*, 1967; Rozental, 1967.
12. On W. Heisenberg: Kuhn *et al.*, 1967; Rozental, 1967.
13. On P. Jordan: Jammer, 1966; Kuhn *et al.*, 1967; Rozental, 1967.
14. On P. Dirac: unpublished interview in the Niels Bohr Arkiv, Copenhagen.
15. This sentence is contained in the interview mentioned in note 14.
16. This information, as well as that on which the rest of this section is based, is taken from Forman, 1971.

References

Books

Bohr, N.: 1958, *Atomic Physics and Human Knowledge*, Wiley, New York.

Bohr, N.: 1961, *Atomic Theory and the Description of Nature*, Cambridge University Press, Cambridge.

Born, M.: 1969, *Physics in my Generation*, Springer, New York.

Born, M.: 1978, *My Life: Recollections of a Nobel Laureate*, Taylor & Francis, London; C. Scribner's Sons, New York.

Courant, R., and Hilbert, D.: 1962, *Methods of Mathematical Physics*, 2 vols, Interscience, New York.

de Broglie, L.: 1964, *The Current Interpretation of Wave Mechanics*, Elsevier, Amsterdam.

de Broglie, L.: 1976, *Recherches d'un demi-siècle*, Albin-Michel, Paris.

Dirac, P. A. M.: 1958, *The Principles of Quantum Mechanics* (4th edn), Clarendon Press, Oxford.

Dirac, P. A. M.: 1974, *The Development of Quantum Mechanics* (Contributi del Centro Linceo Interdisciplinare, No. 4), Accademia Nazionale dei Lincei, Rome.

Dirac, P. A. M.: 1976, 'The Development of Quantum Mechanics' in H. Hora and J. R. Shepanski (eds), *Directions in Physics*, Wiley, Sidney.

Einstein, A.: 1950, Paul Ehrenfest in memoriam, in *Out of My Later Years* (a collection of articles and essays written since 1934), The Philosophical Library, New York.

Einstein, A., and Infeld, L.: 1938, *The Evolution of Physics*, Simon and Schuster, New York.

Feynman, R. P., Leighton, R. B. and Sands, M.: 1965, *Lectures on Physics*, Vol. III, Addison-Wesley, Reading, Mass.

Forman, P. L., and von Meyenn, H.: 1984, *Quantenmechanik und Weimarer Republik*, Vieweg, Braunschweig.

Forman, P. L.: 1971, 'Weimar Culture, Causality, and Quantum Theory, 1918-1927: Adaptation by German Physicists and Mathematicians to a Hostile Intellectual Environment in R. McCormmach (ed.), *Historical Studies in the Physical Sciences*, University of Pennsylvania Press, Philadelphia.

George, A. (ed.): 1953, *Louis de Broglie, physicien et penseur* (Collection *Les savants et le monde*), Albin-Michel, Paris.

Heisenberg W.: 1930, *The Physical Principles of the Quantum Theory*, The University of Chicago Press, Chicago; Dover, New York, 1949.
Heisenberg, W.: 1952, *Philosophic Problems of Nuclear Science*, Fawcett, New York.
Heisenberg, W.: 1959, *Physics and Philosophy*, Harper and Row, New York.
Heisenberg, W., Born, M., Schrödinger, E., and Auger, P.: 1961, *On Modern Physics* (tr. by M. Goodman and J. W. Binns), C. N. Potter, New York; Collier Books, New York, 1962.
Jammer, M.: 1966, *The Conceptual Development of Quantum Mechanics*, McGraw-Hill, New York.
Jordan, P.: 1936, *Anschauliche Quantentheorie*, Springer, Berlin.
Jordan, P.: 1944, *Physics of the Twentieth Century*, Philosophical Library, New York.
Klein, M. J.: 1970, *Paul Ehrenfest*, Vol. I, North-Holland, Amsterdam.
Kuhn, T. S.: 1957, *The Copernican Revolution: Planetary Astronomy in the Development of Western Thought*, Harvard U. P., Cambridge, Mass.
Kuhn, T. S.: 1970, *The Structure of Scientific Revolutions* (2nd edn), The University of Chicago Press, Chicago.
Kuhn, T. S., Heilbron, J. L., Forman, P. L., and Allen, L.: 1967, *Sources for History of Quantum Physics*, The American Philosophical Society, Philadelphia.
Landé, A.: 1965, *New Foundations of Quantum Mechanics*, Cambridge Univ. Press, Cambridge.
Moore, R.: 1966, *Niels Bohr*, Knopf, New York.
Pauli, W.: 1964a, *Collected Scientific Papers*, Interscience, New York.
Pauli, W.: 1964b, *Fisica e Conoscenza*, Boringhieri, Torino.
Planck, M.: 1949, *Scientific Autobiography and Other Papers*, The Philosophical Library, New York.
Rozental, S. (ed.): 1967, *Niels Bohr: His Life and Work as seen by his Friends and Colleagues*, North-Holland, Amsterdam; Wiley, New York.
Schilpp, P. A. (ed.): 1970, *Albert Einstein: Philosopher-Scientist* (3rd edn), Open Court, La Salle.
Schrödinger, E.: 1956, *What is Life? And Other Scientific Essays*, Doubleday, Garden City, N.Y.
Schrödinger, E.: 1957, *Science, Theory and Man*, Dover, New York. Originally published in 1935 under the title *Science and the Human Temperament* (tr. by J. Murphy and H. W. Johnston), Allen and Unwin, London.
Scott, W. T.: 1967, *Erwin Schrödinger*, University of Massachusetts Press, Amherst.
Sommerfeld, A.: 1924, *Atombau und Spektrallinien*, Vieweg, Braunschweig.
Spengler, O.: 1926—1928, *The Decline of the West*, 2 vols, tr. of the German edn (1918—1922), Knopf, New York.

Articles

Ben-Menahem, Y.: 1988, *Struggling with Causality: Schrödinger's Case*, Stanford University, preprint.
Born, M.: 1952, Obituary article for Arnold Sommerfeld, *Obituary Notices of Fellows of the Royal Society* **8** (21), 275.
Ferrero Melgar, M.: 1988, *Found. Phys.* **18**, 357.
Hiley, B. J.: 1977, *Contemp. Physics* **18**, 411.
Landé, A.: 1971, *Found. Phys.* **1**, 191.

Chapter 2

Is Quantum Mechanics a Complete Theory?

Is there something more in the real world than is contained in its description provided by modern physics? A decades-long dispute on this problem saw famous physicists affirmatively answering this question and equally famous physicists erecting or defending a bulwark against the so-called hidden variables. This bulwark had the form of a complicated mathematical theorem. It became useless when it was understood how one can circumvent it. Accordingly we are today free to work on the idea that the affirmative answer to the introductory question is the correct one.

2.1. The Problem of Completeness and of Hidden Variables

Those physicists who did not accept the final version of quantum mechanics disagreed with the philosophy of the Copenhagen and Göttingen schools in particular on one point: they thought that it was possible and useful to *complete* the theory in such a way as to make it causal. Among those physicists were Einstein, Planck, and de Broglie.

In order to understand as clearly as possible the relevance of causality, let us consider the following example. Neutrons are unstable particles, each of which at the end of its individual lifetime decays into a proton, an electron, and an antineutrino. The average lifetime of a large number of neutrons is about 900 s, in the same sense that the average life span of Europeans is about seventy years: Some Europeans live longer than this, others shorter, but when the age-at-death is averaged over millions of people a 70-odd figure results. Thus, individual neutrons can live much shorter (say, 300 s) or much longer (say, 2000 s) than their average lifetime of 900 s. The question then quite naturally arises: How is one to explain these differences? In other words, what are the *causes* underlying the even widely different individual life spans of otherwise identical neutrons?

Today's physics does not provide an understanding of these causes and accepts in fact an acausal philosophy: every decay is a spontaneous process and does not admit a causal explanation. The question about the different individual lifes of similar unstable systems, like neutrons, will according to this line of thought remain forever without answer and should indeed be categorized as a 'nonscientific' question.

Einstein and other opponents of quantum mechanics radically departed from this point of view: they believed instead that our knowledge is presently *incomplete* and that, on becoming more complete, it will naturally lead to an understanding of the individual-lifetime riddle. The causes behind such different individual behaviors are usually called *hidden variables.* One sometimes finds in the literature statements according to which these hidden variables are not measurable in principle. We think that such a point of view is unjustified: if the entities in question exist, they will also finally become measurable — when enough insight would have been gained about their physical nature. Therefore, when we speak about 'hidden variables', the implication is only that they are hidden *now*, in the same sense that atoms have been 'hidden' until *c.* 1910 and pulsars until some years ago. It is, moreover, reasonable to think that, if hidden variables exist, then some time in the future they will also be discovered. When combined with the familiar variables that are presently being used in describing the behavior of atomic and subatomic systems (energy, momentum, etc.), they will then make quantum theory causal. Not only the decay of neutrons, but *all* other quantum processes that currently are considered acausal, including disintegrations generally and all scattering and measurement processes, should in terms of hidden variables find their causal explanation.

Now it is conceivable for quantum mechanics to be an incomplete theory in the same sense as classical thermodynamics. The latter, we know, is successful in predicting the macroscopic equilibrium properties of matter, but is unable to describe such phenomena as thermal fluctuations, Brownian motion, and so on — which can be explained only by taking into account the atomic structure of matter. Could it then be, one may ask, that quantum mechanics likewise is a theory which holds for the average properties of matter, but fails to take into account some features which would explain the varying behavior of its individual constituents? This question has faced quantum mecahnics from the very beginning, and the debate is not settled yet, although some important points have been clarified.

Let us therefore investigate the question in more detail by considering again the decay of neutrons. We assume that there are some — not as yet discovered — variables which determine the different times of decay of neutrons. Two possible types of hidden variables can be imagined:

(1) *Variables related to the internal structure of particles.* For instance, in the discussed example different neutrons would exhibit different lifetimes because they are unlike with respect to some variable internal feature that predetermines the moment of any individual decay. Thus a set of neutrons would resemble a set of bombs which

are externally identical, but set by means of internal timing devices for explosion at different times.

(2) *Variables related to hypothetical fluctuations of the vacuum in small regions surrounding the particles.* Thus a set of neutrons would resemble a set of identical but fragile boats drifting on a stormy sea. Any one of them will break when hit by a wave that is high enough, and this must reasonably happen at random times.

These two types of variables will be called *internal* and *external hidden variables*, respectively. Both types are *local*, in the sense that they are independent at any given time of the *simultaneous* presence and behavior of matter, energy, or vacuum fluctuations outside a small region containing the particle under consideration. Nonlocal hidden variables are also conceivable; they are in fact required if one wants to reproduce *all* the statistical predictions of quantum mechanics, including those pertaining to the correlation of widely separated quantum systems. The problem of nonlocality is however left for future chapters, where it will be discussed at length.

The threat inherent in the idea of hidden variables — which, by declaring quantum mechanics incomplete, implies at the same time a reversal of the philosophical basis of theoretical physics — was immediately understood by physicists of the Copenhagen—Göttingen schools. A welcome (to them) defense against this causal philosophy was provided by John von Neumann, who proposed in 1932 his famous theorem establishing the impossibility of a causal completion of quantum mechanics, *given* a few very general assumptions about the mathematical structure of physical theories (von Neumann, 1955).

Bohr, Born, Pauli, Heisenberg, and Jordan forcefully stressed the importance of this theorem, and the great authority of these physicists, together with the mathematical complexity of von Neumann's theorem, had the practical effect of outlawing the idea of hidden variables.

Bohr publicly aired his opinion on this matter for instance in 1938 when a conference on 'New Theories in Physics' was held in Warsaw (International Institute of Intellectual Cooperation, 1938). In the Proceedings of this conference it is related how von Neumann presented his theorem against hidden parameters and how Bohr expressed his admiration, while pointing out that one of his own papers in more elementary ways arrived at essentially the same conclusions.

Born, too, went on record in support of von Neumann. In a book first published in 1949 he discusses the axioms of von Neumann's theorem and concludes:

The result is that the formalism of quantum mechanics is uniquely determined by these axioms; in particular, no concealed parameters can be introduced with the help of which the indeterministic description could be transformed into a deterministic one. Hence if a future theory should be deterministic, it cannot be a modification of the present one but must be essentially different. How this should be possible without sacrificing a whole treasure of well-established results I leave to the determinists to worry about. (Born, 1964.)

Nobody was ready anyway to accept that quantum mechanics was in need of any modification at a time when the numerous applications and triumphal successes of this theory were all too well known, while its difficulties and ambiguities still lay concealed. The unconditional acceptance of von Neumann's theorem had however the effect of eliminating all scientific dialectics: the philosophy of Copenhagen and Göttingen triumphed almost everywhere and opposition to it was virtually eliminated. (The reasons for this development are discussed in Forman, 1971.) Einstein, who firmly adhered to his rejection of acausality, found himself rather isolated in Princeton, and from 1935 to about 1970 the physicists who worked on the problem of causality were indeed few and far between. Notwithstanding this, important results were obtained by Bohm (1952) and by de Broglie (1960), who managed to do what von Neumann had declared impossible — find a hidden-variable model that does not contradict quantum mechanics in its statistical predictions and, at the same time, provides a causal foundation *for the individual behavior of single quantum systems.* It took some time before Bohm's and de Broglie's results were fully appreciated, in fact until it was understood that von Neumann's theorem, although mathematically correct, could in no way forbid deterministic generalizations of quantum mechanics, for the simple reason that one of its axioms was in general not physically reasonable.

This conclusion will be substantiated in detail in further sections of the present chapter. It will turn out that von Neumann's theorem can really exclude only a very select class of theories, viz., those hidden-variable theories that satisfy its axioms.

2.2. De Broglie's Paradox

The notion concerning the completeness of quantum mechanics implies that nothing exists in objective reality which is not already contained in the quantum mechanical formalism. To be more specific, let us consider a single quantum system, to wit an electron, and neglect its spin for the sake of simplicity. In a nonrelativistic situation the state of the electron is given by its wave function $\phi(x, y, z, t)$, which is a solution of the (time-dependent) Schrödinger equation. Completeness of quantum theory means in this case that *all* one can know about the electron is already contained in the wave function ϕ itself. No additional physical attributes, especially not a spatial localization more precise than that defined by ϕ, can be attributed to the electron if the theory is to be complete. Belief in this type of completeness belonging to quantum mechanics was based on von Neumann's theorem, which will be discussed in Section 2.4 below.

Louis de Broglie noticed that the idea of completeness leads to very paradoxical conclusions when one tries to apply it to concrete physical situations. His famous paradox about the localization of a particle is the following (de Broglie, 1959).

Consider a box B, with perfectly reflecting walls, which can be divided into two compartments B_1 and B_2 by a double sliding partition (Figure 2.1).

Suppose that B contains initially an electron, whose wave function $\phi(x, y, z, t)$ is defined in the volume V of B. The probability density for observing the electron at the point x, y, z at the time t is then given by the squared modulus of $\phi(x, y, z, t)$.

Next, B is divided into the two compartments B_1 and B_2. Part B_1 is then taken to, say, Paris and part B_2 to Tokyo.

The new situation is described quantum mechanically by two wave functions: $\phi_1(x, y, z, t)$, defined in the volume V_1 of B_1, and $\phi_2(x, y, z, t)$, defined in the volume V_2 of B_2.

These two wave functions could simply be equal to the 'old' ϕ when the latter is calculated for the two volumes V_1 and V_2, respectively. Or they could be different from ϕ, in case the process of separating the two boxes itself modifies the state of the electron. Which alternative obtains is however irrelevant; the important thing is that there are now two probabilities P_1 and P_2, both different from zero and adding up to unity, such that P_1 is the probability for finding the electron in box B_1 and P_2 the probability for finding it in box B_2.

Also important is the fact that P_1 can be calculated as the volume integral over V_1 of the squared modulus of ϕ_1 and that P_2 can similarly be calculated

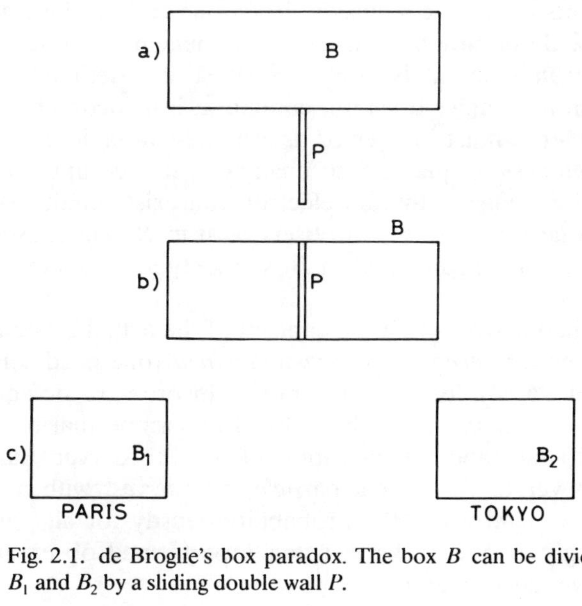

Fig. 2.1. de Broglie's box paradox. The box B can be divided into two compartments B_1 and B_2 by a sliding double wall P.

as the volume integral over V_2 of the squared modulus of ϕ_2. Note that the completeness of the theory forces us to shun statements like: "The electron is actually either in B_1 or in B_2." For, such a statement implies a more precise spatial localization than that provided by the wave functions ϕ_1 and ϕ_2. The only assertion that is compatible with the completeness of quantum mechanics will be: "The particle is in B_1 *and* in B_2."

The situation changes, however, if one makes an observation, because, as is well known, an electron can be *observed* in a well-defined location if one performs a sufficiently accurate position measurement.

If one opens the box in Paris, one will find that either the electron is in B_1, or it is not. In either case one can predict with certainty the outcome of any future observation performed on B_2 in Tokyo: If the electron was present in Paris, it will certainly be found absent in Tokyo, and *vice versa*.

Suppose an observation was performed in Paris at time t_0 and the electron found in B_1, then one is certain that for all $t \geqslant t_0$ the probability P_2 of observing the electron in B_2 in Tokyo is zero. Since P_2 is given by a volume integral of the squared modulus of ϕ_2 over V_2, one must necessarily conclude that the vanishing of P_2 for $t \geqslant t_0$ implies the vanishing of $\phi_2(x, y, z, t)$ for $t \geqslant t_0$. Thus the observation of the electron in Paris changes the wave function in Tokyo, reducing it to zero!

Barring the possibility that an observation in Paris destroys 'part of an electron' in Tokyo and makes it appear in Paris, the natural inclination of most physicists would be to say that the electron observed in Paris at time t_0 *was already there* for $t \leqslant t_0$ and that the wave functions ϕ_1 and ϕ_2 represent only our knowledge prior to observation of the electron's position.

This natural attitude (which corresponds to the philosophical position of *realism*), if pursued to its obvious conclusion, leads one to introduce a new observable parameter λ describing the particle's confinement to either B_1 or B_2. If $\lambda = +1$, the electron is definitely in B_1, if $\lambda = -1$, it is definitely in B_2. Since ordinary quantum mechanics does not contain λ, it is incomplete, and the localization parameter λ must be viewed as a hidden variable. If, on the other hand, the completeness of quantum mechanics is insisted upon, one is lead to the paradoxical conclusion that an electron can exist simultaneously within B_1 and B_2 and that the mere act of observing it in B_1 will cause 'that part' of the electron previously contained in B_2 suddenly to disappear there and to re-appear in B_1.

The conclusion reached above is that the concept of the actual existence in space and time of an electron, *even if very grossly defined* (one needs only to distinguish Tokyo from Paris!), leads to a paradox. In order to defend the completeness of quantum theory one is thus forced to assume that it makes no sense whatsoever to talk about localization of an *un*observed particle. Quantum mechanics never denies that a particle is *observed* with a given localization; in fact it even furnishes the probability density for all possible localizations. Accordingly, if one sticks to actually performed observations, no contradictions are ever encountered.

In the foregoing manner one is compelled to accept a *positivistic* philosophy, in which only arguments involving observations and mathematical schemes are allowed, and renounce objective reality as having no role in scientific reasoning.

All this leads to a rather elementary conclusion: de Broglie's paradox exists only for people who insist on a *realistic* ('particles exist objectively') and *rationalistic* philosophy, in which spacetime is not an illusion of our senses and it is possible to talk about electron localization. From other philosophical perspectives (like that of positivism) no paradox arises at all. It will be seen in subsequent chapters that a similar inference can be drawn from the EPR paradox and other aspects of quantum theory.

Taking refuge in a strictly positivistic philosophy has, however, not been a solution for everybody. Most certainly it was not a solution for Planck, Einstein, de Broglie, and Schrödinger. And even today most working physicists are probably convinced, in the case of de Broglie's paradox, that the electron will be found in Paris in box B_1 only if it was in the same box before the act of observation took place. But in expressing such a conviction they would obviously have conceded that quantum mechanics is *incomplete.* This is after all a natural conclusion: why indeed should any human theory describe physical objects in perfect detail? Our task in the present chapter is in fact to show that *today* we are free to accept the incompleteness of quantum theory. However trivial this conclusion may seem from a *logical* point of view, it must be stressed that it actually represents a most significant milestone in the history of physics, because it has been reached only after forty years of struggle against von Neumann's theorem.

Albert Einstein stated in 1952 his reasons for not believing in the completeness of quantum mechanics. One of them was this:

> The mechanism of quantum theory is built in such a way that the ψ-function of a subsystem of a global system composed of two subsystems is given differently depending on the type of (complete) measurement that one carries out on the other subsystem. This is true also when the two subsystems are spatially separated the one from the other. If the ψ-function described *completely* the real state, this would mean that the measurement on the second subsystem would influence the real state of the first one, and this would correspond to a kind of immediate coupling between spatially separated objects. (George, 1953, p. 9)

The previous argument is merely a simplified version of the EPR paradox, which will be discussed at length in future chapters.

Erwin Schrödinger favored an undulatory description of atomic transitions which would have associated a perfectly causal behavior with atoms and light. After the discovery of his celebrated equation he felt confident

> . . . that the mysterious 'fit and jerk theory' about the jumplike transition from one energy level to another was now ousted. Our wave equation could be expected to describe any changes of this kind as slow and actually *describable* processes. This hope was not informed by personal predilection for continuous description, but if anything by the wish for any kind of description at all of these changes. It was a dire necessity. To produce a coherent train of light waves of 100 cm length and more, as is observed in fine spectral lines, takes a time comparable with the

average interval between transitions. The transition must be coupled with the production of the wave train. Hence if one does not understand the transition, but only understands the 'stationary states', one understands nothing. For the emitting system is busy all the time in producing the trains of light waves, it has no time left to tarry in the cherished 'stationary states', except perhaps in the ground state. (George, 1953, p. 18)

2.3. The Spin-$\frac{1}{2}$ System in Quantum Mechanics

The electron spin was discovered in 1925 by W. Pauli. Considering the structure of complex atoms and the light emitted by them, Pauli recognized that a complete classification and understanding of their line spectra could be achieved only if the following two new hypotheses were introduced:

(1) Besides the usual three quantum numbers introduced by Sommerfeld in 1915 to characterize the electron, a fourth quantum number σ with only two possible values, $\sigma = \pm 1$, has to be used.
(2) In all atoms only one electron can occupy a state defined by definite values of these four quantum numbers.

These assumptions were introduced only formally by Pauli for the sake of understanding and classifying the properties of the emitted light; no real physical foundation for their validity was given. The second assumption became known as the *exclusion principle*, since it excluded the possibility that more than one electron could be found in a state specified by fixed values of the four quantum numbers.

The first assumption received strong support from an experiment, performed by Stern and Gerlach (1922), in which a narrow beam of silver atoms was passed through a strongly inhomogeneous magnetic field and a splitting of the beam into two separate beams was observed. Stern and Gerlach could even measure the magnetic moment of the silver atoms to within 10 percent (they found about 1 Bohr magneton). We now know that what they really measured was the electron's magnetic moment, whose existence in 1922 was not even suspected, since Pauli's discovery of spin came only three years later. The Stern—Gerlach experiment is today explained as follows: an electron can exist in two states, with spin either parallel to the magnetic field or pointing in the opposite direction. Correspondingly, also the magnetic moment can assume two possible orientations. Thus the force exerted by the inhomogeneous magnetic field has two possible values, and the beam of silver atoms, which, for reasons related to their structure, interact with the field essentially as if they were single electrons, split into two divergent beams. Roughly speaking, we can say that there are two types of electrons, those with $\sigma = +1$ and those with $\sigma = -1$, and that the Stern—Gerlach experiment sorts the two types of electrons into separate beams. This experiment thus provides an empirical visualization of the existence of the Pauli quantum number σ.

Soon after the appearance of Pauli's paper, Goudsmit and Uhlenbeck (1926) in The Netherlands tried to understand the physical origin of Pauli's fourth quantum number. Since, according to Sommerfeld's theory, a system possesses as many quantum numbers as degrees of freedom, it seemed that a point particle was incapable of exhibiting more than three quantum numbers. Thus the only way room could be made for the new quantum number was to assume some internal structure. Goudsmit and Uhlenbeck accordingly postulated that the electron was a small rotating sphere and managed to show that this rotation would lead to the needed angular momentum (spin). Although this model was not free of serious difficulties, it was a step forward in the realistic sense.

The problem of electron spin illustrates on a small scale the clash between positivistic and realistic descriptions of nature. Pauli was interested only in describing formally the atomic structure, while Goudsmit and Uhlenbeck (encouraged by Ehrenfest) wanted to understand its real nature. The victory of the positivistic viewpoint in this case is seen in the fact that today everybody accepts and uses the spin formalism, while the concept of a rotating electron has not been developed any further; it is regarded at best as a psychologically satisfying visual aid which cannot be accepted as literally true, because it does not even take into account the wave—particle duality for the electron. One of the consequences of this situation is that, in developing the theory of spin-$\frac{1}{2}$, one must necessarily invoke some abstract, though simple, mathematics. Simple, because the algebra of dichotomic observables is the simplest algebra of quantum theory that still contains all its main concepts (operators, state vectors, eigenvalues, superposition principle, etc.). Abstract, because when one develops the mathematical formalism, it is not known if and in what way every new step corresponds to a property in objective reality.

One fact is clear from the start: the wave function, which is meant to furnish a description as complete as possible of the state of the electron, should depend on the new dichotomic observable σ. Thus we should write for the wave function

$$\psi = \psi(x, y, z, t; \sigma),$$

implying that $|\psi(x, y, z, t; \sigma)|^2$ is the probability density at time t for an electron of type σ. We see that there are two classes of wave functions, those with $\sigma = +1$ and those with $\sigma = -1$. It is found convenient to represent these wave functions with a matrix notation

$$\begin{aligned} \psi(x, y, z, t; +1) &= \psi(x, y, z, t)\begin{pmatrix}1\\0\end{pmatrix}, \\ \psi(x, y, z, t; -1) &= \psi'(x, y, z, t)\begin{pmatrix}0\\1\end{pmatrix}, \end{aligned} \tag{2.1}$$

where ψ and ψ' are now scalar functions of their arguments without any matrix structure.

All the laws of quantum mechanics are linear. This circumstance implies what can be regarded as the most fundamental property of quantum theory, which is often referred to as the *superposition principle*. The latter asserts that if one is given any two wave functions (representing two possible states for the electron under consideration) then an arbitrary linear combination of them will also be a wave function (representing another possible state of the electron), provided only the Born normalization condition is satisfied. This means in particular that a wave function can be obtained from (2.1) by writing

$$\psi_0(x, y, z, t) = c_1\psi \begin{pmatrix} 1 \\ 0 \end{pmatrix} + c_2\psi' \begin{pmatrix} 0 \\ 1 \end{pmatrix} = \begin{pmatrix} c_1\psi \\ c_2\psi' \end{pmatrix}, \tag{2.2}$$

with c_1 and c_2 constants. The straightforward way to introduce Born's normalization condition is to define the *Hermitian conjugate* wave function, given by

$$\psi_0^+(x, y, z, t) = c_1^*\psi^*(1, 0) + c_2^*\psi'^*(0, 1) = (c_1^*\psi^*, c_2^*\psi'^*).$$

The probability density is then assumed to be

$$\rho(x, y, z, t) = \psi_0^+\psi_0 = |c_1|^2|\psi|^2 + |c_2|^2|\psi'|^2,$$

whence, since $|\psi|^2$ and $|\psi'|^2$ are themselves properly normalized probability densities, it follows by integration over all space that $|c_1|^2 + |c_2|^2 = 1$.

Having thus introduced spin states, we next introduce spin observables. In classical physics there is no fundamental difference between states and observables because measurements can always be made in such a way that one can learn from individual systems what their state exactly is. In quantum physics this is no longer possible because the existence of the finite quantum of action h implies that every act of measurement will give rise to a large and unpredictable modification of the state of the system. It is therefore conceptually impossible to identify states and observables, and one must find an independent mathematical description of the observables.

As is well known, quantum mechanics does this by representing observables by linear Hermitian operators which act on the wave functions. For spin states one has a very simple representation of spin observables: They are described by Hermitian 2×2 matrices. The most general spin observable $\mathscr{R}$ is therefore represented by the matrix

$$R = \begin{pmatrix} \alpha_0 & \beta_0 \\ \beta_0^* & \gamma_0 \end{pmatrix}, \tag{2.3}$$

with α_0 and γ_0 real. If one introduces the 2×2 unit matrix I and the Pauli matrices

$$\sigma_1 = \begin{pmatrix} 0 & 1 \\ 1 & 0 \end{pmatrix}, \quad \sigma_2 = \begin{pmatrix} 0 & -i \\ i & 0 \end{pmatrix}, \quad \sigma_3 = \begin{pmatrix} 1 & 0 \\ 0 & -1 \end{pmatrix}, \tag{2.4}$$

it is a simple matter to show that one can write

$$R = \alpha I + \boldsymbol{\beta} \cdot \boldsymbol{\sigma}, \tag{2.5}$$

where $\boldsymbol{\beta} \cdot \boldsymbol{\sigma} = \beta_1\sigma_1 + \beta_2\sigma_2 + \beta_3\sigma_3$ and the new constants α, β_1, β_2, β_3 are related to α_0, β_0, γ_0 through the relations

$$\alpha_0 = \alpha + \beta_3, \qquad \gamma_0 = \alpha - \beta_3, \qquad \beta_0 = \beta_1 - i\beta_2.$$

Quantum mechanics furthermore assumes — and this, too, is well known — that the results of measurements of the observable $\mathscr{R}$ are in all cases identical with the eigenvalues r of the associated matrix R. In order to discover these eigenvalues, we write the eigenvalue equation

$$R \begin{pmatrix} x_0 \\ y_0 \end{pmatrix} = r \begin{pmatrix} x_0 \\ y_0 \end{pmatrix},$$

where $\begin{pmatrix} x_0 \\ y_0 \end{pmatrix}$ is an unknown spin state. On substituting the form (2.5) of R, the foregoing matrix equation resolves into the set of equations

$$(\alpha + \beta_3 - r)x_0 + (\beta_1 - i\beta_2)y_0 = 0,$$
$$(\beta_1 + i\beta_2)x_0 + (\alpha - \beta_3 - r)y_0 = 0,$$

which, being homogeneous, has nonvanishing solutions for x_0 and y_0 only if the determinant formed by their coefficients is zero. From this it follows, via a simple calculation, that

$$r = \alpha \pm |\boldsymbol{\beta}|. \tag{2.6}$$

This important equation will become useful in subsequent paragraphs because it gives the results of all conceivable spin measurements.

To get a better feeling for the contents of Equation (2.6), consider some particular choices for the matrix R: If $R = \sigma_1$, which is the same as taking $\alpha = 0$ and $\boldsymbol{\beta} = \hat{\imath}$, where $\hat{\imath}$ is a unit vector along the x axis, one gets from (2.6) that $r = \pm 1$. Identical conclusions hold for $R = \sigma_2$ or $R = \sigma_3$. If one takes, as a further example,

$$R = \boldsymbol{\sigma} \cdot \mathbf{n},$$

with $\mathbf{n} = (1, 1, 0)$, whence $|\mathbf{n}| = \sqrt{2}$, one gets from (2.6), taken now with $\alpha = 0$ and $\boldsymbol{\beta} = \mathbf{n}$:

$$r = \pm\sqrt{2}.$$

Returning to the general case, we note that measurements of $\mathscr{R}$ on a statistical ensemble of spin-$\frac{1}{2}$ particles, all described by the same spin state $\binom{1}{0}$, will in general give a random sequence of results $\alpha + |\boldsymbol{\beta}|$ (with probability p_1) and of results $\alpha - |\boldsymbol{\beta}|$ (with probability $p_2 = 1 - p_1$).

Quantum mechanics teaches us how to calculate p_1 and p_2. This can be done by starting from two different definitions of the average value of the observable $\mathscr{R}$. One is the experimental definition (weighted average) according to which

$$\langle \mathscr{R} \rangle = p_1(\alpha + |\boldsymbol{\beta}|) + p_2(\alpha - |\boldsymbol{\beta}|). \tag{2.7}$$

The other is the quantum mechanical prediction that, for the spin state $\chi = \binom{1}{0}$, one has

$$\langle \mathscr{R} \rangle = \chi^+ R \chi = \alpha + \beta_3, \tag{2.8}$$

where the last equality is obtained by direct calculation after inserting expression (2.5) for R. Equating the right-hand sides of (2.7) and (2.8) and using the obvious relation $p_1 + p_2 = 1$, one obtains

$$p_1 = \frac{1}{2}\left(1 + \frac{\beta_3}{|\boldsymbol{\beta}|}\right), \quad p_2 = \frac{1}{2}\left(1 - \frac{\beta_3}{|\boldsymbol{\beta}|}\right). \tag{2.9}$$

We have thus obtained all the predictions of quantum mechanics for spin measurements: the outcome for measurements of $\mathscr{R}$ are $\alpha \pm |\boldsymbol{\beta}|$, and their probabilities are p_1 and p_2, respectively, as given above. It could be objected that the spin state considered, χ, is a very special one, but this is not quite true because, considering the most general observable $\mathscr{R}$, one can always use χ as the spin state with a suitable choice of the z axis. Indeed, if the initial state χ' were different from χ, one could perform a rotation of the axes such that the *new* z direction coincided with the direction defined by χ': every spinor χ' can in fact be viewed as an eigenstate with eigenvalue $+1$ of a matrix $\boldsymbol{\sigma} \cdot \mathbf{n}$, for a suitable choice of the unit vector $\mathbf{n}$. The rotation would of course also affect the matrix R, but since we have considered above its *most general* form, our original conclusion holds unchanged.

2.4. A Simple Proof of von Neumann's Theorem

Consider a general spin observable $\mathscr{R}$ corresponding to a linear Hermitian 2×2 matrix R. If one performs a series of N measurements of $\mathscr{R}$ on similarly prepared particles (that is, particles all having the same initial wave function), one will obtain the results

$$\mathscr{R} = r_1, r_2, \ldots, r_N.$$

According to quantum mechanics, everyone of these results will coincide with one of the two eigenvalues of the matrix R.

It is useful to consider also the observable $\mathscr{R}^2$, which is *by definition* that observable which can be measured by measuring $\mathscr{R}$ and taking the square of the result thus obtained. The previous N measurements of $\mathscr{R}$ thus constitute also N measurements of $\mathscr{R}^2$, *viz.*,

$$\mathscr{R}^2 = r_1^2, r_2^2, \ldots, r_N^2.$$

Assume now that the results $r_1, r_2, \ldots, r_N$ are actually determined by a hidden variable λ or, more precisely, by the values that λ had immediately before the measurements. Thus we can write $r_1 = r(\lambda_1)$, $r_2 = r(\lambda_2), \ldots, r_N = r(\lambda_N)$, where $\lambda_1, \lambda_2, \ldots, \lambda_N$ are the values of λ before each individual measurement procedure. If one prepares the particles in such a way that λ is the same before every measurement (say $\lambda = \lambda_0$), then obviously always the same result will emerge:

$$\mathscr{R} = r_1 = r_2 = \cdots = r_N = r(\lambda_0).$$

It follows, for such a fixed-λ ensemble, that

$$\langle \mathscr{R} \rangle = N^{-1} \sum_{i=1}^{N} r_i = r(\lambda_0), \; \langle \mathscr{R}^2 \rangle = N^{-1} \sum_{i=1}^{N} r_i^2 = r^2(\lambda_0).$$

Therefore, the mean-square deviation of the observable $\mathscr{R}$, defined, as in the elementary theory of errors, as the square root of the difference between $\langle \mathscr{R}^2 \rangle$ and $\langle \mathscr{R} \rangle^2$, vanishes; in symbols,

$$\Delta \mathscr{R} = (\langle \mathscr{R}^2 \rangle - \langle \mathscr{R} \rangle^2)^{1/2} = 0.$$

An equivalent but shorter name for $\Delta \mathscr{R}$ is *dispersion*, and this designation will be adopted from here on. We can thus say that there is no dispersion for our observable $\mathscr{R}$ once the hidden variable has a fixed value.

More generally one may consider a situation in which there are several different observables. (As we saw, there is an infinite number of observables already for the spin-$\frac{1}{2}$ particle at rest.) Let $\{\mathscr{R}, \mathscr{S}, \mathscr{T}, \ldots\}$ be the set of *all* the observables pertaining to the particles under consideration. In the spirit of hidden variables, one can say that the results obtained by measuring any of $\mathscr{R}, \mathscr{S}, \mathscr{T}$, etc. are fixed by the values of some hidden variables contained in the set $\{\lambda, \mu, \nu, \ldots\}$. Naturally, the number of hidden variables is in general *not* related to the number of observables. Considering a set of particles in which all the hidden variables have fixed values ($\lambda = \lambda_0$, $\mu = \mu_0$, $\nu = \nu_0$, $\ldots$), every observable must be found, when measured, to have a well-defined value:

$$R = r(\lambda_0, \mu_0, \nu_0, \ldots) = r_0$$

$$\mathscr{S} = s(\lambda_0, \mu_0, \nu_0, \ldots) = s_0,$$

$$\mathscr{T} = t(\lambda_0, \mu_0, \nu_0, \ldots) = t_0,$$

$$\cdots\cdots\cdots\cdots\cdots\cdots$$

It is obviously possible to define the dispersion of the observables $\mathscr{S}, \mathscr{T}, \ldots$ in a way strictly similar to the one used above for $\mathscr{R}$. Repeating the reasoning given above for $\mathscr{R}$ alone, we conclude now that all the observables $\mathscr{R}$, $\mathscr{S}, \mathscr{T}, \ldots$ have zero dispersion. This is obviously a direct consequence of the fact that they all have well-defined (fixed) values. Since these are all the observables of the particles forming the statistical ensemble under investigation, we say that this ensemble is *dispersion-free*.

We need not worry about the fact that different observables might be represented by noncommuting matrices. For, this merely implies that it is not possible to build instruments capable of measuring them simultaneously. But if we have two such observables $\mathscr{R}$ and $\mathscr{S}$, represented by the two noncommuting 2×2 matrices R and S, and just one hidden variable λ with the fixed value λ_0 (the extension of the following arguments to several hidden variables is straightforward), we shall perform on some of the systems measurements of $\mathscr{R}$ and on the remaining systems measurements of $\mathscr{S}$. In the first case one will always find r_0, in the second always s_0. We therefore have $\Delta\mathscr{R} = 0 = \Delta\mathscr{S}$, but this implies no contradiction with the noncommutativity of R and S, because the measurements of $\mathscr{R}$ and $\mathscr{S}$ were performed on different systems. If one believes in hidden variables, one can always think of a given ensemble as a sum of dispersion-free ensembles. The familiar results of quantum mechanics ($\Delta\mathscr{R}\ \Delta\mathscr{S} \geqslant \frac{1}{2}|\langle\psi|[R, S]|\psi\rangle|$) will then hold for the full ensemble but not for its dispersion-free parts. Note that the practical difficulties of preparing dispersion-free ensembles are irrelevant. What is important is that one can logically conceive of any given ensemble as a sum of dispersion-free subensembles.

We may conclude quite generally that the existence of hidden variables implies necessarily the existence of dispersion-free ensembles. It was precisely against the existence of such ensembles that von Neumann's theorem was directed. The time has now come to discuss this theorem. But in the present section we shall restrict ourselves to formulating and proving it for the particular case of spin measurements on spin-$\frac{1}{2}$ particles. General proofs of the theorem will be discussed in Sections 2.6, 2.7, and 2.8 of the present chapter.

Von Neumann's theorem is based on the following three axioms:

N1 *There is a one-to-one correspondence between spin observables and Hermitian* 2×2 *matrices.*

N2 *If the spin observable* $\mathscr{R}$ *corresponds to the matrix* R, *then the observable* $f(\mathscr{R})$ *corresponds to* $f(R)$.

N3 *If* $\mathscr{R}$ *and* $\mathscr{S}$ *are arbitrary spin observables and* a *and* b *real numbers, then the following relation (expressing the linearity of the averages) holds*:

$$\langle a\mathscr{R} + b\mathscr{S}\rangle = a\langle\mathscr{R}\rangle + b\langle\mathscr{S}\rangle.$$

From these axioms von Neumann deduced that *dispersion-free ensembles cannot exist.*

A detailed discussion of the meaning of these axioms is left for Section 2.6 below. Here we mention simply the following definition:

D. *Definition of sum of two observables* $\mathscr{R}$ *and* $\mathscr{S}$. If the two spin observables $\mathscr{R}$ and $\mathscr{S}$ correspond, respectively, to the Hermitian 2×2 matrices R and S, one assumes that $\mathscr{R} + \mathscr{S}$ is the spin observable corresponding to the matrix $R + S$.

Notice that axiom N1 is also intended to convey that the measured values of a spin observable cannot be different from the two eigenvalues of its associated matrix R. As far as N2 is concerned we have included it for historical completeness, but it will not be used in the present simplified proof of the theorem.

Let us suppose a hidden variable λ exists which determines the results of the measurements of all conceivable spin observables. Consider an ensemble in which λ has a fixed value λ_0, i.e., a dispersion-free ensemble. Let $\mathscr{R}$, $\mathscr{S}$, and $\mathscr{T}$ be any three observables and R, S, and T their associated 2×2 matrices. Measurements of these observables on the dispersion-free ensemble will lead necessarily to the conclusion that:

$\mathscr{R}$ has a fixed value $r(\lambda_0)$ which is an eigenvalue of R;

$\mathscr{S}$ has a fixed value $s(\lambda_0)$ which is an eigenvalue of S;

$\mathscr{T}$ has a fixed value $t(\lambda_0)$ which is an eigenvalue of T.

Specializing this general conclusion, let us assume that the observables $\mathscr{R}$, $\mathscr{S}$, and $\mathscr{T}$ correspond, respectively, to the matrices

$$R = \sigma_1, S = \sigma_2, T = \boldsymbol{\sigma} \cdot \mathbf{n}, \tag{2.10}$$

where $\mathbf{n} = (1, 1, 0)$, whence $|\mathbf{n}| = \sqrt{2}$. It follows from (2.10) that $T = R + S$. One therefore concludes, on using definition D, that

$$\mathscr{T} = \mathscr{R} + \mathscr{S}. \tag{2.11}$$

Now, in our dispersion-free ensemble with $\lambda = \lambda_0$, every time one measures $\mathscr{R}$ one finds $r(\lambda_0)$. This means that the ensemble average of $\mathscr{R}$ is also given by $r(\lambda_0)$: $\langle \mathscr{R} \rangle = r(\lambda_0)$. Similarly, one must have $\langle \mathscr{S} \rangle = s(\lambda_0)$ and $\langle \mathscr{T} \rangle = t(\lambda_0)$. Applying now axiom N3, one deduces, from (2.11), $\langle \mathscr{T} \rangle = \langle \mathscr{R} \rangle + \langle \mathscr{S} \rangle$, which implies that $t(\lambda_0) = r(\lambda_0) + s(\lambda_0)$. The latter relation cannot be satisfied, however, since the eigenvalues of $R = \sigma_1$ are ± 1, those of $S = \sigma_2$ are ± 1, and those of $T = \boldsymbol{\sigma} \cdot \mathbf{n}$ are $\pm \sqrt{2}$; and no choice of signs can allow the relation $\pm \sqrt{2} = \pm 1 \pm 1$ to be satisfied (Bell, 1966).

Starting from the axioms of von Neumann's theorem and the existence of dispersion-free ensembles, we have thus reached an absurd conclusion. If one believes axioms N1 and N3 to be correct, one must therefore conclude that dispersion-free ensembles cannot exist. This is the same as saying that a hidden-variable completion of quantum mechanics is impossible.

2.5. The Theorem is not General Enough

If the axioms of von Neumann's theorem were so general that no conceivable physical situation could violate them, we should obviously give up all hope to find a causal completion of quantum mechanics. Fortunately, as we have already anticipated several times, this is not the case. If one superficially inspects axioms N1, N2, and N3, nothing, it seems, could be possibly wrong with them.

About N2 we have already stressed that it was not even needed in the proof of the simplified version of the theorem. Axiom N1 establishes a very simple correspondence between spin observables and 2×2 Hermitian matrices: if this axiom were wrong, one would be forced to construct a theory different from quantum mechanics which makes the correspondence between observables and linear Hermitian operators the very basis of its theoretical formalism. Finally, N3 postulates the linearity property for averages, which is true within quantum mechanics and which seems to be valid also in classical physics. To exemplify this, consider a set of classical particles in an external field, and suppose that these are in motion with different velocities. The general relation $E = T + V$ gives the total energy E of the set as a sum of its kinetic energy T and potential energy V. One may measure the kinetic energies $T_1, T_2, \ldots, T_N$ and potential energies $V_1, V_2, \ldots, V_N$ of the individual particles and calculate $E_1 = T_1 + V_1$, $E_2 = T_2 + V_2, \ldots, E_N = T_N + V_N$. Since

$$\langle E \rangle = N^{-1} \sum_{i=1}^{N} E_i,$$

it is immediately obvious that $\langle E \rangle = \langle T \rangle + \langle V \rangle$. The same result holds if $\langle E \rangle$ is measured directly, for instance by weighing the whole system, while $\langle T \rangle$ and $\langle V \rangle$ are calculated from individual measurements, as above.

Thus, in the present example axiom N3 is valid also within classical physics.

Still it is precisely N3 which fails when one considers other types of measurements.

In order to see this as clearly as possible, we consider a concrete physical model of spin-$\frac{1}{2}$ particles and of spin measurements. This model reproduces all the quantum mechanical predictions for spin measurements (experimental results and probabilities) and therefore does exactly what von Neumann's theorem attempts to forbid, i.e., provide a causal completion of quantum mechanics. Of course, the model itself should not be taken seriously, because, as we shall see, it is for many reasons very different from the reality of subatomic entities: for one thing, it does not take into account the wave—

particle duality. The model is however totally reliable in reproducing the quantum mechanical predictions for spin measurements and therefore constitutes a real counterexample to von Neumann's theorem — which, as we saw, can be formulated also for spin alone. The virtues of the model are therefore methodological and, if one wishes, pedagogical: a concrete set-up reproducing its properties could actually be built in the laboratory. (The present model does not suffer from the drawback of Bell's model (Bell, 1966) that implies unphysical correlations between spin components.)

The properties of our model are embodied in the following four hypotheses:

M1 Spin-$\frac{1}{2}$ particles can be simulated by spinning spheres which are assumed to propagate along the x axis. The intrinsic angular momentum (spin) $\boldsymbol{\lambda}$ of each sphere is assumed in all cases to lie in the plane perpendicular to the direction of motion.

M2 In order to reproduce the probabilistic properties of the quantum mechanical state $\binom{1}{0}$, we assume that an ensemble of spheres, all having the description $\binom{1}{0}$, will exhibit a statistical distribution of spin orientations such that the probability density $\rho(\theta)$ for the spin vector $\boldsymbol{\lambda}$ forming an angle θ with the z direction is given by

$$\begin{aligned} \rho(\theta) &= \tfrac{1}{2}\cos\theta \quad && (-\tfrac{1}{2}\pi < \theta < \tfrac{1}{2}\pi), \\ \rho(\theta) &= 0 && (\text{otherwise}). \end{aligned} \tag{2.12}$$

M3 An instrument designed to measure the quantum mechanical observable $\mathscr{R}$ corresponding to the operator R given in Equation (2.5) works as follows: it measures the *sign* of the projection of $\boldsymbol{\lambda}$ on $\boldsymbol{\beta}$, affixes this to $|\boldsymbol{\beta}|$, and adds α to the result.

M4 One must also reproduce the quantum mechanical predictions of *subsequent* spin measurements on the same spin-$\frac{1}{2}$ particle. In order to do so, one needs to assume that a statistical redistribution of the λ's is generated by the action of the apparatus. The precise assumption is that everytime a new spin observable is measured the λ-vectors are redistributed in the statistical ensemble according to a new density similar to (2.12) but centered around the characteristic direction $\boldsymbol{\beta}$ of the new observable.

The model thus formulated leads to many interesting consequences. First, the procedure of 'measurement' defined in M3 implies that the results obtained will not be different from $\alpha \pm |\boldsymbol{\beta}|$, thus guaranteeing that no disagreement with quantum mechanics can result at this level. Second, as a particular case of the previous result and as a consequence of Equation (2.12), if one measures σ_3 (now $\alpha = 0$ and $\boldsymbol{\beta} = \hat{k}$ (the unit vector along the z axis)), the angle between the $\boldsymbol{\lambda}$'s and $\boldsymbol{\beta}$ will range between $-\frac{1}{2}\pi$ and $\frac{1}{2}\pi$, implying that the projection of $\boldsymbol{\lambda}$ on $\boldsymbol{\beta}$ is never negative. It follows that all measurements will give $\sigma_3 = +1$, a result consistent with the fact that $\binom{1}{0}$, is an eigenstate of σ_3 with eigenvalue $+1$.

Third, it is easy to verify that the probability density (2.12), which is never negative, is properly normalized. We leave the calculation as an exercise to the reader.

Finally, we can calculate the probabilities p_1 and p_2 of finding, respectively, the results $\alpha + |\boldsymbol{\beta}|$ and $\alpha - |\boldsymbol{\beta}|$ when the general observable R is measured. Note in this connection that the hypothesis M3 involves two steps: a true measurement, revealing the sign of $\boldsymbol{\lambda} \cdot \boldsymbol{\beta}$, and a rather mechanical process consisting of the multiplication of this sign into $|\boldsymbol{\beta}|$ and the addition of α to the result. Therefore, p_1 and p_2 coincide, respectively, with the *a priori* probabilities for finding $\boldsymbol{\lambda} \cdot \boldsymbol{\beta} > 0$ and $\boldsymbol{\lambda} \cdot \boldsymbol{\beta} < 0$. In order to establish p_1 one need only calculate the integral of $\rho(\theta)$ over the hatched area σ_0 of Figure 2.2 because, by M2, the $\boldsymbol{\lambda}$'s can lie only in the upper half circle.

Obviously then

$$p_1 = \frac{1}{2} \int_{\sigma_0} \cos\theta \, \mathrm{d}\theta = \frac{1}{2}(1 + \cos\theta_\beta).$$

Remembering that $p_1 + p_2 = 1$, one obtains immediately also

$$p_2 = \tfrac{1}{2}(1 - \cos\theta_\beta).$$

Since

$$\cos\theta_\beta = \frac{\hat{k} \cdot \boldsymbol{\beta}}{|\boldsymbol{\beta}|} = \frac{\beta_3}{|\boldsymbol{\beta}|},$$

these results for p_1 and p_2 coincide exactly with the quantum mechanical probabilities given in Equations (2.9).

It is accordingly clear that our model furnishes results of measurements and probabilities for different results in perfect agreement with quantum mechanics. Still, the outcome of an act of measurement of $\alpha I + \boldsymbol{\beta} \cdot \boldsymbol{\sigma}$ on a

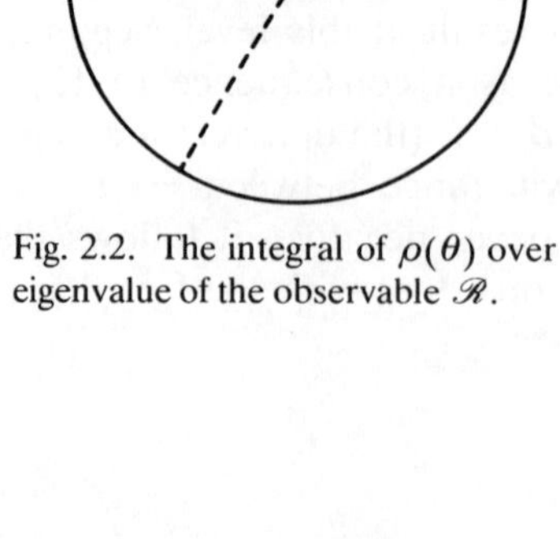

Fig. 2.2. The integral of $\rho(\theta)$ over the hatched area σ_0 gives the probability of the first eigenvalue of the observable $\mathscr{R}$.

given sphere is completely fixed by the value of $\boldsymbol{\lambda}$ for that sphere. Consequently, measurements are causal processes relative to this model, and we have obtained a causal generalization of the quantum mechanical theory of spin-$\frac{1}{2}$ systems.

Our model therefore does exactly what von Neumann's theorem forbids; and this requires that we discuss the validity of von Neumann's axioms within this model. The hidden variable involved here is the vector $\boldsymbol{\lambda}$. A dispersion-free ensemble being one with fixed $\boldsymbol{\lambda}$, let us consider an ensemble of particles all with $\boldsymbol{\lambda}$ pointing along the bisector of the *yz* quadrant and measure on one part of this ensemble the observable σ_2, on a second part the osbervable σ_3, and on a third part the observable $\boldsymbol{\sigma} \cdot \mathbf{m}$ with $\mathbf{m} = (0, 1, 1)$. Note that these observables have been chosen so as to satisfy

$$\sigma_2 + \sigma_3 = \boldsymbol{\sigma} \cdot \mathbf{m}.$$

Since the $\boldsymbol{\lambda}$'s of our dispersion-free ensemble form acute angles with the y and the z axes and since they lie along the $\mathbf{m}$ direction, the result $\alpha + |\boldsymbol{\beta}|$ will be obtained in all three foregoing types of measurements. Therefore, a measurement

$$\text{of } \sigma_2 \text{ always gives } +1 \Rightarrow \langle \sigma_2 \rangle = +1;$$

$$\text{of } \sigma_3 \text{ always gives } +1 \Rightarrow \langle \sigma_3 \rangle = +1;$$

$$\text{of } \boldsymbol{\sigma} \cdot \mathbf{m} \text{ always gives } +\sqrt{2} \Rightarrow \langle \boldsymbol{\sigma} \cdot \mathbf{m} \rangle = +\sqrt{2}.$$

The conclusion is obviously that $\langle \boldsymbol{\sigma} \cdot \mathbf{m} \rangle \neq \langle \sigma_2 \rangle + \langle \sigma_3 \rangle$, and this means that *axiom N3 is not satisfied for our dispersion-free ensemble.*

To formulate and understand the foregoing model is perhaps rather easy, but one should not forget that it took the efforts of physicists like Einstein, de Broglie, Bohm, Bell, and others, working in hostile cultural environments to give us today the possibility of a search for a causal completion of quantum mechanics. The sentiments that these scientists must have harbored against the acausal world outlook are in our opinion well represented by the following statement made by Einstein in 1935:

> I cannot bear the thought that an electron exposed to a ray should by its own free decision choose the moment and the direction in which it wants to jump away. If so, I'd rather be a cobbler or even an employee in a gambling house than a physicist. (Born, 1969, p. 160.)

More complete and rigorous proofs of von Neumann's theorem are presented in the following three sections.

2.6. Von Neumann's Theorem: Assumptions, Definitions, and Results

The basic assumptions used by von Neumann in order to prove his theorem in the general case are the following:

NI *To every linear Hermitian operator in the Hilbert space of a physical system corresponds one and only one observable of that system.*

NII *If the observable $\mathscr{R}$ corresponds to the operator R, then the observable $f(\mathscr{R})$ corresponds to $f(R)$.*

NIII *If $\mathscr{R}$, $\mathscr{S}$, etc., are arbitrary observables and a, b, etc., real numbers, one has, for the expectation values,*

$$\langle a\mathscr{R} + b\mathscr{S} + \cdots \rangle = a\langle \mathscr{R} \rangle + b\langle \mathscr{S} \rangle + \cdots.$$

Concerning assumption NI, note that von Neumann assumed a one-to-one correspondence between observables and operators. Our less stringent assumption is however all he really used, and this leaves open the possibility of describing hidden variables differently from the regular quantum mechanical observables, which are represented by operators.

In order to make the above statements as clear as possible, we recall the following definitions:

D1. *Definition of a function of an observable.* Given an observable $\mathscr{R}$ and a function $f(x)$, we define $f(\mathscr{R})$ in the following operational manner: to measure $f(\mathscr{R})$ one starts by measuring $\mathscr{R}$; if the value r is found for $\mathscr{R}$, one then says that $f(\mathscr{R})$ has the value $f(r)$.

D2. *Definition of a function of an operator.* Given an operator R and a function $f(x)$, the entity $f(R)$ is defined by the well-known development

$$f(R) = \sum_{n=0}^{\infty} \left(\frac{1}{n!}\right) f^{(n)}(0) R^n.$$

The nth power of an operator can be defined starting from the $(n-1)$st power: given the operators R and R^{n-1} and an *arbitrary* vector $|\psi\rangle$ of Hilbert space, one defines

$$|\psi_{n-1}\rangle = R^{n-1}|\psi\rangle,$$

$$|\psi_n\rangle = R|\psi_{n-1}\rangle.$$

Then R^n is an operator such that

$$|\psi_n\rangle = R^n|\psi\rangle.$$

By means of a recurrence procedure one can thus define an arbitrary power of R.

D3. *Definition of the sum of operators.* If the operators R, S, etc., applied to the arbitrary vector $|\psi\rangle$ give rise to the vectors $|\rho\rangle$, $|\sigma\rangle$, etc.:

$$R|\psi\rangle = |\rho\rangle, S|\psi\rangle = |\sigma\rangle, \ldots.$$

then the operator $(R + S + \cdots)$ is defined by

$$(R + S + \cdots)|\psi\rangle = |\rho\rangle + |\sigma\rangle + \cdots$$

for any $|\psi\rangle$.

D4. *Definition of the sum of observables.* If the observables $\mathscr{R}$, $\mathscr{S}$, etc. correspond to the operators R, S, etc., respectively, then $(\mathscr{R} + \mathscr{S} + \cdots)$ is defined to be the observable corresponding to $(R + S + \cdots)$. Such an observable exists by virtue of assumption NI because $(R + S + \cdots)$ is a linear Hermitian operator if R, S, etc., all are.

D5. *Definition of linear combinations of operators and observables.* The definition of the product of a real constant a with an operator R can be obtained as a particular case of definition D2 with $f(x) = ax$. Having so defined aR, bS, etc., one then obtains from D3 the definition of $(aR + bS + \cdots)$. Similarly, one can define $(a\mathscr{R} + b\mathscr{S} + \cdots)$ with the aid of D1 and D4.

In what follows we shall also need the definitions of dispersion-free and homogeneous ensembles:

D6. *Definition of dispersion-free ensembles.* A statistical ensemble is called *dispersion-free* if it requires:

$$\langle \mathscr{R} \rangle^2 = \langle \mathscr{R}^2 \rangle$$

for an arbitrary observable $\mathscr{R}$. (This point was discussed in some detail in Section 2.4 of the present chapter.)

D7. *Definition of a homogeneous ensemble.* A statistical ensemble is called *homogeneous* if every one of its subensembles gives the same expectation value for any particular observable.

Starting from assumptions NI, NII, and NIII above, von Neumann obtained the following results, which together constitute his famous theorem:

R1 *Given a statistical ensemble, a Hermitian operator ρ exists such that, for an arbitrary observable $\mathscr{R}$, one has*

$$\langle \mathscr{R} \rangle = \mathrm{Tr}(\rho R),$$

where R is the operator corresponding to $\mathscr{R}$.

R2 *The operator ρ is positive-definite, meaning that*

$$\langle \eta | \rho | \eta \rangle \geqslant 0$$

for an arbitrary state $|\eta\rangle$.

R3 *The operator ρ is unique for a given statistical ensemble.*
R4 *There exist no dispersion-free ensembles.*
R5 *A statistical ensemble is homogeneous if and only if a normalized vector $|\phi\rangle$ exists such that*

$$\rho = P_\phi,$$

where $P_\phi = |\phi\rangle\langle\phi|$.

In the following section we shall prove the above results.

2.7. General Proof of von Neumann's Theorem

Given a complete orthonormal set of vectors $\{|n\rangle\}$ and an arbitrary operator R, one can always write

$$R = \sum_{m,n} |n\rangle R_{nm} \langle m|, \tag{2.13}$$

where

$$R_{nm} = \langle n|R|m\rangle. \tag{2.14}$$

Assume R to be Hermitian and thus to correspond to an observable $\mathscr{R}$. In this case one has

$$R^*_{mn} = \langle m|R|n\rangle^* = R_{nm},$$

whence

$$\begin{aligned} \mathrm{Re}(R_{mn}) &= \mathrm{Re}(R_{nm}), \\ \mathrm{Im}(R_{mn}) &= -\mathrm{Im}(R_{nm}). \end{aligned} \tag{2.15}$$

From (2.13) and (2.15) it then follows that

$$\begin{aligned} R &= \sum_n |n\rangle\langle n|R_{nn} + \sum_{m<n} \{|n\rangle\langle m|R_{nm} + |m\rangle\langle n|R_{mn}\} \\ &= \sum_n |n\rangle\langle n|R_{nn} + \sum_{m<n} \{|n\rangle\langle m| + |m\rangle\langle n|\}\,\mathrm{Re}(R_{nm}) \\ &+ \sum_{m<n} \{i|n\rangle\langle m| - i|m\rangle\langle n|\}\mathrm{Im}(R_{nm}). \end{aligned}$$

On defining the Hermitian operators

$$U_{nn} = |n\rangle\langle n|, \tag{2.16}$$

$$V_{nm} = |n\rangle\langle m| + |m\rangle\langle n|, \tag{2.17}$$

$$W_{nm} = i(|n\rangle\langle m| - |m\rangle\langle n|), \tag{2.18}$$

this outcome can be written as

$$R = \sum_{n} U_{nn} R_{nn} + \sum_{m<n} \{ V_{nm} \operatorname{Re}(R_{nm}) + W_{nm} \operatorname{Im}(R_{nm}) \}. \tag{2.19}$$

By assumption NI there are observables $\mathscr{U}_{nn}$, $\mathscr{V}_{nm}$, $\mathscr{W}_{nm}$ corresponding to the Hermitian operators U_{nn}, V_{nm}, W_{nm}. From the assumed nature of the correspondence between observables and Hermitian operators, (2.19) then implies that

$$\mathscr{R} = \sum_{n} \mathscr{U}_{nn} R_{nn} + \sum_{m<n} \{ \mathscr{V}_{nm} \operatorname{Re}(R_{nm}) + \mathscr{W}_{nm} \operatorname{Im}(R_{nm}) \}. \tag{2.20}$$

In fact, if (2.20) were not valid, there would be two observables corresponding to R: one would be $\mathscr{R}$ and the other, because of (2.19), the right-hand side of (2.20).

From (2.20), on applying NIII, one gets

$$\langle \mathscr{R} \rangle = \sum_{n} R_{nn} \langle \mathscr{U}_{nn} \rangle + \sum_{m<n} \{ \operatorname{Re}(R_{nm}) \langle \mathscr{V}_{nm} \rangle + \operatorname{Im}(R_{nm}) \langle \mathscr{W}_{nm} \rangle \}. \tag{2.21}$$

It is easy to prove that one can equivalently write

$$\langle \mathscr{R} \rangle = \sum_{n} R_{nn} \langle \mathscr{U}_{nn} \rangle + \sum_{m<n} \{ \tfrac{1}{2} R_{nm} (\langle \mathscr{V}_{nm} \rangle - i \langle \mathscr{W}_{nm} \rangle) + \tfrac{1}{2} R^*_{nm} (\langle \mathscr{V}_{nm} \rangle + i \langle \mathscr{W}_{nm} \rangle) \},$$

which, in terms of the quantities defined by

$$\rho_{nn} = \langle \mathscr{U}_{nn} \rangle, \tag{2.22a}$$

$$\rho_{mn} = \tfrac{1}{2} (\langle \mathscr{V}_{nm} \rangle - i \langle \mathscr{W}_{nm} \rangle), \tag{2.22b}$$

$$\rho_{nm} = \tfrac{1}{2} (\langle \mathscr{V}_{nm} \rangle + i \langle \mathscr{W}_{nm} \rangle), \tag{2.22c}$$

gives rise to

$$\begin{aligned}
\langle \mathscr{R} \rangle &= \sum_{n} \rho_{nn} R_{nn} + \sum_{m<n} \rho_{mn} R_{nm} + \sum_{m<n} \rho_{nm} R^*_{nm} \\
&= \sum_{n} \rho_{nn} R_{nn} + \sum_{m<n} \rho_{mn} R_{nm} + \sum_{m<n} \rho_{nm} R_{mn} \\
&= \sum_{n} \rho_{nn} R_{nn} + \sum_{m<n} \rho_{mn} R_{nm} + \sum_{n<m} \rho_{mn} R_{nm} \\
&= \sum_{n} \rho_{nn} R_{nn} + \sum_{m \neq n} \rho_{mn} R_{nm} \\
&= \sum_{m,n} \rho_{mn} R_{nm}.
\end{aligned} \tag{2.23}$$

If one finally defines an operator ρ such that

$$\rho_{mn} = \langle m | \rho | n \rangle, \tag{2.24}$$

this outcome can obviously be written as

$$\langle \mathscr{R} \rangle = \mathrm{Tr}(\rho R). \tag{2.25}$$

The first point (R1) of von Neumann's theorem has thus been proved.

We next prove that ρ must be positive-definite (point R2).

Consider any state $| \phi \rangle$ such that $\langle \phi | \phi \rangle = 1$ and let

$$P_\phi = | \phi \rangle \langle \phi |. \tag{2.26}$$

If $\mathscr{P}_\phi$ is the observable corresponding to P_ϕ, then, by assumption NII, $\mathscr{P}_\phi^2$ corresponds to P_ϕ^2. One can thus write

$$\begin{aligned} \langle \mathscr{P}_\phi^2 \rangle &= \mathrm{Tr}(\rho P_\phi^2) = \mathrm{Tr}(\rho P_\phi) \\ &= \langle \phi | \rho | \phi \rangle \end{aligned}$$

because $P_\phi^2 = P_\phi$.

But $\mathscr{P}_\phi^2$ is a positive-definite quantity and its average value is thus certainly nonnegative, i.e., $\langle \mathscr{P}_\phi^2 \rangle \geqslant 0$. Consequently

$$\langle \phi | \rho | \phi \rangle \geqslant 0. \tag{2.27}$$

Finally, we show that ρ is unique for a given statistical ensemble. In fact, if one had two different statistical operators ρ and ρ', Equation (2.25) would require that

$$\mathrm{Tr}(\rho R) = \mathrm{Tr}(\rho' R)$$

for an arbitrary Hermitian operator R. Taking, in particular, $R = P_\phi$ with P_ϕ defined in (2.26), one would get

$$\langle \phi | \rho | \phi \rangle = \langle \phi | \rho' | \phi \rangle.$$

But, since $| \phi \rangle$ is an arbitrary (normalized) vector, it follows from the above relation that $\rho = \rho'$. [Actually this conclusion is a consequence of Equation (2.29) below, taken together with $R = \rho - \rho'$ and $| m \rangle = | \phi \rangle$: Because $\langle \phi | (\rho - \rho') | \phi \rangle = 0$, Equation (2.29) implies that $\langle n | (\rho - \rho') | \phi \rangle = 0$, whence $(\rho - \rho') | \phi \rangle = 0$, since $| n \rangle$ is arbitrary. But, seeing that $| \phi \rangle$ is also arbitrary, this means that $\rho - \rho' = 0$.] Hence ρ is unique. A similar derivation can be used to show that two statistical ensembles having equal average values for *all* observables are described by the same ρ.

The next point we need to demonstrate is the nonexistence of dispersion-free ensembles. If such ensembles existed one would have

$$\langle \mathscr{R}^2 \rangle = \langle \mathscr{R} \rangle^2 \tag{2.28}$$

for an arbitrary observable $\mathscr{R}$. If R is the corresponding operator, it would follow from (2.28) that

$$\mathrm{Tr}(\rho R^2) = \{\mathrm{Tr}(\rho R)\}^2.$$

In particular, for $R = P_\phi$ with $|\phi\rangle$ again arbitrary,

$$\mathrm{Tr}(\rho P_\phi) = \{\mathrm{Tr}(\rho P_\phi)\}^2$$

since $P_\phi^2 = P_\phi$. Therefore

$$\langle\phi|\rho|\phi\rangle = \{\langle\phi|\rho|\phi\rangle\}^2,$$

whence

$$\langle\phi|\rho|\phi\rangle = 0 \text{ or } 1.$$

Suppose now that $|\phi'\rangle$ and $|\phi''\rangle$ are two vectors such that $\langle\phi'|\phi'\rangle = 1 = \langle\phi''|\phi''\rangle$. One can always vary $|\phi\rangle$ continuously from $|\phi'\rangle$ to $|\phi''\rangle$. In so doing, $\langle\phi|\rho|\phi\rangle$ is also varied continuously. But, since this matrix element can only be 0 or 1, one concludes that it will always remain zero or always unity, the case where it alternates between these two values being excluded. In other words, the statistical operator of a dispersion-free ensemble satisfies necessarily one of the two following conditions:

(a) $\langle\phi|\rho|\phi\rangle = 0$ for all vectors $|\phi\rangle$, whence $\rho = 0$. (This follows from Equation (2.29) below with $R = \rho$ and $|m\rangle = |\phi\rangle$.)
(b) $\langle\phi|\rho|\phi\rangle = 1$ for all vectors $|\phi\rangle$, whence $\rho = 1$. (This follows from Equation (2.29) below with $R = \rho - 1$ and $|m\rangle = |\phi\rangle$.)

The case (b) above can readily be excluded. In fact, $\rho = 1$ does not represent a dispersion-free ensemble, because it is not true that $\mathrm{Tr}(R^2) = \{\mathrm{Tr}(R)\}^2$ for an arbitrary operator R.

Only the case $\rho = 0$ is left, and it gives

$$\langle\mathscr{R}\rangle = \mathrm{Tr}(0) = 0$$

for an arbitrary observable. At most this can represent the vacuum state, since no physical system can have a zero expectation value for all its observables. We have therefore proved that no *physical* ensemble can be dispersion-free.

Finally, statement R5 of von Neumann remains to be proved.

We begin by establishing first that *if a statistical ensemble is homogeneous, then a normalized vector $|\phi\rangle$ exists such that $\rho = P_\phi$* $\equiv |\phi\rangle\langle\phi|$. Let ρ be the statistical operator of an ensemble that is homogeneous by hypothesis. Of course, $\rho \neq 0$, otherwise the ensemble would not represent physical systems, as we just saw. The statement $\rho \neq 0$ equivalently asserts that at least one vector $|f_0\rangle$ different from zero exists such that $\rho|f_0\rangle \neq 0$. It is easy to see that, for a vector of this kind, one has $\langle f_0|\rho|f_0\rangle > 0$. In fact, we already know that

$\langle f_0|\rho|f_0\rangle \geqslant 0$ since ρ is positive-definite. But the Schwarz inequality

$$|\langle n|R|m\rangle|^2 \leqslant \langle n|R|n\rangle\langle m|R|m\rangle, \tag{2.29}$$

which is valid for arbitrary $|n\rangle$, $|m\rangle$, and R, when applied to the case $R = \rho$, $|m\rangle = |f_0\rangle$, reveals that $\langle f_0|\rho|f_0\rangle = 0$ would correspond to $\langle n|\rho|f_0\rangle = 0$. And this result for arbitrary $|n\rangle$ would in turn imply $\rho|f_0\rangle = 0$, contrary to hypothesis.

From $\langle f_0|\rho|f_0\rangle > 0$ it follows that the operators ρ_1' and ρ_2', defined by

$$\rho_1' = \frac{\rho|f_0\rangle\langle f_0|\rho}{\langle f_0|\rho|f_0\rangle}, \tag{2.30}$$

$$\rho_2' = \rho - \frac{\rho|f_0\rangle\langle f_0|\rho}{\langle f_0|\rho|f_0\rangle}, \tag{2.31}$$

are positive-definite. In fact, given an arbitrary $|f\rangle$, one has

$$\langle f|\rho_1'|f\rangle = \frac{\langle f|\rho|f_0\rangle\langle f_0|\rho|f\rangle}{\langle f_0|\rho|f_0\rangle} = \frac{|\langle f|\rho|f_0\rangle|^2}{\langle f_0|\rho|f_0\rangle} \geqslant 0$$

and

$$\langle f|\rho_2'|f\rangle = \langle f|\rho|f\rangle - \frac{|\langle f|\rho|f_0\rangle|^2}{\langle f_0|\rho|f_0\rangle}$$

$$= \frac{\langle f|\rho|f\rangle\langle f_0|\rho|f_0\rangle - |\langle f|\rho|f_0\rangle|^2}{\langle f_0|\rho|f_0\rangle} \geqslant 0,$$

the last inequality being a consequence of (2.29).

We have therefore proved that one can write

$$\rho = \rho_1' + \rho_2', \tag{2.32}$$

where ρ_1' and ρ_2' are positive-definite operators. In general, however, $\mathrm{Tr}(\rho_1') \neq 1$ and $\mathrm{Tr}(\rho_2') \neq 1$, so that ρ_1' and ρ_2' are not density operators. It is therefore useful to introduce two operators ρ_1 and ρ_2, such that

$$\rho_1' = \alpha\rho_1 \quad \text{and} \quad \rho_2' = \beta\rho_2 \quad (\alpha > 0, \beta > 0),$$

which do have a unit trace: $\mathrm{Tr}(\rho_1) = \mathrm{Tr}(\rho_2) = 1$. One can then write

$$\rho = \alpha\rho_1 + \beta\rho_2, \tag{2.33}$$

and, if the trace on both sides of (2.33) is taken, the outcome is

$$\alpha + \beta = 1. \tag{2.34}$$

Notice that, if we set

$$|\phi\rangle = \frac{\rho|f_0\rangle}{\|\rho|f_0\rangle\|}, \tag{2.35}$$

then $\langle\phi|\phi\rangle = 1$. It furthermore follows from (2.30) that $\rho_1' = c|\phi\rangle\langle\phi|$, whence $\rho_1 = |\phi\rangle\langle\phi|$ and $c = \alpha$ because $|\phi\rangle\langle\phi|$ has unit trace. This shows that there exists a statistical ensemble E_1 for which ρ_1 is the density operator, to wit, a pure sample with state vector $|\phi\rangle$.

Is it possible to find an ensemble corresponding to the density operator ρ_2? von Neumann did not answer this important question. In his theorem he seems simply to *assume* that for every positive-definite operator of unit trace, such as ρ_2, there exists a statistical ensemble having that operator as density operator. In any event, in order to proceed, we shall assume, with von Neumann, that there is such a one-to-one correspondence between statistical ensembles and unit-trace positive-definite operators. Let then E_2 be the ensemble corresponding to ρ_2. If one forms with E_1 and E_2 a larger ensemble E' in such a way that E_1 contains the fraction α, and E_2 the fraction β of the systems comprised by E' (this is possible because of (2.34)), it becomes immediately apparent that $\alpha\rho_1 + \beta\rho_2 = \rho$ (see (2.33)) is the density operator of E'. But ρ is also the density operator of our initial ensemble E, which is homogeneous. Since E and E' are described by the same density operator, they are indistinguishable experimentally and E' must be homogeneous too. This implies that for all observables one must have

$$\langle \mathscr{R} \rangle_{E'} = \langle \mathscr{R} \rangle_{E_1} = \langle \mathscr{R} \rangle_{E_2},$$

whence it follows, as we saw, that $\rho = \rho_1 = \rho_2$ and thus

$$\rho = |\phi\rangle\langle\phi|, \tag{2.36}$$

which shows that both E' and E are pure samples.

The second half of the von Neumann's statement R5 avers: *If a vector* $|\phi\rangle$ *exists such that* $\rho = P_\phi \equiv |\phi\rangle\langle\phi|$, *then the statistical ensemble described by* ρ *is homogeneous.* In order to prove this assertion, we divide the ensemble E described by ρ into two subensembles E' and E'' with statistical operators ρ_1 and ρ_2, respectively; so that, if E' and E'' contain the fractions α and β, respectively, of the systems in E, one necessarily has

$$\rho = \alpha\rho_1 + \beta\rho_2. \tag{2.37}$$

We then establish in succession the following results: if $\langle\phi|f\rangle = 0$, then (i) $\rho_1|f\rangle = 0$, (ii) $\rho_1|\phi\rangle = c|\phi\rangle$, and (iii) $\rho_1 = \rho_2 = \rho$. The last result obviously implies that the statistical ensemble is homogeneous, given the arbitrariness of the splitting of E into E' and E''.

The proof of (i) starts from the observation that, for any vector $|f\rangle$ orthogonal to $|\phi\rangle$,

$$\rho|f\rangle = |\phi\rangle\langle\phi|f\rangle = 0.$$

The fact that ρ_1 and ρ_2 are positive-definite implies:

$$0 \leqslant \alpha\langle f|\rho_1|f\rangle \leqslant \alpha\langle f|\rho_1|f\rangle + \beta\langle f|\rho_2|f\rangle = \langle f|\rho|f\rangle = 0.$$

Therefore, $\langle f|\rho_1|f\rangle = 0$, which, on substitution into Equation (2.29) with

$R = \rho_1$, $|m\rangle = |f\rangle$, and $|n\rangle$ arbitrary, gives $\langle n|\rho_1|f\rangle = 0$. The vector $|n\rangle$ being arbitrary, this means that

$$\rho_1|f\rangle = 0. \tag{2.38}$$

To prove (ii), note that what we have just shown is that any vector $|f\rangle$ perpendicular to $|\phi\rangle$ is also perpendicular to $\rho_1|n\rangle$, whatever $|n\rangle$ might be. Consider therefore a complete set of states

$$|f_1\rangle, |f_2\rangle, \ldots, |f_n\rangle \ldots,$$

with $\langle f_i|f_j\rangle = \delta_{ij}$ and $|f_1\rangle \equiv |\phi\rangle$. Then

$$\rho_1|n\rangle = \sum_i |f_i\rangle\langle f_i|\rho_1|n\rangle = \sum_i |f_i\rangle c_n\delta_{i1} = c_n|f_1\rangle = c_n|\phi\rangle.$$

In particular, for $|n\rangle = |\phi\rangle$, one has

$$\rho_1|\phi\rangle = c|\phi\rangle. \tag{2.39}$$

To establish the validity of step (iii), note first that an arbitrary vector $|h\rangle$ can be written

$$|h\rangle = |\phi\rangle\langle\phi|h\rangle + |h'\rangle,$$

with

$$\langle h'|\phi\rangle = 0.$$

Therefore,

$$\rho_1|h\rangle = \rho_1|\phi\rangle\langle\phi|h\rangle + \rho_1|h'\rangle = \rho_1|\phi\rangle\langle\phi|h\rangle;$$

the last step here is justified because, $|h'\rangle$ being orthogonal to $|\phi\rangle$, one has $\rho_1|h'\rangle = 0$. By applying (2.39), one gets

$$\rho_1|h\rangle = c|\phi\rangle\langle\phi|h\rangle = cP_\phi|h\rangle = c\rho|h\rangle$$

since $\rho = P_\phi$. Because $|h\rangle$ is arbitrary, we deduce immediately that $\rho_1 = c\rho$. Furthermore, since ρ and ρ_1 are statistical operators, they satisfy $\mathrm{Tr}(\rho) = \mathrm{Tr}(\rho_1) = 1$. Hence, $c = 1$ and thus $\rho_1 = \rho$. It follows moreover from (2.37) that

$$\rho_2 = \beta^{-1}(\rho - \alpha\rho_1) = \beta^{-1}(1-\alpha)\rho = \rho$$

because $\alpha + \beta = 1$. We have thus proved that

$$\rho_1 = \rho_2 = \rho.$$

This result, in conjunction with Equation (2.36), concludes the proof of R5 and therefore of von Neumann's entire theorem.

The proof of R4 is obviously the most important part of this theorem. The conclusion that dispersion-free ensembles cannot exist clearly rules out hidden variables, for the latter imply the former, as we have seen in the present chapter. More accurately we should say that *hidden variables satisfying von Neumann's axioms* are ruled out.

The last point, R5, of von Neumann's theorem implies that homogeneous ensembles do exist, being those (and only those) represented as 'states' in quantum mechanics. In fact, saying that the density operator of an ensemble is $\rho = |\phi\rangle\langle\phi|$ is the same as claiming that there is a unique state vector $|\phi\rangle$ describing the ensemble. Thus the quantum-mechanical description of homogeneous ensembles is the most general one possible that accords with the general assumptions NI, NII, and NIII. Therefore, while item R4 rules out hidden variables, R5 reaffirms quantum mechanics. If one accepts von Neumann's axioms, it is consequently impossible to doubt the completeness of quantum mechanics.

It should however be stressed that, as has been known since 1952, assumption NI is not true. In fact, the so-called *superselection rules* (Wick *et al.*, 1952) provide examples of Hermitian operators that are not associated with any observables. To illustrate, consider the superselection rule for electric charge. If $|q_1\rangle$ and $|q_2\rangle$ are two *different* eigenstates of the charge operator for which all other quantum numbers are equal, then one can construct, for example, the vector

$$|\eta\rangle = \frac{1}{\sqrt{2}}(|q_1\rangle + |q_2\rangle)$$

and the operator

$$P_\eta = |\eta\rangle\langle\eta|.$$

Since P_η is Hermitian, one would expect it to represent an observable $\mathscr{P}_\eta$, and a measurement of $\mathscr{P}_\eta$ would necessarily leave the system in the state $|\eta\rangle$ whenever the result $\mathscr{P}_\eta = 1$ is found. But nobody knows how to produce, or has ever seen, a state like this: observations of whatever kind always result in *well-defined* values of electric charge. This means that the state $|\eta\rangle$ does not describe any physical reality and the operator P_η does not correspond to any measurable physical quantity. More recent proofs of von Neumann's theorem, like the one we shall examine in the following section, do not suffer from weaknesses of this sort and thus apply also to the realistic situation in which superselection rules exist.

2.8. Jauch and Piron's Theorem

A newer and more complete proof of von Neumann's theorem was given by Jauch and Piron (Jauch and Piron, 1963; Piron, 1963). They start by observing that the measurement of any observable quantity can be reduced to the determination of a series of yes—no experiments. Such yes—no experiments (also called *propositions*) are taken as the basic entities for building a mathematical formalism. The latter should, of course, be general enough to include quantum mechanics and its conceivable hidden-variable generalizations. According to Jauch and Piron, this generality is ensured by the fact that the propositions of a physical system have a logical structure quite independent of the particular fact that in quantum mechanics they are represented by projection operators.

Let the set of all the propositions of a physical system be denoted by $\mathscr{L}$. Within $\mathscr{L}$ a partial ordering is given by the implication (a true) $\Rightarrow$ (b true), where a and b are any two of its propositions. The foregoing implication is briefly indicated by writing $a \leqslant b$. The notation $a < b$ is also equivalent to (a true) $\Rightarrow$ (b true), but with the added condition that a and b are necessarily different propositions.

The first two axioms of Jauch and Piron are given by:

JPI $a \leqslant b$ *and* $b \leqslant a \Rightarrow a = b$.
JPII $a \leqslant b$ *and* $b \leqslant c \Rightarrow \alpha \leqslant c$.

With any pair of propositions a and b of $\mathscr{L}$ one can associate two other propositions of $\mathscr{L}$:

$a \wedge b$ (a and b are simultaneously true);

$a \vee b$ (either a or b is true).

The next two axioms are, in practice, formal definitions of $a \wedge b$ and $a \vee b$. They are, with x and y denoting propositions of $\mathscr{L}$:

JPIII $x \leqslant a \wedge b$ *is always equivalent to* $x \leqslant a$ *and* $x \leqslant b$.
JPIV $a \vee b \leqslant y$ *is always equivalent to* $a \leqslant y$ *and* $b \leqslant y$.

In this scheme it would be absurd for all possible propositions to be simultaneously true, while trivially some proposition must be true. The definitions of the absurd proposition ϕ and of the trivial proposition I are therefore

$$\phi \equiv \bigwedge_{a_i \in \mathscr{L}} a_i, \quad I \equiv \bigvee_{a_i \in \mathscr{L}} a_i.$$

With every proposition a is associated another proposition a' signifying 'non a'. Concering a', one assumes:

JPV $(a')' = a$.
JPVI $a \vee a' = I, a \wedge a' = \phi$.
JPVII $(a \vee b)' = a' \wedge b'$.

In the foregoing discussion there is nothing that distinguishes classical physics from quantum mechanics. A characterization of classical propositions was established by Birkhoff and von Neumann (1938) who showed that classical systems obey a *distributive law* which may be stated in two equivalent forms:

$$(\mathrm{D})\quad \begin{cases} a \wedge (b \vee c) = (a \wedge b) \vee (a \wedge c), \\ a \vee (b \wedge c) = (a \vee b) \wedge (a \vee c). \end{cases}$$

When any three propositions a, b, and c of a physical system satisfy (D), we say that the system is *classical*; and the logical structure obtained by adding (D) to the JP postulates is that of a *Boolean* algebra.

The characterization of quantum mechanical systems is much less direct; it cannot be achieved simply by allowing some propositions of $\mathscr{L}$ to violate the law (D). In fact, the basic feature of quantum mechanical observables is that sometimes they fail to commute. And noncommutativity is a property of *two* operators, e.g., projection operators corresponding to our yes—no propositions, while (D) is formulated for three propositions.

The way Jauch and Piron go about solving this problem is first, to introduce the concept of compatibility. Two propositions a and b are said to be *compatible* if and only if they satisfy

(C) $(a \wedge b') \vee b = (b \wedge a') \vee a.$

The shorthand notation $a \leftrightarrow b$ is used to signify this compatibility property. Relation (C) has exactly the properties one would associate with measurements that can be performed simultaneously without disturbing each other: in particular it implies that the projection operators P_a and P_b, which in quantum mechanics correspond to the propositions a and b, respectively, commute. The axiom characteristic for quantum systems thus is:

JPVIII *$a \leqslant b$ implies $a \leftrightarrow b$.*

A lattice satisfying axioms JPI through JPVIII is called a *generalized proposition system*. Such a system admits superselection rules, which are introduced as follows.

We say that a proposition $x \in \mathscr{L}$ *belongs to the center* Z *of* $\mathscr{L}$ if it is compatible with every other proposition in $\mathscr{L}$. The center always contains ϕ and I. If the center contains nothing else, it is characterized as *trivial*; and a lattice $\mathscr{L}$ is *coherent* when it has a trivial center. If $\mathscr{L}$ has a nontrivial center, one calls it *reducible* and superselection rules are said to exist. This definition is easily understood if one recalls that an observable admitting superselection rules commutes with every other observable.

Next, Jauch and Piron introduce the concept of *state*, which they interpret as a probability function $\omega(x)$ defined for all $x \in \mathscr{L}$. Such a function is assumed to satisfy:

S1 $0 \leqslant \omega(a) \leqslant 1.$

S2 $\omega(\phi)=0, \omega(I)=1$.
S3 *If* $a \leftrightarrow b$, *then* $\omega(a)+\omega(b)=\omega(a \wedge b)+\omega(a \vee b)$.
S4 *If* $\omega(a)=\omega(b)=1$, *then* $\omega(a \wedge b)=1$.
S5 *If* $a \neq \phi$, *then a state exists such that* $\omega(a) \neq 0$.

It is not difficult to see that quantum mechanical states satisfy S1 through S5. Whether these conditions are so general that any conceivable hidden-variable theory must satisfy them is a very important question, which we shall discuss later.

Two states ω_1 and ω_2 are different if a proposition a exists such that $\omega_1(a) \neq \omega_2(a)$. If $\omega_1 \neq \omega_2$ and $\lambda_1 > 0, \lambda_2 > 0, \lambda_1 + \lambda_2 = 1$, then

$$\omega(a) = \lambda_1\omega_1(a) + \lambda_2\omega_2(a)$$

defines a new state different from ω_1 and ω_2. It is, in fact, easy to see that such a ω satisfies S1 through S5 if ω_1 and ω_2 do. A state ω of this type is known as a *mixture*. If a state is not a mixture it is called *pure*. The *dispersion* of a state ω on a proposition a is defined by

$$\sigma(a) = \omega(a) - \omega^2(a). \tag{2.40}$$

A state ω is said to be *dispersion-free* if one has $\sigma(a) = 0$ for all $a \in \mathscr{L}$. For a dispersion-free state one has either $\omega(a) = 0$ or $\omega(a) = 1$. This means that any proposition a is either certainly false (which means that the eigenvalue of P_a is zero) or certainly true (which means that the eigenvalue of P_a is unity). In either case, the observable corresponding to P_a has a fixed value, which is the very definition of a dispersion-free state. It is easy to see that every mixture has a nonvanishing dispersion. Therefore a dispersion-free state is always pure.

LEMMA I. *If a proposition system* $\mathscr{L}$ *admits hidden variables and if* $\omega(a)=\omega(b)$ *for all states, then* $a=b$.

Proof. Given the proposition $x = a \wedge (a \wedge b)'$, one can show that $x = \phi$. In fact, by S5, if $x \neq \phi$, a state exists such that $\omega(x) \neq 0$. But since every state is a mixture of dispersion-free states if hidden variables exist — as we are assuming — a dispersion-free state ω_0 must exist such that $\omega_0(x) \neq 0$, which implies that $\omega_0(x) = 1$. Because $x = a \wedge (a \wedge b)' \leqslant a$ and $\omega_0(x) = 1$, it follows that $\omega_0(a) = 1$. But, by hypothesis, $\omega_0(a) = \omega_0(b)$. Hence $\omega_0(a) = \omega_0(b) = 1$. By S4 it then follows that $\omega_0(a \wedge b) = 1$ and therefore $\omega_0((a \wedge b)') = 0$. On the other hand, $x = a \wedge (a \wedge b)' \leqslant (a \wedge b)'$. Consequently, $\omega_0(x) = 1$ implies $\omega_0((a \wedge b)') = 1$. We thus arrive at a contradiction, which means that

$$x = \phi = a \wedge (a \wedge b)'.$$

But $\phi \vee (a \wedge b) = a \wedge b$ (ϕ is always false). Hence

$$(a \wedge (a \wedge b)') \vee (a \wedge b) = a \wedge b.$$

On the other hand, it is not difficult to see that

$$(a \wedge (a \wedge b)') \vee (a \wedge b) = a.$$

Therefore $a = a \wedge b$. It follows that $a \leqslant b$. The foregoing reasoning can easily be repeated under interchange of a and b, introducing $y \equiv b \wedge (b \wedge a)'$, and so on. One then finds $y = \phi$ and $b \leqslant a$. Since $a \leqslant b$ and $b \leqslant a$, JPI finally leads to $b = a$.

LEMMA II. *If a proposition system $\mathscr{L}$ admits hidden variables, then, for any pair of propositions a, b $\in \mathscr{L}$ and any state ω,*

$$\omega(a) + \omega(b) = \omega(a \wedge b) + \omega(a \vee b). \tag{2.41}$$

Proof. Since by hypothesis, $\mathscr{L}$ admits hidden variables, every state ω is of the form

$$\omega(a) = \sum_i \lambda_i \omega_{0i}(a),$$

where $\lambda_i > 0$ (all i), $\Sigma_i \lambda_i = 1$, and the ω_{0i} are dispersion-free states. Since (2.41) is a linear relation, it is true for all regular states if it is true for all dispersion-free states. It therefore suffices to prove (2.41) for dispersion-free states.

For any dispersion-free state ω_0 and any two propositions a and b, there are four possibilities:

$$\begin{aligned} &\omega_0(a) = 0, \qquad \omega_0(b) = 0,\\ &\omega_0(a) = 1, \qquad \omega_0(b) = 0,\\ &\omega_0(a) = 0, \qquad \omega_0(b) = 1,\\ &\omega_0(a) = 1, \qquad \omega_0(b) = 1. \end{aligned} \tag{2.42}$$

Notice that if

$$\omega_0(a') + \omega_0(b') = \omega_0(a' \vee b') + \omega_0(a' \wedge b')$$

then

$$1 - \omega_0(a') + 1 - \omega_0(b') = 1 - \omega_0(a' \vee b') + 1 - \omega_0(a' \wedge b')$$

or, by JPVII,

$$\begin{aligned} \omega_0(a) + \omega_0(b) &= 1 - \omega_0((a \wedge b)') + 1 - \omega_0((a \vee b)')\\ &= \omega_0(a \wedge b) + \omega_0(a \vee b). \end{aligned}$$

Consequently, if (2.41) holds for a', b' it necessarily holds for a, b. The last two lines in (2.42) can thus be reduced to the first two, since $\omega_0(a) = 0$, $\omega_0(b) = 1$ implies $\omega_0(a') = 1$, $\omega_0(b') = 0$ and $\omega_0(a) = \omega_0(b) = 1$ implies

$\omega_0(a') = \omega_0(b') = 0$. It is thus sufficient to prove Lemma II for the first two cases of (2.42), provided this can be done for arbitrary a and b.

If $\omega_0(a) = \omega_0(b) = 0$, it follows that $\omega_0(a') = \omega_0(b') = 1$. By applying S4, one gets $\omega_0(a' \wedge b') = 1$. Therefore,

$$\omega_0(a \vee b) = 1 - \omega_0((a \vee b)') = 1 - \omega_0(a' \wedge b') = 0.$$

Furthermore,

$$\omega_0(a \wedge b) \leqslant \omega_0(a) = 0 \Rightarrow \omega_0(a \wedge b) = 0,$$

because '$a \wedge b$ true' implies 'a true', which necessarily means that $\omega_0(a \wedge b) \leqslant \omega_0(a)$. Relation (2.41) is then valid as all its terms are zero.

If $\omega_0(a) = 1$, $\omega_0(b) = 0$, we can reason as follows. Obviously

$$\omega_0(a \vee b) \geqslant \omega_0(a) = 1 \Rightarrow \omega_0(a \vee b) = 1,$$

because 'a true' implies '$a \vee b$ true', which necessarily means that $\omega_0(a) \leqslant \omega_0(a \vee b)$. Furthermore,

$$0 = \omega_0(b) \geqslant \omega_0(a \wedge b) \Rightarrow \omega_0(a \wedge b) = 0.$$

Therefore, (2.41) can be written $1 + 0 = 1 + 0$, which is true.

THEOREM. *If the proposition system $\mathscr{L}$ admits hidden variables, then, for any two propositions $a, b \in \mathscr{L}$, one has $a \leftrightarrow b$.*

Proof. Given an arbitrary state ω, one can write, by thrice applying Lemma II,

$$\begin{aligned}
\omega((a \wedge b') \vee b) &= \omega(a \wedge b') + \omega(b) - \omega((a \wedge b') \wedge b) \\
&= \omega(a \wedge b') + \omega(b) \\
&= \omega(a) + \omega(b') - \omega(a \vee b') + \omega(b) \\
&= 1 + \omega(a) - \omega(a \vee b') \\
&= \omega(a) + \omega((a \vee b')') \\
&= \omega(a) + \omega(a' \wedge b) \\
&= \omega(a) + \omega(a' \wedge b) - \omega((a' \wedge b) \wedge a) \\
&= \omega((a' \wedge b) \vee a).
\end{aligned}$$

We have thus proved that, for all states,

$$\omega((a \wedge b') \vee b) = \omega((b \wedge a') \vee a).$$

Since $\mathscr{L}$ admits hidden variables, it follows from Lemma I that

$$(a \wedge b') \vee b = (b \wedge a') \vee a,$$

which is the very definition of $a \leftrightarrow b$.

The foregoing theorem rules out hidden variables, because it is well known that incompatible propositions do exist for a quantum mechanical system. For instance, the propositions

$a \equiv$ an electron has a coordinate in the interval $(x, x + \Delta x)$

and

$b \equiv$ an electron has momentum in the interval $(p, p + \Delta p)$ are not compatible if $\Delta x\, \Delta p < \hbar$.

Notice that no hypothesis about the center of $\mathscr{L}$ was necessary in order to prove the last theorem. Accordingly, the center is allowed to be nontrivial, and, as a consequence, the physical system can admit superselection rules. In this sense, therefore, the theorem of Jauch and Piron is more general than that of von Neumann, which is not valid if superselection rules are present.

In order to assess the generality of Jauch and Piron's theorem, we must discuss its axioms. Those designated as JPI through JPVII supply the mathematical foundations of the propositional calculus and should be general enough to cover the logical structure of hidden variables. Axiom JPVIII is a very natural assumption, as it is inconceivable that b could be true every time a is true unless a and b are compatible. Only axioms S3 and S4 remain as candidates for possible criticism, since S1, S2, and S5 are beyond suspicion, given their very simple and natural structure.

Axiom S3 is valid for classical physics as well as for compatible observables of quantum mechanics. We illustrate this with two examples. Classically, consider the throwing of two dice. Let a be the proposition 'the outcome is between 3 and 7' and b the proposition 'the outcome is between 5 and 10'. Then $a \wedge b$ stands for 'the outcome is between 5 and 7' and $a \vee b$ for 'the outcome is between 3 and 10'. Thus, if ω_i is the probability of the outcome i, with $3 \leqslant i \leqslant 10$, then we have

$$\omega(a) = \sum_{i=3}^{7} \omega_i, \ \omega(b) = \sum_{i=5}^{10} \omega_i, \ \omega(a \wedge b) = \sum_{i=5}^{7} \omega_i, \ \omega(a \vee b) = \sum_{i=3}^{10} \omega_i .$$

Relation S3 is obviously satisfied.

Quantum-mechanically, consider an electron in the xy plane. Let a be the proposition 'the electron is in the region R_1' and b the proposition 'the electron is in the region R_2'. These are certainly compatible propositions, as they should. If $R = R_1 \cup R_2$ and $R' = R_1 \cap R_2$, then $a \wedge b$ states that the electron is in R' and $a \vee b$ that it is in R. The probabilities $\omega(a)$, $\omega(b)$, $\omega(a \wedge b)$, and $\omega(a \vee b)$ are given by integrals of the squared modulus of the electron wave function over the regions R_1, R_2, R', and R, respectively. It is a simple matter to check that S3 is indeed satisfied.

The least that can be said about S3 is that one has every reason to expect that it will be valid for hidden-variable models.

Thus there remains only S4 to discuss. Notice that S4 is not an axiom but a consequence of S3 if $a \leftrightarrow b$. In fact, if $\omega(a) = \omega(b) = 1$ and $a \leftrightarrow b$, it follows from S3 that $\omega(a \wedge b) + \omega(a \vee b) = 2$, whence $\omega(a \wedge b) = \omega(a \vee b) = 1$. Consequently, $\omega(a) = \omega(b) = 1 \Rightarrow \omega(a \wedge b) = 1$, for $a \leftrightarrow b$.

One could be surprised at the possibility $\omega(a) = \omega(b) = 1$ if a and b are incompatible. This is actually possible, although rare, because in quantum mechanics a and b are represented by the projection operators P_a and P_b on the subspaces H_a and H_b. If $a \not\leftrightarrow b$, then $[P_a, P_b] \neq 0$ and H_a and H_b are different subspaces. Nevertheless, $H_a \cap H_b$ need not be empty, and, if the state vector of the physical system is in $H_a \cap H_b$, then both propositions a and b are certainly true. As an example note that a particle with momentum $p = 0$ has angular momentum $L = 0$, even though momentum and angular momentum do not commute. It is known that S4 is always valid for quantum mechanical states. Is it necessarily valid also for the states of any conceivable hidden-variable theory? The negative answer to this question (Bell, 1966) is best illustrated with a specific counterexample.

For spin-$\frac{1}{2}$ measurements, Jauch and Piron's propositions are represented by projection operators, which can be of three types: the zero operator ϕ, the unit operator I, and operators of the form $P(\hat{\alpha}) = \frac{1}{2} + \frac{1}{2}\boldsymbol{\sigma} \cdot \hat{\boldsymbol{\alpha}}$, where the $\boldsymbol{\sigma}$'s denote, as usual, Pauli matrices.

Consider then the following three propositions:

a = the spin is certainly along the $+x$ axis,

b = the spin is certainly along the $+y$ axis,

$a \wedge b$ = the spin is certainly along the $+x$ *and* $+y$ axes.

Quantum mechanically we associate with these propositions three projection operators:

$$a \Rightarrow \tfrac{1}{2} + \tfrac{1}{2}\sigma_x,$$

$$b \Rightarrow \tfrac{1}{2} + \tfrac{1}{2}\sigma_y,$$

$$a \wedge b \Rightarrow \phi.$$

The association of ϕ with $a \wedge b$ is a dire necessity, for, in the two-dimensional case, there is no vector that corresponds to fixed values of σ_x *and* σ_y, and any projection operator of the form $\frac{1}{2} + \frac{1}{2}\boldsymbol{\sigma} \cdot \boldsymbol{\alpha}$ would *always* project onto some vector. The hidden-variable model presented in Section 2.5 gives rise to dispersion-free states for which

$$\omega_0(a) = \omega_0(b) = 1 \quad \text{and} \quad \omega_0(a \wedge b) = 0,$$

as is easy to verify. The result $\omega_0(a \wedge b) = 0$ is a consequence of the measurement procedures in that model: one measures the sign of the projection of $\hat{\lambda}$, multiplies this by $|\boldsymbol{\beta}|$ (which is zero), and adds α (also zero). The result $a \wedge b = 0$ is then always found for every conceivable dispersion-free ensemble. Axiom S4 of Jauch and Piron is therefore unduly restrictive, just as was the case with NIII of von Neumann.

2.9. The Debate on Impossibility Proofs

In the preceding sections we have examined the two most ambitious impossibility proofs against the existence of hidden variables and shown them to be inadequate to their task. In fact, a rather simple hidden-variable model can be constructed that does not satisfy every axiom of either of the examined proofs. That this sort of finding invariably obtains for *every* impossibility proof ever proposed has been convincingly demonstrated in books (d'Espagnat, 1965; Belinfante, 1973; Jammer, 1974), review articles (Mugur-Schächter, 1964; Bell, 1966; Bohm and Bub, 1966; Capasso *et al.*, 1970; Ballentine, 1970), and other technical papers (Blokhintzev, 1953; Ochs, 1972), so that the matter in question can now be viewed as conceptually settled. The argument that clinches this conclusion is the mere fact that hidden-variable models *do exist* that are capable of reproducing *all* quantum mechanical predictions for essential physical applications of the theory; where by 'essential' we also mean that impossibility proofs hold for these applications even when they are considered in isolation from the theory as a whole. Such is the case for the hidden-variable model pertaining to a spin-$\frac{1}{2}$ particle at rest that was discussed in Section 2.5 of the present chapter. Such, too, was the case for the important pilot-wave model of Louis de Broglie (1927) and the model of David Bohm (1952), both of which will be discussed in a future chapter.

Impossibility proofs were also proposed by Salomon (1933), Destouches-Février (1945), Gleason (1957), Misra (1967), Kochen and Specker (1967), and Gudder (1968). None of them has withstood the test of time. It is interesting also to recall a qualitative argument put forward by Heisenberg in a 1935 lecture at Vienna University (Heisenberg, 1952, pp. 45—58). In Heisenberg's opinion the statistical nature of quantum theory is fundamentally different from that of classical thermodynamics. The statistical character of the latter theory expresses our lack of knowledge as to the state occupied by the system that is being studied, while, in quantum theory, ignorance about the results of future experiments performed on this system can be compatible with a knowledge of its state as complete as possible. There could apparently be no objection to an assumption that, say, a radium atom possesses hitherto unknown variables that accurately determine the time and direction of emission of an alpha particle. Heisenberg believed, however, that such an assumption could *with the aid of a more detailed analysis* be shown to imply the incorrectness of those fundamental statements of quantum theory which had led to so many successful predictions. His argument goes as follows: When experiments are performed in atomic physics, one has to draw a line somewhere between the measuring apparatus (treated classically) and the measured system (represented quantum theoretically by a wave function). Such a procedure does not involve a contradiction, because the classical description of macroscopic bodies can be obtained as a limiting case from quantum mechanics. However, the necessity of expressing the properties of a quantum system in classical terms compels one to take into account an essential disturbance, uncontrollable in principle, of the system by the measuring apparatus. This disturbance, which stems from the finite value of the quantum of action (Planck's constant h), inescapably demands a statistical description of the system under study. Therefore, argued Heisenberg, the only place where hidden variables could appear would be at the interface between apparatus and system, which is where the uncontrollable disturbance arises. Since, however, this boundary can be displaced in such a way that the earlier interface becomes due for a quantum treatment, one would be able to create a quantum mechanical object that exhibit also hidden variables. But this, Heisenberg concluded, would generate "a contradiction between the logical (*gesetzmässigen*) consequences of the new properties and the connections of quantum theory". Although Heisenberg did little to justify this conclusion, he applied it, to the example of α-particle emission, which is worth discussing because it illuminates the lack of cogency of his entire argument.

Let alpha particles, emitted by a nucleus, be reflected at a diffraction grating in clearly defined directions, depending on the particles' (presumably sharply defined) energies; these directions are determined by the *whole* grating. Now, if there existed hidden variables of the radium atom that would allow one to predict in which direction any alpha particle is emitted, then it would also be possible to predict precisely the point of impact of the latter on the diffraction grating. But such precise knowledge would imply a local interaction with the grating which could not contribute elsewhere to the creation of the diffraction pattern: diffraction could thus not occur, contrary to observation. The objection to Heisenberg's impossibility proof is that it rests on an oversimplified classical picture of a particle. A hidden-variable model of a dual, wave—particle structure, like that proposed by de Broglie and Bohm and discussed in our next two chapters, would obviously invalidate Heisenberg's argument.

One should not forget that one of the most important 'impossibility proofs' was provided by Bohr's *complementarity*, which can be seen in two ways. Firstly, as a positive statement, and it can then be considered a formulation of wave-particle duality: for this reason the presentation of complementarity is left for the next chapter, where also its limitations will be stressed. Secondly, as a negative statement, since it tries to forbid a description of physical phenomena simultaneously in causal terms and in space and time. Here it will merely be remarked that the hidden-variable model of spin measurements discussed in Section 2.5 above has the following features:

(i) It provides a causal description of quantum spin measurements.
(ii) It provides also a description of these measurements in space and time.
(iii) It describes all conceivable spin measurements as predetermined by objective properties of the measured system, even though there are observables that cannot be measured at the same time ('incompatible observables').

Therefore, that model does twice what complementarity tries to forbid. The first achievement (causality in space and time) is perhaps not very relevant to a direct application of complementarity to spin measurements, but the second one (simultaneous predetermination of all spin components) definitely is. Therefore also the interdictions arising from complementarity can be overcome (Selleri, 1985).

References

Books

Belinfante, F. J.: 1973, *A Survey of Hidden-Variables Theories*, Pergamon, Oxford.
Born, M.: 1964, *Natural Philosophy of Cause and Chance*, Dover, New York.
Born, M.: 1969, *Physics in My Generation*, Springer, New York.
de Broglie, L.: 1960, *Non-Linear Wave Mechanics*, Elsevier, Amsterdam.
d'Espagnat, B.: 1965, *Conceptions de la physique contemporaine*, Hermann, Paris.
Forman, P. L.: 1971, 'Weimar Culture, Causality, and Quantum Theory, 1918—1927: Adaptation by German Physicists and Mathematicians to a Hostile Intellectual Environment' in R. McCormmach (ed.), *Historical Studies in the Physical Sciences*, University of Pennsylvania Press, Philadelphia.
George, A. (ed.): 1953, *Louis de Broglie, physicien et penseur* (Collection *Les savants et le monde*), Albin-Michel, Paris.
Heisenberg, W.: 1952, *Philosophic Problems of Nuclear Science*, Fawcett, New York.
Jammer, M.: 1974, *The Philosophy of Quantum Mechanics*, Wiley, New York.
Mugur-Schächter, M.: 1964, *Étude du caractère complet de la théorie quantique*, Gauthier-Villars, Paris.
New Theories in Physics (Conference Proceedings, Warsaw 1938): 1939, International Institute of Intellectual Cooperation, Paris.
von Neumann, J.: 1955, *The Mathematical Foundations of Quantum Mechanics*, Princeton University Press, Princeton, N.J.

Articles

Ballentine, L. E.: 1970, *Rev. Mod. Phys.* **42**, 358.
Bell, J. S.: 1966, *Rev. Mod. Phys.* **38**, 447.
Birkhoff, G., and von Neumann, J.: 1938, *Ann. Math.* **37**, 823.
Blokhintsev, D. I.: 1953, *Sowjetwissenschaft* (*Naturwiss. Abteilung*) **6**, 545.
Bohm, D.: 1952, *Phys. Rev.* **85**, 166 and 180.
Bohm, D., and Bub, J.: 1966, *Rev. Mod. Phys.* **38**, 470.
Capasso, V., Fortunato, D., and Selleri, F.: 1970, *Riv. Nuovo Cimento* **2**, 149.

de Broglie, L.: 1927, *J. Phys. Radium* **8**, 225.
de Broglie, L.: 1959, *J. Phys. Radium* **20**, 963.
Destouches-Février, P.: 1945, *C. R. Acad. Sci.* **222**, 553 and 587.
Gleason, A. M.: 1957, *J. Math. Mech.* **6**, 885.
Goudsmit, S., and Uhlenbeck, G. E.: 1926, *Nature* **117**, 264.
Gudder, S. P.: 1968, *Rev. Mod. Phys.* **40**, 229.
Jauch, J. M., and Piron, C.: 1963, *Helv. Phys. Acta* **36**, 829.
Kochen, S., and Specker, E. P.: 1967, *J. Math. Mech.* **17**, 59.
Misra, B.: 1966, *Nuovo Cimento* **47**, 841.
Ochs, W.: 1972, *Nuovo Cimento* **10B**, 172.
Pauli, W.: 1925, *Z. Phys.* **31**, 765.
Piron, C.: 1963, 'Axiomatique quantique', Thesis, Université de Lausanne.
Salomon, J.: 1933, *J. Phys.* (*Paris*) **4**, 34.
Santos, E.: 1988, 'The search for hidden variables in quantum mechanics', in Selleri, F. (ed.), *Quantum Mechanics versus Local Realism. The Einstein, Podolsky, and Rosen Paradox*, Plenum, New York, London.
Selleri, F.: 1985, 'Bohr's complementarity in the light of modern researches in the foundations of quantum theory', in *Rivista di storia della scienza* **2**, 545.
Stern, O., and Gerlach, W.: 1922, *Z. Phys.* **9**, 349.
Wick, G. C., Wightman, A. S., and Wigner, E. P.: 1952, *Phys. Rev.* **88**, 101.

Chapter 3

The Wave—Particle Duality

Some experiments seem to imply that an electron can simultaneously occupy different regions of space; others again that it can be localized in a single tiny region. The same is true for photons, protons, neutrons, mesons, etc. Every microscopic object therefore seems to have a dual nature, which involves an extended phenomenon having the form of a wave and a localized phenomenon that behaves like a particle.

What does all this mean? Well, a consensus was never reached among the founders of modern physics on how one should *answer* this question, and thus people agreed for a long time simply not to *ask* it. Times are changing, though, and conditions appear ripe for taking a fresh look at the same question. We begin by discussing several different interpretations of the wave—particle duality along with some recent experimental evidence.

3.1. Duality for Photons

The classical tradition has been to consider the world to be an association of observable objects . . . moving about according to definite laws of force, so that one could form a mental picture in space and time . . . . It has become increasingly evident in recent times, however, that nature works on a different plan. Her fundamental laws do not govern the world as it appears in our mental picture in any very direct way, but instead they control a substratum of which we cannot form a mental picture without introducing irrelevancies.

So begins the preface to the first edition (1930) of Dirac's *The Principles of Quantum Mechanics*, the most important book on the subject (Dirac, 1958). A similar point of view has been expressed by almost all textbooks written ever since, so that the renouncement of a spacetime understanding of atomic phenomena has become the dominant attitude among physicists today.

It is the purpose of the first two sections of the present chapter to discuss the fundamental experimental evidence on which the foregoing negative conclusions are based. In Section 3.1 we deal with photons and present the twofold evidence which indeed seems to point to the impossibility of understanding in spacetime terms the behavior of atomic objects; Section 3.2 analyzes the same type of evidence for neutrons, with similar conclusions. The remaining sections of the present chapter review the most important interpretations of duality that historically were proposed by the founders of quantum theory.

Extensive observations on the self-interference of photons were made by Lajos Jánossy at the Central Research Institute of Physics in Budapest in the fifties. His final experiment used an interferometer of the Michelson type and monochromatic photons coming from the 5461 Å (green) line of mercury (Jánossy and Naray, 1958). The experiment had the features shown in Figure 3.1).

1. *Arm-length of about 14 m.* A photon can be viewed as a packet of length comparable with the coherence length of a spectral line, which in the case of the green mercury line corresponds to a linear dimension of about 1 m. If the dimensions of the interferometer were smaller than this 'size' of the photon, one might account for the interference phenomenon by imagining that the photon 'floods' the entire apparatus and thus is acted on simultaneously by both mirrors at the extremities of the apparatus' arms. But this picture obviously fails when the interferometer has an arm-length much larger than the photon's coherence length, which will certainly be the case for arms 14 m long.

For this reason Jánossy and Naray decided to build an improved version of a previous interferometer which had an arm length of 10 cm and which

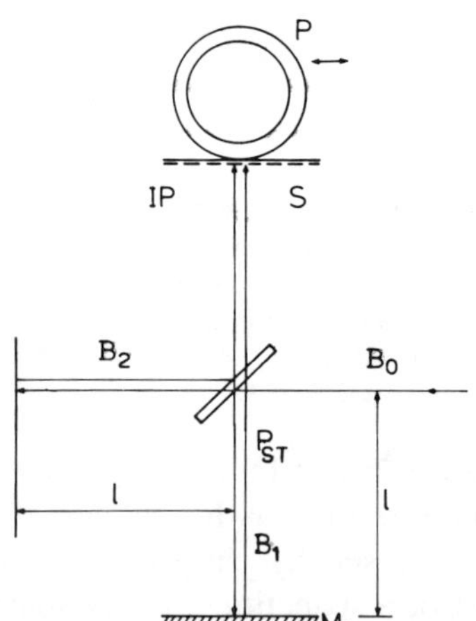

Fig. 3.1. Set up of the Jánossy—Naray experiment. A low-intensity source sends beam B_0 toward a semitransparent mirror. Emerging beams B_1 and B_2 are reflected back by totally reflecting mirrors M_1 and M_2, respectively. The interference pattern IP is formed on screen S with a slit behind which a photodetector P is placed. Arm length about 14 m, longitudinal coherence length of the photonic wave packets less than 1 m. (Redrawn from Jánossy and Naray, 1958.)

already had been used to establish that the interference pattern did not depend on light intensity.

2. *Underground tunnel cut in rock.* In order to achieve the necessary stability, the apparatus was built in an underground tunnel cut in rock 30 m under the surface of the earth. But even the presence and activity of a human being in the tunnel could disturb the regular functioning of the apparatus. To solve this problem, the experimenters installed remote controls, observing the interference pattern from outside by help of a periscope. After all instrumental adjustments had been carried out, it was found necessary to leave the apparatus undisturbed for a few days, during which interval all mechanical tensions relaxed and temperature equilibrium was reached. Only then could a successful series of measurements be carried out. The delicacy of this beautiful experiment depended on the fact that, in order to measure an interference pattern point by point, a total time of about an hour was required. During this time it was necessary to have *fringe stability*, which in practice meant that no fringe shift larger than 1/10 of a fringe could take place. This *desideratum* in turn required the arms of the apparatus to remain constant within a length of the order of $\lambda/20$, λ being the radiation wavelength. Since a variation of this size could be generated by a temperature change as small as 10^{-5} K, all sources of thermal and mechanical disturbances had obviously to be eliminated.

In the experimental setup depicted in Figure 3.1 an incoming beam (B_0) is split by a semitransparent mirror into a transmitted component (B_2) and a reflected component (B_1). Both beams are reversed on reflection at mirrors M_2 and M_1, respectively, and thus made to interact once again with the semitransparent mirror, which splits each of them into two parts. Only the transmitted part of B_1 and the reflected part of B_2 are of interest, since, propagating along a common path, they give rise to an interference pattern on the plane of a slit. The interference detecting device *IP* of Figure 3.1 consists of a plane (shown as a black line) with a slit on which the interference pattern (shown as a broken line) is formed. Behind the slit a photomultiplier *P*, capable of detecting photons which pass through the slit, is placed. The photomultiplier plus slit was displaced by small steps along a horizontal track, and thus the rate of incident photons was counted at various points of the fringe system. The foregoing description of the apparatus is naturally a simplified one, as many aspects which are of technical but not conceptual relevance has been omitted (monochromator, lenses, light attenuator, electronics, etc.).

3. *Low photon intensity.* Since a photon takes 10^{-7} s to travel 30 m, Jánossy and Naray, who used an apparatus of this path-length, required an intensity much lower than 10^7 photons s^{-1} to ensure that only one photon was present in their apparatus at any given time. In fact, they used an intensity as low as 10^4 photons s^{-1}, a figure that is obviously small enoguh for the stated purpose.

Jánossy and Naray end their paper with the following statement:

As the result of our measurements we have therefore to conclude that *interference phenomena are perfectly normal even at such low intensities where at one time on average less than one photon is to be found in the arrangement; this is true even if the dimensions of the arrangement greatly exceed the coherence length of the photons giving rise to the patterns.* (*Ibid.*)

An attempt to understand this result in spacetime terms would naturally assume that when the photon wave packet arrives at the semitransparent mirror it is split into two parts, which follow different trajectories, are reflected by different mirrors, and finally recombine in the region where the interference pattern is observed. This proposed explanation seems however to be invalidated by another experiment.

If one allows a low-intensity photon beam to fall on, and be split into two parts by a semitransparent mirror, coincidences should be registered by two photomultipliers placed in the paths of the transmitted and reflected beams if the energy—momentum of the incident photon were really divided into two parts. However, accurate measurements performed by Clauser (1974) and, as a byproduct of a more complex experiment, by Dagenais and Mandel (1978), revealed no such coincidences. The essential aspect of Clauser's experiment, from our present point of view, is that two phototubes viewed photons on opposite sides of a dielectric beam splitter (which acts just like a semitransparent mirror). Their readings contradicted, to a high degree of statistical accuracy, the predictions of any classical or semiclassical theory in which the probability of photon detection is assumed proportional to the classical intensity. The Clauser experiment therefore refuted the idea that a photon wave packet is simply a classical electromagnetic field (uniformly carrying energy-momentum) which is divided by the beam splitter into two identical parts. The same conclusion can be drawn from the experiment by Dagenais and Mandel.

The foregoing experiments, involving a pair of photomultipliers, are consistent with the idea that a single photon which interacts with a beamsplitter chooses at random between reflection and transmission. The Jánossy—Naray experiment, by contrast, suggests that the *same* photon interacting with the *same* beamsplitter is partly reflected and partly transmitted. These two disparate answers, it was thought, one could not possibly reconcile.

The traditional strategy among physicists for escaping from the paradox of dualism has been to stick to the mathematical apparatus of quantum theory, which, in conjunction with its Copenhagen—Göttingen interpretation, has led to accurate predictions for all known experiments, including those discussed in the present section. Against those people who want to understand the behavior of atomic entities *also* in spacetime terms, it has been objected that space, time, and causality are merely conceptions of the classical (macroscopic) domain and that their extrapolation to atomic dimensions is not permissible. The same prohibition should, in this view point, extend to the distinction between broadly defined trajectories with *macroscopic* separation, such as $M—M_1—M$ and $M—M_2—M$ in Figure 3.1, where M is the semitransparent mirror. The question concerning the photon's interaction with the semitransparent mirror and the more general question dealing with its propagation within the experimental apparatus would, according to this point of view, be considered meaningless and essentially nonscientific; not only because one tries, by the very asking of these questions, to extrapolate 'classical conceptions' to the quantum domain, but also because the former are necessarily formulated for the unobserved photon. The well-known recipe, '*do not ask any question about what is not directly observed*', works in this case equally well as we saw that it did for de Broglie's paradox in the previous chapter. Also in this instance, it is therefore possible to invoke the foregoing recipe to escape from a paradoxical situation. The price one has to pay is however high and almost amounts to the abolition of objective reality.

3.2. Duality for Neutrons

When an x-ray enters a crystal as a wave packet having dimensions larger than interatomic distance of 10^{-8} cm, one would expect its behavior to be very different from that of a neutron under the same conditions, since x-rays interact with electric charges and therefore 'see' a medium which is replete with dispersing elements (atoms and their constituents, electrons and nuclei), while neutrons interact only with nuclei and therefore see a medium in which only one part in 10^5 is not empty space. It turns out however that this difference in the way the medium is 'perceived' is unimportant, and a neutron beam behaves in a crystal very like an x-ray beam of equal de Broglie wavelength does: all the laws of crystal optics relating to reflecting, transmission, diffraction, and dispersion remains valid for neutrons as well.

To understand why this happens, it is necessary to realize that the essential feature determining the optical laws for x-rays is the coordinate-space arrangement of atoms in the crystalline lattice: it is their regular geometrical ordering which, starting from the elementary spherical waves that emanate from the single atoms, produces the general macroscopic (diffracted, transmitted, etc.) x-ray wave. Thus, although neutron waves interact only with nuclei, since the latter have the same geometrical ordering as the atoms, the resulting diffracted or transmitted waves are geometrically indistinguishable from x-ray waves.

The first neutron interferometer was built and operated in 1974 by Rauch, Treimer, and Bonse at the Austrian Nuclear Institute in Vienna (Rauch *et al.*, 1974). The interferometer is constructed with a silicon crystal which is required to be essentially perfect: it must be a single crystal, about 10 cm long, and completely free of dislocations and other defects in its regular atomic structure. For example, one can start from a right parallelepiped about 8 cm long and 5 cm in diameter and cut away part if it, leaving three slabs connected by the remainder of the crystal (see Figure 3.2). The slabs are then about 5 mm thick and 3 cm apart.

The cutting of the slabs is an operation which must be done with great care: in a typical interferometer the distance Δx between the crystal plates and the thickness t of the slabs are, respectively,

$$\Delta x = (2.729\,36 \pm 0.000\,09)\ \text{cm},$$

$$t = (0.439\,54 \pm 0.000\,08)\ \text{cm}.$$

Distances in the interferometer are therefore exact to within 10^{-4} cm or about 10 000 atomic layers. This precision is achieved by cutting the original crystal with optically controlled diamond tools and by polishing the surfaces.

When a beam of neutrons strikes the first slab in the interferometer at an angle θ with the normal to the surface (Bragg angle) of magnitude 20—30°, it is scattered by planes of atoms *perpendicular* to the face of the crystal. This kind of scattering, called *Laue scattering,* gives rise to two beams: a transmitted one, at the Bragg angle θ, and a diffracted one, at the same angle but on the opposite side of the scattering planes. In other words, the emerging beams form a V whose vertex lies in the first slab. At the second slab each of these beams is again Laue-scattered, the four emerging beams forming a W whose vertices lie in the second slab. The geometry of the apparatus is such that the two external beams of this W do not have any further interaction with the interferometer (see Figure 3.3). The important event occurs when the two internal beams of the W converge in a point on the face of the third slab: there each of them is again Laue-scattered and gives rise to two beams emerging from the third slab. The pair of V's are, however, spatially superimposed in such a way that each of the two resulting beams is the sum of a transmitted component and of a diffracted component.

Let ψ_R and ψ_L be the wave functions describing the beams that arrive at the *third* slab from the right and left, respectively. If we consider only cases in which neutrons do actually arrive at the third slab (and neglect therefore

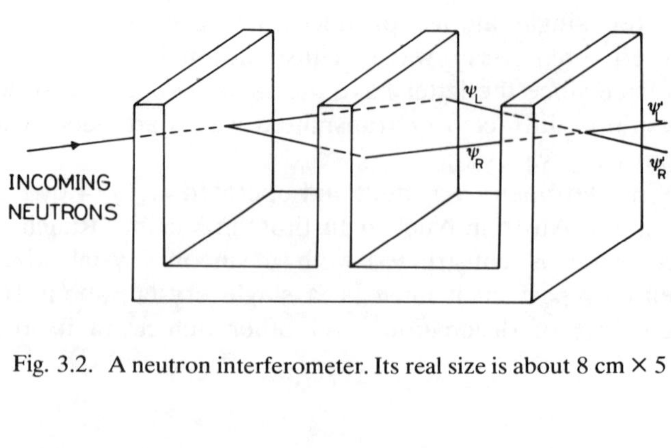

Fig. 3.2. A neutron interferometer. Its real size is about 8 cm × 5 cm × 5 cm.

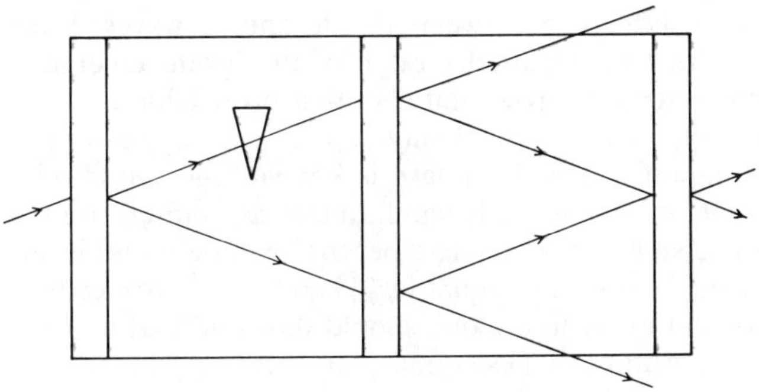

Fig. 3.3. Paths of the neutronic wave packets inside the interferometer. A phase shifter is inserted on one of these paths.

the cases where neutrons travel along the external beams of the W), we can assume that there is an overall normalization

$$|\psi_R|^2 + |\psi_L|^2 = 1.$$

At the third slab, each wave is resolved into a transmitted wave T and a diffracted wave D: ψ_R gives rise to ψ_R^T and ψ_R^D, while ψ_L produces ψ_L^T and ψ_L^D. The coherence of this phenomenon guarantees that ψ_R, ψ_R^T, and ψ_R^D have the same phase, and likewise for ψ_L, ψ_L^T, and ψ_L^D.

Furthermore, the amplitudes of diffracted and transmitted waves arising from either ψ_R or ψ_L are taken to be equal for simplicity; conservation of probability then requires that

$$|\psi_R^T| = |\psi_R^D| = \frac{1}{\sqrt{2}}\,|\psi_R|,$$

and

$$|\psi_L^T| = |\psi_L^D| = \frac{1}{\sqrt{2}}\,|\psi_L|.$$

Now, contained in the V that emerges from the third slab, there are two waves, which we call ψ'_R and ψ'_L and which result from the physical spatial superposition (and therefore the algebraic sum) of two different waves

$$\psi'_R = \psi_L^T + \psi_R^D$$

and

$$\psi'_L = \psi_R^T + \psi_L^D.$$

It is possible to show that

$$|\psi'_R|^2 = \tfrac{1}{2}(|\psi_R|^2 + |\psi_L|^2 + 2|\psi_R|\,|\psi_L|\cos\alpha)$$

and

$$|\psi'_L|^2 = \tfrac{1}{2}(|\psi_R|^2 + |\psi_L|^2 - 2|\psi_R|\,|\psi_L|\cos\alpha), \qquad (3.1)$$

where α is the phase difference between the R and L waves. From this result, one sees, first, that the squared moduli of the beams emerging from the neutron interferometer add up to unity, so that probability is conserved, and furthermore that the moduli individually depend on the relative phase of the right and left components. If this phase is known, one can check if the famous probability law of Born is really valid: In fact ψ'_R and ψ'_L are spatially separated, and the probability that a single neutron will be found in the left-hand or the right-hand path should equal $|\psi'_L|^2$ and $|\psi'_R|^2$, respectively. On repeating the experiment many times, one should thus find that the emerging neutron *fluxes* are proportional to these same quantities.

Actual experiments have been done with monochromatic neutron beams emerging from a nuclear reactor. Monochromaticity is achieved by letting neutrons of different velocities, but with the same direction of motion, impinge on a crystal at some suitable angle. Dispersion takes place, just as in the case of light, and different wavelengths (thus, different neutron velocities) emerge at different angles. On selecting with suitable absorbers only the neutrons that emerge at some definite angle, one thus obtains approximately monochromatic neutrons. The beam so produced is allowed to enter the neutron interferometer. Typically it represents a flux of about 100 neutrons s^{-1}, which corresponds to an average time interval $\Delta t \simeq 10^{-2}$ s, but interferences have been observed with average intervals as large as a fifth of a second. The thermal neutrons emerging from a reactor normally have a velocity of 2×10^5 cm s^{-1}, which corresponds to a de Broglie wavelength of 2×10^{-8} cm. Every neutron thus traverses the entire interferometer in a time

$$\tau \simeq 8 \text{ cm}/2 \times 10^5 \text{ cm s}^{-1} = 4 \times 10^{-5} \text{ s}.$$

The probability for simultaneously having *two* neutrons in the apparatus is approximately given by $\tau/\Delta t$ and therefore is of the order of 10^{-3}—10^{-4}. It should furthermore be noticed that the coherence length of the neutron wave packet has been measured and found to be around 5×10^{-5} cm.

One can therefore conclude that, in the large majority of the cases, a neutron wave interferes with itself. All the experiments performed thus far (Bonse and Rauch, 1979), using the most diverse physical means for varying the relative phase α (phase-shifting materials, magnetic fields, the terrestrial gravitational field, the Earth's rotation, etc.), have produced results that are consistent with Equation (3.1) to a high degree. Thus Born's statistical postulate (particle distribution over space given by the squared modulus of the wave function) has been fully confirmed.

One is obviously tempted to infer that, in the case of a single neutron, something propagates over both possible paths within the interferometer, as if the neutron were split into two 'parts' by the first slab. But this conclusion is apparently contradicted by a second experiment which one can do by putting neutron counters immediately behind the first slab: every neutron is then revealed only by a single counter. Since neutron counters are in the final analysis sensitive to incoming energy-momentum, we can now conclude that energy-momentum (and therefore mass) always take only one of the two paths that are possible across an interferometer slab. One sees here repeated the situation encountered for photons in the first section. Nature seems to give contradictory answers to the *same* physical question posed by different experimental setups.

The question is now the following: *How does a single neutron behave when it interacts with the slab of an interferometer*? If one uses a complete interferometer, with neutron counters put behind the third slab and with some phase-shifting device, one is led to the conclusion that the neutron *is split* by each slab and that it propagates partially over each of the two paths diverging from the interaction point. But if one uses only a single silicon slab backed by neutron counters, one is forced instead to conclude that the neutron *is not split* by the slab, but that it chooses at random between the two trajectories emerging from the slab. Accurate measurements, moreover, easily show that the neutron observed exclusively on one of the two paths is just a regular neutron with all its well-known properties: mass, magnetic moment, spin, β instability, and so on.

The problem identified in the foregoing manner has been known since the birth of quantum mechanics and could appropriately be called a *paradox*. It has pushed the majority of modern physicists to consider a spacetime description of atomic entities as an impossibility. Also in the present case, obviously, the considerations of the closing part of the previous section can be repeated, in particular about the fact that a paradox can be evaded if one is willing to pay the price of abolishing all reference to objective reality when it is not directly observed.

3.3. Einstein's Discovery of Duality

Einstein's discovery of the equivalence between mass and energy (Einstein, 1905a) could have brought about a tremendous unification of the scientific world-view. Since energy and momentum are always conserved, while mass can be created and destroyed, the old idea that matter is irreducible had to be replaced with the idea that everything is really composed of energy-momentum. This allowed one, at least in principle, to understand in a unified way matter itself, light, electromagnetic and gravitational fields, and so forth, briefly all objects of physical investigation. If this unification in practice did not take place, it was for good reasons, starting from the fact that Einstein himself was forced in the same year 1905 to admit the existence of something that was real (in the sense that it was assumed to propagate causally in space and time) but contained neither matter nor energy-momentum. This admission was a necessary consequence of Einstein's dualistic picture of the electromagnetic radiation (Einstein, 1905b).

Difficulties for the classical theory of radiation were found in 1902 by Lenard and in 1903 by Ladenburg in experimental studies of the photoelectric effect. They were led in particular to three conclusions: (1) photoelectricity was not transmitted by atoms and therefore very probably was transmitted by electrons (which had been discovered by J. J. Thomson in 1897); (2) when a given metal was irradiated, electrons were ejected only if the radiation had a frequency above a certain minimum ν_0; (3) the kinetic energy of the ejected photoelectrons was independent on the light intensity. These conclusions appeared impossible to understand on the basis of the classical representation of light. Therefore the idea gained acceptance that electromagnetic energy was not uniformly distributed over the wavefront. In a lecture at Yale University in 1903, J. J. Thomson made reference to 'bright specks on a dark background' as being the correct representation of the wavefront. (This story is told, e.g., by Jammer, 1966.)

Also, Einstein started to work on the same idea. He was of course well aware of the evidence in favor of a continuous picture of the radiation wave, but thought that such evidence reflected only temporal averages (over long-lasting observations) of radiation field quantities and that one could therefore not exclude the appearance of new phenomena for instantaneous values of these quantities, such as those entering into the description of absorption and emission processes. Einstein's basic idea was that the energy of an electromagnetic wave of frequency ν came in discrete units of value $h\nu$, each of which occupied only a very small region of space; these were the *quanta* or particles of radiation. Einstein announced his discovery in the following words:

> According to the hypothesis that I want here to propose, when a ray of light expands starting from a point, the energy does not distribute on ever increasing volumes, but remains constituted of a finite number of energy quanta localized in space and moving without subdividing themselves, and unable to be absorbed or emitted partially. (Einstein, 1905a)

This was the birth of wave—particle duality. The particles served at the very least as carriers of the energy belonging to the radiation field, while the electromagnetic wave constituted an extended entity surrounding the particles. The existence of particles was essential for an understanding of the photoelectric effect, while the presence of the waves allowed one to comprehend the interference and diffraction of electromagnetic radiation. Waves and particles were strictly associated with one another: For instance, the very definition of particle energy ($E = h\nu$) contains the frequency of the wave. From this beautiful idea Einstein obtained his famous equation for the photoelectric effect

$$\tfrac{1}{2}mv^2 = eV = h\nu - \phi,$$

where e, m, and v are the electric charge, mass, and velocity of the expelled electron, V is the minimum potential difference capable of stopping the electron, and ϕ denotes the metal's *work function.* Besides explaining Lenard's observations, Einstein could make very important predictions: the ratio of observable quantities $eV + \phi$ and ν had to be independent of the nature of the metal and the intensity or frequency of the light; furthermore, the ratio had to equal the constant h, which earlier was precisely evaluated by Planck. Millikan stressed much later that these were indeed very bold hypotheses, because in 1905 there was no way of knowing "whether the quantity h to which Planck had already assigned a numerical value appeared at all in connection with photoelectric discharge" (Millikan, 1947, p. 238). Nine years were to pass before Millikan himself would verify the validity of Einstein's predictions.

Einstein's dualistic idea had at first very little success among other physicists. He kept working on it, however, and in 1909 showed that Planck's formula for blackbody radiation implied that the particle and wave aspects of light gave *additive* contributions to the mean square fluctuation of energy in any small volume. In the same year he proposed to treat light quanta as singularities surrounded and guided in their motion by a continuous wave-phenomenon.

In 1914, Millikan used light of different frequencies and measured, as a function of ν, the potential difference V capable of stopping completely the photoelectric current. A plot of eV versus ν did in fact produce a straight line whose slope was exactly equal to Planck's constant. In the same year, Mayer and Gerlach studied small particles of metal dust suspended in an electric field. The particles were illuminated and the liberation of an electron from any one of them was observed as a sudden acceleration. It was noticed that the liberation of photoelectrons was sometimes instantaneous, in sharp contradiction with what one would expect from a continuous-wave description of electromagnetic radiation (Mayer and Gerlach, 1914).

Einstein wrote a new paper on dualism in 1917 (Einstein, 1917). In it, he considered a gas of molecules interacting through the emission and the absorption of electromagnetic radiation; he assumed for the molecules the existence of discrete energy states, capable of transforming into each other through emission of absorption of electromagnetic radiation; he postulated that possible processes were stimulated emission under the influence of a radiation field, absorption of the same type, and emission 'without excitation from external causes'; and, finally, he assumed that different energy states occurred in the molecular gas with a frequency determined by the canonical distribution over states of statistical mechanics. Among the results Einstein obtained on the basis of these assumptions was a very simple derivation of Planck's formula which in modern times is still repeated in many textbooks. But Einstein considered more important another result which is nowadays almost forgotten.

When radiation is emitted or absorbed by a molecule, it imparts to the latter a momentum (except if one is dealing with a spherically symmetric wave). The maximum momentum is obtained when all of the energy E is emitted in one direction, and in this case it equals E/c. Einstein could prove that the experimentally established Maxwell distribution of molecular velocities was obtained only under the hypothesis that all the energy was emitted in (or absorbed from) a unique direction in every individual radiation-matter interaction process. That the Maxwell distribution should be obtained only on properly taking into account the velocities communicated to the molecules by radiation emission, or absorption, can be understood if one recalls that the Maxwell distribution was originally derived from a consideration of molecular *collisions* alone and that it is verified experimentally. Since matter is always emitting and absorbing radiation, this observation means that incorrect results would obtain unless the right treatment for the radiation-molecule interaction is used.

In his final paragraph Einstein stressed that the following points could be considered as 'fairly certainly proved':

> If a radiation bundle has the effect that a molecule struck by it absorbs or emits a quantity of energy $h\nu$ in the form of radiation (ingoing radiation), then a momentum $h\nu/c$ is always transferred to the molecule. For an absorption of energy, this takes place in the direction of propagation of the radiation bundle, for an emission, in the opposite direction. (*Ibid.*)

Einstein considered the previous result as the *most important conclusion of his paper*, obviously because it threw light on the real nature of the electromagnetic radiation.

A more spectacular proof of the directional character of quanta was given by Compton in 1923. He found that, in the scattering of x-rays from matter, a change of frequency occurred and was able to derive a theoretical formula in complete agreement with his experimental data. Compton wrote:

> The present theory depends essentially upon the assumption that each electron which is effective in the scattering, scatters a complete quantum. It involves also the hypothesis that the quanta of radiation are received from definite directions and are scattered in definite directions. The experimental support of the theory indicates very convincingly that a radiation quantum carries with it directed momentum as well as energy. (Compton, 1923.)

Compton's experiment suggested that corpuscles were indeed present in the electromagnetic radiation. Much stronger evidence for such a presence was furnished in 1925 by Compton and Simon who observed recoil electrons from x-ray scattering in a cloud chamber. If an unobserved corpuscle is scattered, one can calculate from energy and momentum conservation the direction along which it will move. The same corpuscle could undergo a second scattering from an electron, and the resulting electron track should have its starting point on the calculated trajectory of the unseen corpuscle (Figure 3.4). Compton and Simon observed 18 such double scatterings,

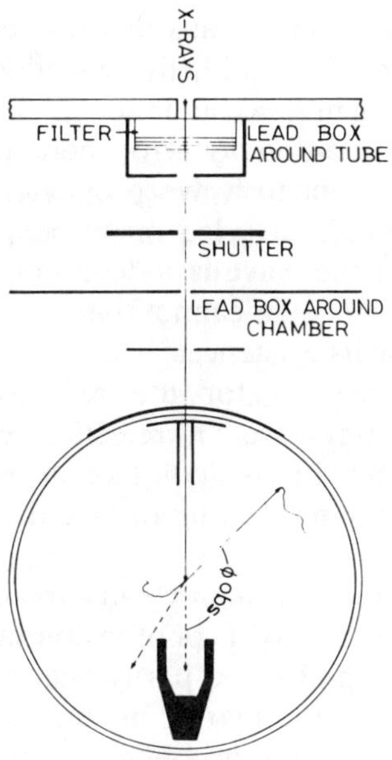

Fig. 3.4. Cloud chamber event of the Compton—Simon experiment. An electron is set in motion in the chamber by collision with incoming x-ray. Observed electron trajectory allows one to calculate trajectory of unseen photonic particle after collision (energy-momentum conservation). If corpuscular hypothesis for photonic particle is correct, the starting point of an eventual secondary electron will always be on the calculated photonic trajectory.

which was, within statistical errors, what they expected from the known cross section for x-ray scattering on electrons. Discussing this experiment, Compton wrote two years later: "... we can find no interpretation of the scattering except in terms of the deflection of corpuscles or photons of radiation" (Compton, 1927).

Einstein's objective dualism associated wave and corpuscle in such a way that the particle properties, energy E and momentum $\mathbf{p}$, and the wave properties, frequency ν and wave number vector $\mathbf{k}$, were related by

$$E = h\nu \text{ and } \mathbf{p} = \hbar\mathbf{k}.$$

As is well known, de Broglie extended these relations to material particles (electrons, neutrons, etc.), so that they appear today as expressing a very fundamental property of nature.

A problem immediately coming to mind within Einstein's philosophy is the following: if the localized particle carries all the energy and momentum, in which sense can the wave be considered real? This problem was felt so acutely by Einstein that he referred to these waves as *Gespensterfelder* (ghost fields): an object without energy and momentum is in fact unable to exert a pressure when impinging on a material surface, which means that it does not have that quality that makes us call something real. Still, the equations of quantum theory describe this wave as propagating in space and time.

The difficulties associated with the concept of an *empty wave* (here and in the following this expression is taken as equivalent to 'wave propagating in space and time but devoid of energy-momentum') have led many people to discard the idea as a scientific impossibility: if the wave is indeed empty, it cannot induce changes in physical objects; hence it cannot be observed; therefore it does not make any sense to postulate its existence.

This argument is probably one of the reasons for the rejection of Einstein's duality by the large majority of physicists, a rejection which displeased Einstein so much that he could write: "I must look like an ostrich hiding always his head in the relativistic sand for not having to face the ugly quanta" (quoted by de Broglie, 1979).

Aside from this difficulty, Einstein's formulation of duality can explain the experimental evidence rather naturally: in the case of a photon interacting with a beamsplitter, the wave is forced to split and to be partly transmitted and partly reflected. If phototubes are placed on the two paths, one will be able to detect only energy-momentum, in other words, only the particle, which is forced by the beamsplitter to choose between reflection and transmission. No coincidences should therefore take place, as observed. In the case of the Jánossy—Naray experiment, the waves propagate over both paths until they recombine just before the interference detecting device. The total wave so generated would then act on the particle and force it to go with greater probability to regions where the amplitude is large. The counting of many particles would in this way allow one to reconstruct the interference pattern, which would be present although unobservable, even in the case of a single photon.

3.4. De Broglie's Duality

Einstein's idea of dualism was generalized by Louis de Broglie. He regarded the double nature of radiation (corpuscles and waves) as a well-established phenomenon and proposed to extend this duality to electrons and other particles. His very interesting reasoning was as follows (de Broglie, 1925): Consider a particle of rest mass m_0 and thus rest energy m_0c^2. All forms of energy are reducible to motion. But relativity teaches us that the rest mass is also some form of energy. Therefore, some form of motion should be associated with it. Since a particle is a localizable structure which can be brought to rest, the motion related to the mass must take place within a very small region. It should then be some kind of rotation or vibration; in any event, a periodic motion with some associated frequency ν_0.

Then how can ν_0 be determined? Physically one expects ν_0 to increase with energy. Furthermore, Planck's relation for photons, $E = h\nu$, has the convenient feature that E and ν transform in the same way, so that $E = h\nu$ can be true in all frames of reference. De Broglie therefore assumed very naturally that $E = h\nu$ holds also for electrons, where E is the total energy. In the particle's rest frame, the Planck relation becomes $\nu_0 = m_0c^2/h$.

The trouble with the foregoing description is that, relative to a frame moving with velocity v, there is not one but two different frequencies.

The first frequency is obtained if one notices that an observer in the moving frame, because of time dilation, sees the internal periodic motion slowed down according to the formula

$$\Delta t_1 = \frac{\Delta t_0}{(1-\beta^2)^{1/2}},$$

where $\beta = v/c$, and Δt_0 and Δt_1 represent the same time interval measured in rest frames of the particle and the observer, respectively. Now, the internal rotation (or vibration) of the particle is an objective property of the particle itself, and the stationary and moving (relative to the particle) observers must count the same number n of full cycles in the time intervals Δt_0 and Δt_1, respectively. Therefore,

$$\nu_1 = \frac{n}{\Delta t_1} = \left(\frac{n}{\Delta t_0}\right)(1-\beta^2)^{1/2} = \nu_0(1-\beta^2)^{1/2}. \tag{3.2}$$

Thus in the moving system one has $\nu_1 < \nu_0$.

The second frequency appears when one notices that h is relativistically invariant, so that the transformation of $\nu_0 = m_0c^2/h$ can be effected simply by transforming m_0c^2. This gives

$$\nu_2 = \frac{E}{h} = \frac{m_0c^2}{h(1-\beta^2)^{1/2}} = \frac{\nu_0}{(1-\beta^2)^{1/2}}. \tag{3.3}$$

Obviously one has $\nu_2 > \nu_0$, whence $\nu_2 \neq \nu_1$.

De Broglie solved the dilemma posed by the presence of two different frequencies in the moving system by attributing ν_1 to the particle's *internal* motion and ν_2 to an undulatory phenomenon *external* to the particle, but associated with it. To be specific, the transformation law leading from ν_0 to ν_2 is identical with the one leading from m_0c^2 to E: in other words, the frequency ν_2 transforms like the fourth component of a 4-vector. A well-known physical entity that involves a 4-vector having a frequency as fourth component is the (invariant) phase of a plane wave, given by

$$F = \mathbf{k}\cdot\mathbf{x} - \nu_2 t,$$

where $\mathbf{k}$ is the wave-number vector and ν the frequency. Accordingly, de Broglie in this way associated a wave of frequency ν_2 with a particle of mass m_0.

But now notice that, in the particle rest frame, $\beta = 0$ and therefore $\nu_1^0 = \nu_0 = \nu_2^0$, from (3.2) and (3.3). Thus the wave and the internal periodic motion have the same frequency. This is an important physical conclusion: It looks as if the internal motion generates the wave and, naturally, imparts to it the internal frequency. If this conclusion is valid, every observer should in his frame find that the *internal motion and the wave are always in phase.* (A cork that bobs up and down in the water and thus generates circular waves provides a good analogy. In any moving frame whatsoever one will see a crest when the cork is up and a valley when it is down.) But since $\nu_1 \neq \nu_2$, it is not obvious that this condition can be satisfied. If G denotes the periodic particle property, assumed to execute simple harmonic vibrations, and ψ the displacement describing the wave, then, postulating one-dimensional wave propagation for simplicity, one will have:

$$G = G_0 \sin 2\pi\nu_1 t, \qquad \psi = \psi_0 \sin 2\pi\nu_2 \left(t - \frac{x}{u}\right), \tag{3.4}$$

where u is the phase velocity of the wave with frequency ν_2 in the moving system. If the particle is assumed to pass through the origin at time zero, then one sees that G and ψ will initially ($t = 0$) have the same (vanishing) phase at the location ($x = 0$) of the particle. In general, the 'particle's phase' at time t is

$$\nu_1 t = \frac{\nu_1 x}{\beta c}, \tag{3.5}$$

where x denotes the particle's coordinate at time t, since the particle moves with the velocity βc. Simultaneously, the 'wave's phase' at the location of the particle is

$$\nu_2 \left(\frac{x}{\beta c} - \frac{x}{u}\right). \tag{3.6}$$

On equating expressions (3.5) and (3.6), while taking formulas (3.2) and (3.3) into account, one can easily show that

$$u = \frac{c}{\beta}. \tag{3.7}$$

The two oscillations therefore remain in phase if and only if the wave velocity is c/β, which obviously corresponds to a superluminal propagation. This outcome, however, does not constitute a real drawback of the theory, since one can show that the wave under study cannot be used to transmit signals with superluminal velocity.

After having, in the above rigorous manner, causally linked particle and wave, de Broglie was able to deduce a highly significant result. The wavelength λ, the frequency ν_2, and the phase velocity u are obviously related via

$$\nu_2 \lambda = u. \tag{3.8}$$

On substituting in this equation ν_2 from (3.3) and u from (3.7), one gets

$$\lambda = \frac{h}{m_0 v/(1-\beta^2)^{1/2}} = \frac{h}{p}, \tag{3.9}$$

which is the famous de Broglie result that relates wavelength to particle momentum.

Another important result obtains if one considers a nonmonochromatic wave, for instance a wave given by

$$\psi = \int d\nu_0 \rho(\nu_0) \sin 2\pi(\nu_2 t - kx), \tag{3.10}$$

where ν_2 and ν_0 are related via Equation (3.3) and ρ is some weight-function. As is well known, the *group velocity* of ψ is given by

$$v_g = \frac{d\nu_2}{dk}. \tag{3.11}$$

Considering the two equations

$$k = \frac{\nu_2}{u} = \frac{\nu_0 \beta}{c(1-\beta^2)^{1/2}} \quad \text{and} \quad \nu_2 = \frac{\nu_0}{(1-\beta^2)^{1/2}},$$

one can quickly calculate v_g in the following way:

$$v_g = \frac{d\nu_2}{dk} = \frac{d\nu_2}{d\beta} \bigg/ \frac{dk}{d\beta} = \beta c = v \tag{3.12}$$

Equation (3.12) has a very simple physical interpretation: the wave packet and the particle travel with the same velocity and remain physically associated with one another at all times if they were so initially. This result could hardly be anticipated with monochromatic waves, since the phase equality between G and ψ existed between a particle in motion with velocity $v < c$ and a wave train propagating with velocity $c/\beta > c$. That is, the phase equality existed while the waves were, so to speak, rapidly sweeping over the particle!

Other interesting conclusions of de Broglie were the following:

(1) A photon can be treated as the vanishing-mass case of a particle in his theory. Inded, in the limit $m_0 \to 0$, Einstein's relations $E = h\nu$ and $p = h\nu/c$ follow from de Broglie's formalism. A great unification between particles and electromagnetic radiation was thus achieved, since both came to be viewed as corpuscles embedded in a wave phenomenon.

(2) Consideration of a bound electron leads to the inference that its stable

states are those, and only those, for which the de Broglie wave is stationary (corresponding to orbits whose length ℓ is an integral multiple of the wavelength: $\ell = n\lambda$); otherwise the wave interferes destructively with itself, while its frequency along with the particle's energy become ill defined and varying in time. The condition $\ell = n\lambda$ can be transformed exactly into the Bohr—Sommerfeld quantization rule.

(3) On the basis of a natural generalization of his theory to the case in which an external potential was acting on the wave—particle system, de Broglie was able to show that Fermat's principle for wave propagation could be transformed exactly into Maupertuis' variational principle for particle propagation. In this way one could fully understand the physical reasons for the striking similarity of the two principles.

Waves and particles were for de Broglie both objectively existing in space and time. Indeed, in his Nobel lecture he stated: "The electron . . . must be associated with a wave, and this wave is no myth; its wavelength can be measured and its interferences predicted" (de Broglie, 1929).

In 1927 Davisson and Germer provided a check on the validity of de Broglie's description of material particles by studying experimentally the scattering of electrons by nickel crystals (Davisson and Germer, 1927). The grating constant of the latter was known to be 2.15 Å, and an application of Bragg's formula to the observed diffraction peaks in electron scattering allowed Davisson and Germer to obtain the electron wavelength $\lambda = 1.65$ Å. The agreement between this figure and the number calculated with the aid of de Broglie's formula $\lambda = h/mv$ (for nonrelativistic electrons) brilliantly confirmed de Broglie's theory in its most important aspect.

In the same year 1927, George Paget Thomson (son of J. J. Thomson, the famed discoverer of the electron) completed his experiments involving the diffraction of electrons by thin films (Thomson, 1928). His empirical data, too, were in complete agreement with de Broglie's theory.

The dualistic theory of matter was applied by its discoverer in many physical situations: in a 1977 paper, de Broglie discussed the Pfleegor—Mandel (1967) experiment, in which interferences between the beams emitted by two different lasers were observed under conditions where the probability for having more than one quantum of energy $h\nu$ outside the lasers was negligibly small. He expressed his ideas in the following way (de Broglie, 1977):

> In agreement with classical conceptions, for me a particle is a very small object which is constantly localized in space and a wave is a physical process which propagates in space. . . .

Later he added:

> The particle is a very small region of high concentration of energy which is embodied in the wave in which it constitutes some kind of singularity generally in motion.

With these ideas, the explanation of Pfleegor and Mandel's observations becomes elementary. Lasers 1 and 2 (Figure 3.5) continuously emit two undulatory beams, which intersect and interfere with one another in the region *R*, there giving rise to interference maxima (shown as black stripes in Figure 3.5). If a photon is emitted from one of the two lasers (arrow in Figure 3.5), it will be guided by the wave to reach with greater probability those parts of *R* where the wave amplitude is larger. When many photons are allowed to fall on a screen *S* inserted into the region *R*, the interference pattern produced by the wave will be revealed by the absorbed-photon distribution. From this viewpoint, the particles, by their observed statistical distribution over space, provide only the means for detecting interference in regions where such interference exists. A similar explanation holds for the experiments discussed in the first two sections of the present chapter.

The many problems posed by this interpretation to Heisenberg's uncertainty relations, the indistinguishable nature of photons, the arbitrary normalization of a probabilistic wave, and so on, are considered solved by de Broglie with his *theory of the double solution* (de Broglie, 1956).

There remains, of course, the fundamental problem of the nature of the directly unobservable wave. On this point, a revolutionary thinker such as de Broglie prefers to be rather conservative and to assume that *almost* all energy-momentum is associated with the particle, but that a tiny fraction of it — so minute as to have escaped all means of observation thus far — is smeared out over the wave, as it were. De Broglie believes that this idea can, e.g., explain the cosmological redshift without invoking any Doppler effect: the condition for observing the light from a distant galaxy is that its quanta evade any absorption while traveling from the source to our instruments. But even if this condition is satisfied, argues de Broglie, it is very likely that the accompanying *wave* will be slightly absorbed by interstellar matter, which is

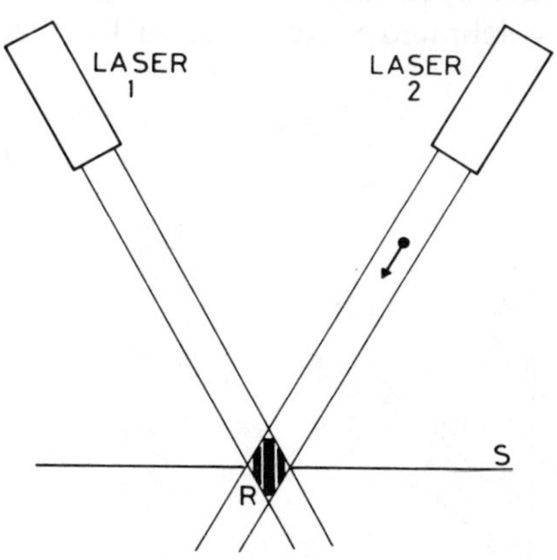

Fig. 3.5. Set up of the Pfleegor—Mandel experiment. Photonic particles are guided by the wave to go with larger probability to the regions of constructive interference (black stripes).

known to be present in space. Such absorption would result in a small energy loss for the dual wave—particle system and therefore in a slight reddening of the light.

Furthermore, since the lengthier the path described by light the longer the duration of its interaction with interstellar matter, one reaches the conclusion that the redshift should increase with distance, as is indeed observed.

A proposal which *in principle* might allow laboratory tests of the idea that a small fraction of energy-momentum is associated with the wave was advanced by de Broglie (1971) in connection with the diffraction of a monochromatic light beam at a small hole. It is known that the diffraction pattern in this case exhibits a central maximum and some secondary lateral maxima. By using suitable optical devices (Figure 3.6), it is possible to eliminate these secondary maxima. This phenomenon, usually called *apodization*, at least in the French literature, results when a plate P of varying thickness is placed between the light source and a hole in a screen S: the light falling on S then varies in intensity over the circular region of the hole, and this variation has the effect of eliminating the secondary maxima of the diffraction pattern. To explain the occurrence of this phenomenon even when the incident light has a very low intensity ('one photon at a time'), one needs to assume that the wave associated with every photon undergoes a partial absorption in the plate P. In de Broglie's picture this means that the energy of the dual wave—particle system should diminish slightly, and the challenge was therefore posed to find a method that would reveal this effect.

3.5. Schrödinger's Waves

De Broglie's undulatory theory was developed further by Erwin Schrödinger (1926) who was able to deduce his celebrated wave equation for the wave function $\psi(\mathbf{x}, t)$:

$$i\hbar \frac{\partial \psi}{\partial t} = -\frac{\hbar^2}{2m} \nabla^2 \psi + V\psi. \tag{3.13}$$

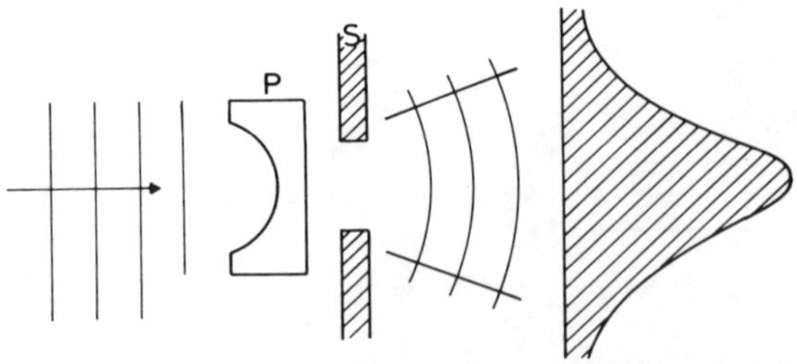

Fig. 3.6. De Broglie's 'apodization' experiment. The variable-thickness plate P is used to eliminate the secondary maxima of the diffraction figure.

Here $\hbar$ is Planck's constant divided by 2π, m the electron mass, ∇^2 the Laplacian operator, and $V = V(\mathbf{x})$ the potential acting on the electron. An important consequence of this equation is that the volume integral extended over all space of $|\psi|^2$ does not depend on time. This gives rise to a conservation law, which was assumed to be that for electric charge by Schrödinger himself and that for total probability in Born's probabilistic interpretation (Born, 1926).

Many authors stress the essential continuity between the ideas of de Broglie and those of Schrödinger, but there are also in fact significant differences between the two viewpoints involved. The most important element of continuity is of course implicit in the discovery of the wave equation itself, since this equation provides a rigorous dynamical foundation for the wave aspect of elementary entities that de Broglie first discovered. On the other hand, a fundamental difference resides in the fact that Schrödinger did not accept the presence of particles, but reduced both light and matter to purely undulatory phenomena.

Philosophically, Schrödinger's ideas were often on the realistic side: for instance, in a 1926 paper he mentioned the "... de Broglie—Einstein undulatory theory, according to which a moving corpuscle is nothing but the foam on a wave radiation in the basic substratum of the universe" (quoted in Jammer, 1966, p. 57). In a paper demonstrating the equivalence of matrix mechanics and wave mechanics, Schrödinger referred directly to 'ether waves'. These quotations should clarify what Schrödinger had in mind, even though his four papers on wave mechanics do not discuss the physical origin of the waves. This omission was intentional; Schrödinger wrote in the first paper of his series:

> It is, of course, strongly suggested that we should try to connect the function ψ with some *vibration process* in the atom, which would more nearly approach reality than the electronic orbits, ... I originally intended to found the new quantum conditions in this more intuitive manner, but finally gave them the above neutral mathematical form .... (Schrödinger, 1926a.)

Anyway, Schrödinger's contributions were *objectively* on the realistic side as regards the ease with which waves can be visualized and the causal connections existing between the processes involved (e.g., Schrödinger viewed electromagnetic waves as causally linked to the frequency *beats* of atomic waves). This circumstance was immediately recognized by physicists of the realistic ilk — and by those of the opposite conviction, as well. Thus, Planck wrote Schrödinger, in a letter dated 24 May 1926, "You can imagine the interest and enthusiasm with which I plunge into the study of these epoch-making works" (Przibram, 1967, p. 6), and invited him immediately to give seminars in Berlin on the subject. Similarly, Einstein wrote, in a letter dated 16 April 1926, "Professor Planck pointed your theory out to me with well justified enthusiasm, and then I studied it too, with the greatest interest ..." (*ibid.*, p. 23).

The reactions in Göttingen and Copenhagen to Schrödinger's theory were very different. In a letter to Pauli, Heisenberg wrote: "The more I ponder about the physical part of Schrödinger's theory, the more disgusting it appears to me" (Jammer, 1966, p. 272). Born, too, understood the realistic philosophy underlying wave mechanics, as witness this remark in his Nobel acceptance talk: "Schrödinger thought that his wave theory made it possible to return to deterministic classical physics." Born pioneered the probabilistic interpretation (Born, 1926) of quantum waves, whereby the whole of wave mechanics was reinterpreted in probabilistic terms. He started from a real difficulty in wave mechanics: Schrödinger wanted to interpret electrons as *just waves* (without any associated particles) and to attribute the observed corpuscular properties of electrons to *wave packets*, i.e., to the confinement of wave disturbances to small regions of space. But this view gave rise to several difficulties. First, wave packets usually spread during their propagation, and electron *localization* (i.e., the property allowing an electron to manifest its presence within a small region of space) is thereby lost. Schrödinger was aware of this objection and tried to counter it with the example of the harmonic oscillator, for which the wave packet remains concentrated during the passage of time and thus localization is preserved. Heisenberg showed, however, that the harmonic oscillator was a very special problem and that in practically all other cases wave *dispersion* took place.

The second type of difficulty with Schrödinger's approach arose from its assumption that electric charge was dispersed over the entire electronic 'cloud'. To be precise, Schrödinger postulated that the charge $-e$ of a single electron filled space with a density

$$q(\mathbf{x}, t) = -e|\psi(\mathbf{x}, t)|^2.$$

Consequently, elements of charge were supposed to be spread throughout the whole region in which the wave function ψ was different from zero. Schrödinger was well aware also of the dilemma this conclusion posed; he wrote:

> The original wave-mechanical model of the hydrogen atom is not self-consistent. The electronic cloud effectively shields the nuclear charge towards the outside, making up a neutral whole, but is inefficient inside; in computing its structure, its own field that it will produce must not be taken into account, only the field of the nucleus. (George, 1953, p. 22.)

The third type of difficulty pertained to interference and diffraction phenomena involving electrons. If electrons were nothing but waves, we should, in diffraction experiments, detect a single electron simultaneously in different places; in other words, all localization would be lost, contrary to experience. Schrödinger thought that this difficulty could be solved by introducing the following idea:

> One must, of course, give up thinking of, e.g., an electron as of a tiny speck of something

moving within the wave train along a mysterious unknowable path. One must regard the 'observation of an electron' as an *event* that occurs within a train of de Broglie waves when a contraption is interposed in it, which, by its very nature cannot but answer by discrete responses: A photographic emulsion, a luminescent screen, a Geiger counter. And one must, to repeat this, hold on to the wave aspect throughout. (*Ibid*., p. 26.)

The foregoing idea is very simple, but untenable today in light of recent experimental evidence, for instance, that arising from neutron interferometry. Roughly speaking, Schrödinger assumed that an electron actually has an extended structure and that any of its *observed* localizations (e.g., a spark on a metallic screen) is only due to the *very nature of the employed instrument*, which cannot, by assumption, react otherwise. If this picture were correct for photons and neutrons, it would imply: (1) in the case of photon experiments involving two semitransparent mirrors, along with two counters, either both counters or none should register a photon arrival, depending on whether or not half a wave packet is capable of triggering a 'localized detection' in the phototube; (2) in the case of neutrons that have undergone Laue-scattering in a silicon slab, again either both neutron counters or none should signal 'neutron detection'. Both predictions are however sharply contradicted by abundant experimental evidence, as should be clear from the empirical facts presented in the first two sections of this chapter (only single counts occur).

In the same year, 1926, that Schrödinger's equation was published, Born advanced the hypothesis that quantum waves were only *probability waves* and, specifically, that the probability density for finding a particle at the point (x, y, z) at time t was given by the squared modulus of its wave function. Thus Born wrote: "The motion of particles conforms to the laws of probability, but the probabiltiy itself is propagated according to the law of causality" (quoted in Jammer, 1966, p. 285). In practice, Born's assumption was very successful in correctly predicting results for particle scattering experiments. Born's waves have often been regarded as observable *only* through the measurement of $|\psi|^2$ as a particle probability density. But, in fact, their best interpretation is in terms of the 'virtual waves' introduced by Bohr, Kramers, and Slater (Bohr *et al*., 1924), which, in Heisenberg's words, are "something standing in the middle between the idea of an event and the actual event, a strange kind of physical reality just in the middle between possibility and reality" (Heisenberg, 1958, p. 41).

Schrödinger's reaction to Born's interpretation of wave mechanics was very negative, as one would expect. In a 1927 letter to Planck, he wrote:

What seems most questionable to me in Born's probability interpretation is that . . . the most remarkable things come forth naturally: the probabilities of events that a naive interpretation would consider to be independent do not simply multiply when combined, but instead 'the probability amplitudes interfere' in a completely mysterious way (namely, just like my wave amplitudes, of course). In a brand new article by Heisenberg even my much smiled at wave packets are said to have finally found their suitable interpretation as 'probability packets'. The first is especially comical. It can also be expressed in this way: the Born probability (more correctly its square root) is a two-dimensional vector; its addition is to be carried out vectorially. The multiplication is still more complicated, I believe. Well, as God wills; I keep quiet. That is, if one really *must*, I too will become accustomed to such things. (Przibram, 1967, p. 20.)

Schrödinger was forced to wage a difficult battle in Copenhagen, where Bohr had invited him to lecture in the fall of 1926. To quote Heisenberg, Born asked Schrödinger "not only to lecture on his wave mechanics, but also to stay in Copenhagen so long that there would be adequate time to discuss the interpretation of quantum theory" (Rozental, 1967, p. 103). Heisenberg also describes the intensity of the Bohr—Schrödinger debate:

. . . though Bohr was an unusually considerate and obliging person, he was able in such a discussion, which concerned epistemological problems which he considered to be of vital importance, to insist fanatically and with almost terrifying relentlessness on complete clarity in all arguments. He would not give up, even after hours of struggling, before Schrödinger had admitted that his interpretation was insufficient, and could not even explain Planck's law. Every attempt from Schrödinger's side to get round this bitter result was slowly refuted point by point, in infinitely laborious discussions. It was perhaps from overexertion that after a few days Schrödinger became ill and had to lie abed as a guest in Bohr's home. Even here it was hard to get Bohr away from Schrödinger's bed . . . . (*Ibid.*, p. 103.)

Heisenberg concludes:

Schrödinger finally left Copenhagen rather discouraged, while we at Bohr's Institute felt that at least Schrödinger's interpretation of quantum theory, an interpretation rather too hastily arrived at using the classical wave theories as models, was now disposed of . . . . (*Ibid.*, p. 104.)

Schrödinger's mathematically alluring theory contained several features that clearly rendered invalid the picture of a purely wave-like electron. Some of these features were discussed above; another one worth mentioning is the formulation of wave mechanics in 'configuration space', i.e., an often fictitious space whose dimensionality equals the number of degrees of freedom of the system under study. In this formulation, the quantum wave that describes two 'particles' is theoretically imagined to propagate in a six-dimensional space. But no derivation of this conjecture from a more physical foundation was ever furnished by Schrödinger.

Max Born's probabilistic interpretation managed to overcome Schrödinger's difficulties, as it was related to a less ambitious program. One often reads in textbooks that, for Born, quantum waves were merely a mathematical expedient useful *only* for calculating probabilities. But such assertions fail to tell the full story, considering that Born could write:

The question of whether the waves are something 'real' or a function to describe and predict phenomena in a convenient way is a matter of taste. I personally like to regard a probability wave, even in $3N$-dimensional space, as a real thing, certainly as more than a tool for mathematical calculations. . . . Quite generally, how could we rely on probability predictions if by this notion we do not refer to something real and objective? (Born, 1964, p. 105.)

Therefore, at least the problem posed by wave propagation in configuration space continues to exist also for Born's waves if these are conceived as something more than just a mathematical tool for the calculation of probabilities.

Other difficulties with quantum waves can be overcome, at least in principle, as will be shown in the next chapter.

3.6. Bohr's Complementarity

Niels Bohr's papers up to 1926 treated the electromagnetic field in purely undulatory, i.e., essentially classical, terms. If in his 1913 paper on the quantization of the hydrogen atom (Bohr, 1913) the nature of the electromagnetic field is basically left unexplored, his 1918 paper on the *correspondence principle* (Bohr, 1918) invokes classical electrodynamics not only in the region of high quantum numbers but also in that of intermediate and low numbers. In this way, intensities, polarizations, and selection rules for the emitted radiation were correctly calculated.

The success of these calculations was greater than could have been anticipated. For instance, the selection rules were immediately deducible if one assumed that spherical waves were emitted — in defiance of all the evidence which at that time seemed to indicate a localized structure of radiation quanta: photoelectric measurements and Einstein's calculation of molecular velocity distributions. In any event, the fortunate outcome of the semiclassical calculations helps to understand why Bohr was not willing to accept Einstein's photonic picture of the electromagnetic field. Also the famous 1924 paper by Bohr, Kramers, and Slater treats the electromagnetic field as continuous. The story associated with this publication is very interesting.

Slater, who was convinced by Compton's results that particles were indeed present in electromagnetic radiation, thought of combining this fact with the undulatory properties of radiation by postulating a 'virtual' radiation field that was continuously being created by 'virtual' atomic oscillators. The virtual field was supposed to be strictly linked to the particles, carrying them along with itself during its motion. A paper stating what Slater had in mind in early 1924 appeared only in 1925. As he wrote there,

> The essential feature was the emission of the field before the ejection of the corpuscle: that is, during the stationary state before the transition . . . . (Slater, 1925.)

With these ideas in his mind, Slater went to Copenhagen, hoping to develop his theory in collaboration with Bohr and Kramers. But this is what happened instead:

> I had gone to Copenhagen with the idea that the field of oscillators would be used to determine the behavior of photons, which I preferred to regard as real entities, satisfying

conservation as we now know that they did, and I wished to introduce probability only insofar as the waves determined the probability of the photon's being at a given place at a given time. Bohr and Kramers opposed this view so vigorously that I saw that the only way to keep peace and get the main part of the suggestion published was to go along with them with the statistical idea. (Quoted by van der Waerden, 1967, p. 13.)

And so it came that Einstein's picture was dropped and a completely different paper was written. In the latter, the electromagnetic field is described as continuous everywhere (thus allowing neither localized nor extended quanta of energy), but 'virtual'. Roughly speaking, this means that the field has to be introduced if correct calculations are to be performed, but that, all the same, it is not real in the usual sense. Such a field is after all not too different from the empty wave of Einstein except for the absence of associated particles (see Figure 3.7).

In the Bohr—Kramers—Slater (BKS) theory an atom in its fourth stationary state (we enumerate states in order of increasing energy), for instance, emits continuously three virtual spherical waves with frequencies $\nu_i = (E_4 - E_i)/h$, where $i = 3, 2, 1$. These waves, together with any other waves that might emanate from nearby atoms, give rise to an electromagnetic field which, through its overall intensity, determines a transition probability

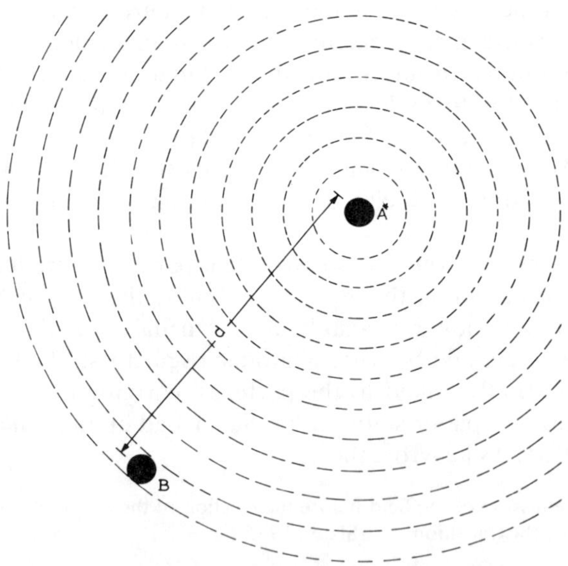

Fig. 3.7. In the BKS theory an excited atom A^* emits a spherical virtual wave (and nothing else). A second atom B under the action of this wave has a probability of jumping to a higher energy level. No time correlation between A^* de-excitation and B excitation is expected. In Einstein's picture the second event is instead predicted to take place a time d/c after the first one.

(in Einstein's 1917 sense) for our initial atom to different stationary states. The jump of this atom from the given stationary state to another takes place under violation of energy (and momentum) conservation: in fact, the atomic energy changes while no energy is transferred to the radiation field, which by hypothesis, is virtual and thus carries no energy. BKS themselves wrote of this:

> . . . we abandon on the other hand any attempt at a causal connection between the transitions in distant atoms, and especially a direct application of the principles of conservation of energy and momentum, so characteristic of the classical theories. (Bohr *et al.*, 1924.)

Nonetheless, these authors claimed that energy conservation would still be valid as a *statistical* concept. Also, in the BKS picture there are no 'spontaneous' transitions: all transitions are stimulated, albeit by a virtual field.

A further prediction of the BKS theory is that no time correlation exists between atomic events. To understand this point, consider a gas of identical atoms, half of them in the state E_1 and half in the state E_2 ($> E_1$). Einstein's photon theory requires any atom A making a transition from E_2 to E_1 to emit a photon, which travels through space and eventually excites a second atom B from E_1 to E_2. Thus, there is clearly a time relation between the events A and B, calculable from the distance AB and the speed of light. In the BKS theory, however, such a relation does not obtain: no energy-quanta are propagated through space, since only a continuous field exists, which determines only a *transition probability* for individual atoms.

It was precisely this prediction which led Bothe and Geiger (1925) and Compton and Simon (1925a, b) to carry out experiments which disproved the BKS theory. These authors used the Compton scattering of x-rays, which in the BKS theory would only be a continuous beam of virtual, short-wave radiation. Thus, in place of atom A making a transition between two stationary states, we have an electron that is suddenly scattered (according to a probability law similar to the one holding for atoms), and, in place of atom B, we have the discharge of a counter due to ionization processes. Bothe and Geiger found a very sharp time correlation between electron scattering and x-ray absorption, and thereby conclusively established that the BKS theory could not provide a correct description of the physical world.

The foregoing experimental results motivated Bohr to reconsider his totally negative opinion toward wave—particle duality. The outcome was the celebrated *principle of complementarity*, which summarized Bohr's peculiar way of accommodating a picture of the world in his dualism. It was presented in 1927, twenty-two years after Einstein had started to work on the problem.

Returning from a skiing holiday in Norway, Bohr was told by Heisenberg about his discovery of the uncertainty relations. At first Bohr was rather dissatisfied with Heisenberg's paper, objecting not only to certain statements of the first version, which were incorrectly founded, but also to the fact that wave—particle dualism was not the central fact.

Further, he had probably already grown familiar, while he was in Norway, with the concept of complementarity which would make it possible to take the dualism between the wave and particle picture as a suitable starting point for an interpretation. This concept of complementarity fitted well the fundamental philosophical attitude which he had always had, and in which the limitations of our means of expressing ourselves entered as a central philosophical problem. (Rozental, 1967, p. 106.)

After several weeks of discussion, Heisenberg and Bohr concluded that they really were talking about the same thing and that the uncertainty relations could be considered as a special case of the more general complementarity principle.

In fact, Heisenberg's discovery of the uncertainty relations was an important bridge between the formalism of quantum theory and the physical world. But it did not provide answers to all the relevant questions, such as: What is the composition of the electromagnetic field, particle or waves? And how can one take into account the undulatory properties of the electrons which de Broglie had introduced?

Bohr faced such questions directly and found a 'solution' revolving around the assumption that we cannot hope to solve this kind of contradiction. Rosenfeld, one of Bohr's pupils, wrote in this context:

While the great masters [Planck, Einstein, Born, and Schrödinger] were vainly trying to eliminate the contradictions in Aristotelian fashion by reducing one aspect to another, Bohr realized the futility of such attempts. He knew that we have to live with this dilemma . . . and that the real problem was to refine the language of physics so as to provide room for the coexistence of the two conceptions. (Quoted in Landé, 1965, pp. 147—148.)

Obviously, to live with a dilemma is something fundamentally different from solving it. One cannot fail to notice at this point the close resemblance between Bohr's renunciation of any solution to the fundamental problem of atomic physics and the irrationalistic philosophies of Kierkegaard and Høffding, to which Bohr had been exposed since his youth and by which he had been strongly influenced. For, Kierkegaard's 'qualitative dialectic' consisted of the statement that contradictions in life and nature are rigid and impossible to overcome, coupled to sharp attacks against the rationalism of Hegel and his conviction that contradictions can always be synthesized (i.e., resolved) at a higher level.

Bohr's complementarity can be introduced in the following manner. (See Bohr, 1961, for the original presentation of this idea.) Any experimenter lives in a macroscopic world, and conceptions typical of this condition, such as causality, space, and time, are deeply impressed on the human mind. But it is not necessary and, according to Bohr, it is in fact not true that conceptions even of such a general nature should have an unlimited applicability in the domain of microphenomena. The key to a correct understanding of this fundamental point is the existence of the quantum of action h.

Let us examine the experimental meaning of causality and of spacetime. Causality, for Bohr, is what obtains in any process that takes place according to well-defined rules. These rules are in practice those of energy-momentum conservation. Therefore, an experimenter who wishes to check the rigorous validity of the causality law must perform infinitely precise measurements of energy and momentum. The relations $\Delta E = 0$ and $\Delta p = 0$ imply however, because of Heisenberg's relations, that $\Delta t = \infty$ and $\Delta x = \infty$, respectively, which mean that absolutely no localization in time and in space is achievable during the measurements that are by hypothesis required. This complete lack of localization leads Bohr to infer that space and time in practice do not exist. (It should be noted that here, as in many other places, Bohr makes an important concession to the positivistic philosophy, assuming that what is not observed does not exist.) One therefore concludes that any attempt at establishing causality will preclude the observation of space and time.

Conversely, one may wish to observe spacetime localizations. An ideal measurement would now imply $\Delta x = 0$, whence $\Delta p = \infty$. This means that an arbitrarily large momentum is exchanged between observational apparatus and atomic system. Furthermore, such an exchange is *in principle* impossible to determine 'if the measuring apparatus has to serve its purpose', as Bohr shows in detail in several concrete situations. Under these conditions, it is obviously impossible to verify the law of causality (i.e., to check energy-momentum conservation). One thus infers that observation of space and time is incompatible with any experimental verification of the validity of the law of causality. Consequently, it makes no (positivistic) sense to even talk about such a law when spacetime observations are being performed.

Bohr concludes from the above considerations that there is a *complementary relation* (rigid mutual exclusion) between spacetime coordination and causality: These two ideas can never be employed simultaneously.

A similar conclusion is reached when one considers the conceptions of particle and wave. When studying causality (e.g., in the photoelectric effect or the Compton effect), one discovers that the conservation of energy and momentum 'finds its adequate expression just in the light quantum idea put forward by Einstein'. Conversely, the conceptual tool that is suitable for predicting possible points P of localization of an observed system after a prior localization in P', is the wave function, whose squared modulus gives the probability density for observing the system at different points P. It is naturally impossible to know in which point the localization will manifest itself. Therefore, the evolution of the system appears as intrinsically stochastic, which is the same as saying once more that causality does not hold. It should furthermore be stressed that one is here speaking an undulatory language. There accordingly also exists a mutual exclusion between particle and wave, which must thus be considered complementary descriptions of atomic systems. The complementarity interpretation of the fundamental experiments on duality presented in the first two sections of this chapter will be discussed at the end of the next section, along with their interpretation based on Fock's 'relativity to the observational means'.

We should emphasize that the (Bohr) wave, which represents one of the complementary aspects of any atomic entity (the other one being the particle), is a successor to the BKS virtual wave and philosophically very akin to the probability wave that Born in 1926 introduced into quantum mechanics. In fact, as was stated above, the wave is applicable to the investigation of spacetime localizations. But, since strict localization precludes the possibility for checking the causality law (i.e., energy-momentum conservation), one is faced once more, as in the BKS paper, with a wave that does not generate energy-conserving transitions. Nonetheless, the wave can modify the probabilities of atomic transitions, and this is a feature that we shall visit in the next chapter.

A closing remark on complementarity: recent research on this principle has shown that the difference between the wave and particle aspects is less sharp than Bohr imagined and that undulatory phenomena are easily observable — even in cases where one is 99 percent sure that particle propagation has occurred (Wootters and Zurek, 1979; Bartell, 1980). In the present author's opinion, these more recent findings tend objectively to build a bridge between Bohr's complementarity and the Einstein—de Broglie version of dualism.

It should also be noted that the well-known interpretation of quantum waves in probabilistic terms, proposed by Max Born in 1926 and now universally accepted, was essentially a reformulation and a generalization of the ideas first advanced by Bohr, Kramers, and Slater. This historical continuity was stressed particularly by Heisenberg who wrote:

> The probability wave of Bohr, Kramers, and Slater . . . was a quantitative version of the old concept 'potentia' in Aristotelian philosophy. It introduced something standing in the middle between the idea of an event and the actual event, a strange kind of physical reality just in the middle between possibility and reality. Later, when the mathematical framework of quantum theory was fixed, Born took up this idea of the probability wave and gave a clear definition of the mathematical quantity in the formalism . . . . (Heisenberg, 1958, p. 41.)

In this manner the 'virtual' waves of BKS became the familiar wave functions of quantum theory.

Returning to Bohr's complementarity, it is interesting to note that, according to Rosenfeld, Bohr did not see a complete symmetry between the particle and the wave aspects. *In the case of light* he seems to have ascribed physical meaning more to the wave than to the particle. In fact, Bohr

> . . . attached reality, i.e., reality as it was defined by him, to those aspects that could be directly observed in certain limiting circumstances by direct macroscopic observation. And, of course, in the case of radiation it is clear that direct observation in the limiting case of small values of $h\nu/kT$ gives the usual classical wave description of Hertz and Maxwell. As to the photon or the light quantum concept, introduced by Einstein, Bohr regarded it as a useful but an

auxiliary concept, one which he later called symbolical, meaning thereby that it was not an aspect of the radiation phenomena which could be directly observed as such. . . . Bohr always made this distinction between the two aspects of radiation. (Rosenfeld, 1973, p. 252.)

The latter statement is essentially repeated but also made more precise when Rosenfeld later writes:

. . . the situation between the complementary aspects of light, the wave aspect which is in direct correspondence with the classical description, and the concept of photon which is symbolical, which is only an expression for the exchange of energy and momentum between matter and radiation, is not paralleled, as de Broglie thought, by the wave and particle aspects of matter, but it is just the other way around. For matter, the aspect which is in correspondence with classical observation is the particle aspect, of course; whereas the wave aspect is a symbolical one . . . . (*Ibid.*, p. 260.)

3.7. Fock's Relativity with Respect to the Means of Observation

In 1957 the Soviet physicist Vladimir A. Fock, on visit in Copenhagen, several times discussed with Bohr the interpretation of quantum mechanics. Their disagreements did not stem from the physical applications of the theory, but rather from some statements that appeared in Bohr's papers and could be considered as important concessions to the philosophy of positivism. After some exchanges of opinion, Fock wrote a paper with the title 'My Answer to Professor Niels Bohr' (Fock, 1964) and gave it to the Danish physicist who examined it with the greatest care. On many points Bohr finally agreed with Fock, and this is reflected in papers he authored in the last part of his life. (An interesting discussion of the Bohr—Fock controversy of which I could not find any English equivalent is contained in Tagliagambe, 1972.) Bohr's modified views are contained, for instance, in a 1959 paper, published in a Russian journal, and they were exposed in seminars held in Moscow during 1961.

Fock's paper contained four main objections to Bohr's early ideas on the foundations of quantum physics.

Fock's first objection revolves around Bohr's insistence on the negative content of quantum mechanical conceptions, a content which puts severe limitations on classical descriptions as a consequence of the uncertainty relations. Some of these limitations are of course real, but Fock rejects the idea that there should be a profound difference between the mathematical symbols of classical physics, which represent something real, and the mathematical symbols of quantum mechanics, which should according to Bohr be purely abstract entities. In fact, Fock sees no difference in the role of mathematics in classical and in quantum physics. *In particular the wave function ψ represents something real, as it allows one to predict the evolution of probabilities*, which reflects something real and concrete.

The second objection concerns Bohr's frequent statement that causality was no longer true in quantum physics. Fock replies to this that only Laplacian determinism has broken down, whereas 'simple causality', as contained in the statement that well-defined natural laws do exist, remains valid. It is therefore an error to exclude the term 'causality' from atomic physics: one should stress instead that some type of probabilistic causality still reigns.

The third objection attacks the use of the complementarity principle made by Bohr. One should emphasize that complementarity expresses limitations imposed by Heisenberg's relations only on the *classical* description of phenomena. As far as a quantum mechanical description is concerned, no limitation at all is implied. The properties of atomic objects are electric charge, mass, spin, type of wave equation, and so on. Experimental probability distributions can be calculated with an ever increasing precision, while we penetrate with our understanding into the nature of such phenomena and make use of more sophisticated quantum mechanical theories.

Fock's final objection deals with Bohr's idea of the existence of an 'uncontrollable interaction' between the measuring apparatus and the object under study. Fock concurs with the view that a boundary has to be drawn between what is regarded as the quantum mechanical object under investigation, on the one hand, and the measuring apparatus to be described classically, on the other. However, *where* exactly to place the boundary has to be an arbitrary decision. Therefore, considering the finite physical interaction between observed system and measuring agent, one certainly is allowed to study also the latter simply by shifting the boundary closer to the observer. It is therefore possible in principle to study any physical process taking place in the measuring apparatus by means of quantum theory: for instance, a theory of photographic emulsions is feasible. The circumstance that somewhere one has to draw a 'gnoseologic' boundary, does not entail a renunciation of the physical description of parts of the apparatus.

The above four objections do not explicitly contain Fock's idea of relativity with respect to the means of observation, which was introduced by him in the sixties (see, e.g., Fock, 1978) and can be viewed as a consequence of his third objection. In developing this new idea, Fock starts from the philosophy of knowledge, in which one must distinguish the subject with his consciousness, his mental faculties, and his perceptions, from the object of the external world studied by him. In classical physics, the subject can study the object without influencing in any way its properties and its evolution. The classical observer is, in Fock's words, a 'spy' of the external world, which can perceive without being perceived. In quantum physics this 'spying' becomes impossible, because the existence of the *quantum of action h* implies a finite inter*action* between atomic system and measuring apparatus. The observer is, however, a classical macroscopic object: his reasoning and his fundamental concepts are necessarily those of macroscopic physics (space, time, energy, momentum, and so on). The apparatuses, which are merely amplifiers of the observer's perceptive abilities, must also be described classically, in the sense that they can only convey informations about spacetime, energy-momentum, and so on. A classical language is thus used to describe the results of measurements. But a classical language cannot be used without limitations, because one is now measuring microscopic objects with quantum properties. Fock believed that such limitations were precisely expressed by the Heisenberg relations

$$\Delta x\, \Delta p_x \geqslant \hbar, \Delta y\, \Delta p_y \geqslant \hbar, \Delta z\, \Delta p_z \geqslant \hbar,$$

in which Δx, Δy, Δz are the linear dimensions of the region in coordinate space containing the atomic object under study and Δp_x, Δp_y, Δp_z similarly characterize a region in momentum space of the same object. An accurate description of quantum measurements is, according to Fock, the following: *The means of observation must be described in classical terms but with due regard to the uncertainty relations.*

A micro-object is revealed in its interaction with an instrument. For instance, the track of a charged particle can be rendered visible by the condensation of tiny droplets around the trail of ions it leaves in the supersaturated vapor of a cloud chamber or by the development of the latent image created by a similar trail in a photographic emulsion. The macroscopic manifestations of the interactions of atomic objects with measuring instruments are the basic elements from which all understanding and all theoretical systematization are fashioned. Measurements of position and momentum must always satisfy Heisenberg relations, the latter expressing an absolute limitation on the way all experimental setups function. It is thus possible to build an instrument capable of localizing a particle in space with near-perfect precision, but the same equipment can reveal next to nothing about the particle's momentum. Conversely, an instrument designed to perfectly define a particle's momentum will fail to localize the object in any way.

One can therefore accept Bohr's proposal and call *complementary* those properties (such as corpuscular or undulatory appearance) which manifest themselves in their pure forms only under mutually exclusive sets of experimental conditions, while in other experiments assuming only their imperfect, modified forms (e.g., an object's incomplete localization in coordinate and momentum space, respectively, which is mandated by the uncertainty relations).

According to Fock, there is no sense in considering complementary properties simultaneously (in their pure forms), and thus no contradiction can arise from wave—particle duality.

This way of considering duality makes each aspect of a micro-object arise exclusively in conjunction with a certain class of measuring apparatuses. Thus one observes the particle aspect only when apparatuses capable of producing a perfect localization in momentum space are used. Similarly, the wave aspect becomes apparent only in experimental setups that produce perfect localization in coordinate space (while destroying any localization in momentum space). Accordingly, the two aspects exist relative to different classes of apparatuses.

It is of considerable interest to apply the complementarity idea to the interpretation of the contradictory experimental evidence discussed in the first two sections of the present chapter. Considering, for instance, the experiments on photons, Bohr would start by stressing that the Jánossy–Naray experiment and the one employing counters are mutually incompatible: one can choose to detect photons behind the beamsplitter, but, in so doing, one would *ipso facto* destroy any possibility to observe interference fringes; alternatively, one can elect to observe interference, but only at the cost of surrendering in principle any possibility to know along which path the photon has propagated. The two experiments appear therefore as complementary in Bohr's sense: they exclude each other and cannot be performed simultaneously. Nevertheless, only the information gathered by performing both experiments will enable us to complete the catalog of properties governing the behavior of photons. Any questions about the behavior of the unobserved photon are to be flatly turned aside: remember that Bohr advocated the use of the term 'reality' only for describing *experiments*, i.e., human acts of observation. Note also that detection of the interference fringes introduces a wave picture of reality and that waves are not localized — consistent with the fact that it is impossible to tell over which path the photon propagates. Detecting the photon on one of the two paths, amounts to a check that energy is conserved, because the photons are found to be single chunks of energy $h\nu$, as is known to be true also for the incident light. But this finding, as Bohr always stressed, goes along with the corpuscular picture, a fact again consistent with the observed localization of the photon.

For some period of time, Bohr would have said that neither the wave picture nor the particle picture represents the objective reality of a photon, because these conceptions are classical in nature and can only serve as makeshifts in the quantum domain. It seems however probable that Bohr did not always consistently hold this negative opinion. His virtual wave of 1924 was an aspect that had a classical analog and was therefore considered 'more real' by him. Bohr's acceptance of Fock's 1957 criticisms of complementarity points in the same direction, as Fock also stressed that the ψ function represented 'something real'.

Apropos of Fock's 'relativity with respect to the means of observation', it is interesting to note that this conception applies in a very straightforward manner to the experiments discussed in the first two sections of the present chapter. Experimental setups such as Jánossy and Naray's are designed to actualize interference fringes, and, relative to such apparatuses, the photon discloses its wave aspect. Conversely, apparatuses capable of localizing photons, such as those involving counters along with a beamsplitter, elicit the particle aspect of radiation while curbing its potential to interfere. One can thus say that, relative to the first (second) class of instruments, the photon behaves as a wave (particle). The question that remains unanswered concerns the true nature of the photon: Is it *really* a wave or a particle or a combination of wave and particle or something entirely different but unknowable 'in classical terms'? Fock does not give a direct answer to this question, but one can guess from his papers that he probably favored the latter alternative.

3.8. Heisenberg Beyond Complementarity

Heisenberg wrote in 1958:

> . . . the concept of complementarity introduced by Bohr into the interpretation of quantum theory has encouraged the physicists to use an ambiguous rather than an unambiguous language, to use the classical concepts in a somewhat vague manner in conformity with the principle of uncertainty, to apply alternatively different classical concepts which would lead to contradictions if used simultaneously. In this way one speaks about electronic orbits, about matter waves and charge density, about energy and momentum, etc., always conscious of the fact that these concepts have only a very limited range of applicability.
>
> When this vague and unsystematic use of the language leads into difficulties, the physicist has to withdraw into the mathematical scheme and its unambiguous correlation with the experimental facts. (Heisenberg, 1958, p. 179.)

It is interesting to notice that this withdrawal 'into the mathematical scheme' has become a mass phenomenon in contemporary theoretical physics and that Bohr's complementarity and the other forms of duality are half-forgotten. The previous statement thus shows that Heisenberg was a very perceptive forecaster of a historical process which has led the community of physicists from the awkward language of complementarity to the clear-cut instructions of a mathematical formalism. It can also be said that Heisenberg moreover advocated such a development. Analyzing the use of descriptive concepts in atomic physics, he insisted that they were often vague and that this vagueness was similar to that characterizing the notion of 'temperature' in reference to a single atom. As is well known, temperature is a statistical concept which makes sense only when applied to a lump of matter containing billions of atoms; for a single atom, the same concept is not defined, inasmuch as one cannot correlate it with any mechanical property of this individual system. In a similar way, according to Heisenberg, all the classical concepts become in the atomic domain just as well or just as little defined as the 'temperature of an atom'. They are in fact correlated with statistical expectations, behaving something like tendencies and possibilities, and can best be described with the notion of 'potentia' of Aristotelian philosophy:

> So the physicists have gradually become accustomed to considering the electronic orbits, etc., not as reality but rather as a kind of 'potentia'. The language has already adjusted itself, at least to some extent, to this true situation. But it is not a precise language in which one could use

the normal logical patterns; it is a language that produces pictures in our mind, but together with them the notion that the pictures have only a vague connection with reality, that they represent only a tendency toward reality. (*Ibid.*, p. 181.)

The way out of this dismal situation was for Heisenberg either the above-quoted 'withdrawal into the mathematical scheme' of quantum theory or the development of a different, precise language that follows definite logical patterns in complete conformity with the same scheme.

The modifications of classical logic advocated in the latter connection by Birkhoff and von Neumann and by von Weizsäcker are cited with approval by Heisenberg. In classical logic one of the two statements '*A* is *B*' or '*A* is not *B*' is necessarily true: '*tertium non datur*.' In quantum logic — which, be it noted, has developed at a remarkable pace in recent times (Beltrametti and Cassinelli, 1976) — particularly this 'law of the excluded middle' has to be modified.

Werner Heisenberg is probably the physicist who has done most to encourage a withdrawal into the mathematical certainties of quantum physics. One of his first papers, written in 1925 in collaboration with Kramers, already had a most mathematical formulation. The comments, which Heisenberg later made about his paper are very revealing of his approach:

> . . . I found in the formulae, which were the results of my collaboration with Kramers, a mathematics which in a certain sense worked automatically, independently of all physical models.
>
> This mathematical scheme had for me a magical attraction, and I was fascinated by the thought that perhaps here could be seen the first threads of an enormous net of deep-set relations. (Rozental, 1967, p. 98.)

About this attitude in those years Heisenberg also wrote:

> I was perhaps, already at that point, more prepared than Bohr to leave the models and take the step over to mathematical abstraction. (*Ibid.*)

This step was taken in the summer of 1925 when Heisenberg discovered matrix mechanics.

Guided by a daring combination of experimental facts, symmetry requirements, and mathematical intuition, Heisenberg erected an important pillar of modern theoretical physics. The outcome of his inspired effort left however very little to physical intuition, since the 'position' of a particle had become an infinite-dimensional matrix of complex numbers. This shortcoming was underlined by explicit statements of a strong philosophical flavor; this one, for example:

> . . . it seems sensible to discard all hope of observing hitherto unobservable quantities, such as the position and period of the electron . . . . (van der Waerden, 1967, p. 262.)

A conclusion of this kind gave however rise to very serious difficulties when it was tried to apply the theoretical formalism to physical situations: Not only had experiments been performed which strongly suggested that 'trajectory' was a meaningful concept for atomic particles (notably the Bothe—Geiger and Compton—Simon experiments), but the quantum theorists themselves could not avoid making reference to position and momentum in their intuitive meaning, when discussing concrete physical situations. These contradictions were fully realized by Bohr and Heisenberg. Indeed their writings of this period often contain the word 'paradox', and, according to Heisenberg, much time was spent in the last part of 1926 to conceive ideal experiments "each displaying the paradoxes in a more clear-cut way" (Rozental, 1967, p. 105). But at first the theorists had little success, since, as Bohr and Heisenberg had to agree, it was not clear how the mathematical formalism of quantum theory was to be used even for the simple case of an electron track in a cloud chamber.

It is interesting to understand the reasoning by which Heisenberg discovered the uncertainty relations. On the one hand he trusted fully the mathematical formalism developed since 1925. Discussing the papers by Dirac and Jordan, he wrote:

> ... Dirac and Jordan developed the transformation theory ... and the completion of this mathematical formalism again confirmed us (i.e., Bohr and Heisenberg) in our belief that there was no more to change in the formal structure of quantum mechanics, and that the remaining problem was to express the connection between mathematics and experiment in a way free of contradictions. (*Ibid.*, p. 105.)

On the other hand he held the idea that *nature had to conform* to the formalism of quantum theory. In fact, Heisenberg recalled that in February 1927, during a walk under the stars in a park in Copenhagen,

> The obvious idea occurred to me that one should postulate that nature allowed only experimental situations to occur which could be described within the framework of the formalism of quantum mechanics. (*Ibid.*)

Once more we see that Heisenberg gave priority to the mathematical formalism over considerations about objective reality. And once again he made a great discovery which has shaped theoretical physics in the twentieth century and which has led to almost incredible successes in the interpretation of atomic phenomena. Even so, it is fair to say that nobody really understands, to this day, *why* quantum theory works and that epistemology still lags far behind physics.

There are some considerations that will make for a fuller understanding of Heisenberg's discovery of uncertainty relations:

1. Heisenberg's description of the interaction between electromagnetic radiation and the electron implied a rejection of the Bohr—Kramers—Slater theory and at least a partial acceptance of Einstein's picture of light. It can thus be said that the realization that Einstein had not been wrong with his

theory of photoelectricity was a prerequisite for the discovery of the uncertainty relations.

2. Strong statements of a philosophical nature were made by Heisenberg also in his 'uncertainty' paper. He stated, for instance, that, if one wants to talk of the position of an object,

> he has to describe an experiment by which the 'position of an electron' can be measured; otherwise this term has 'no meaning at all'. (Jammer, 1966, p. 328.)

A page later he stated that

> the path comes into existence only when we observe it. (*Ibid.*, p. 329.)

In the closing part of his paper, Heisenberg wrote:

> In view of the intimate connection between the statistical character of the quantum theory and the imprecision of all perception, it may be suggested that behind the statistical universe of perception there lies hidden a 'real' world ruled by causality. Such speculations seem to us — and this we stress with emphasis — useless and meaningless. For physics has to confine itself to the formal descriptions of the relations among perceptions. (*Ibid.*, p. 330.)

3. There is an element of arbitrariness in Heisenberg's conclusions. The fact that one cannot *measure* position and momentum of an electron simultaneously and with great precision does not necessarily mean that such quantities do not have any physical meaning. It is only within a positivistic or an idealistic philosophy that one can draw such a conclusion. The validity of the uncertainty relations still allows one to calculate position and momentum *in the past* with any desired accuracy. About this fact Heisenberg wrote in 1930:

> Then for these past times $\Delta x\ \Delta p$ is smaller than the usual limiting value, but this knowledge of the past is of a purely speculative character, since it can never . . . be used as an initial condition in any calculation of the future progress of the electron and thus cannot be subjected to experimental verification. It is a matter of personal belief whether such a calculation concerning the past history of the electron can be ascribed any physical reality or not. (Heisenberg, 1949, p. 20.)

Heisenberg's own 'personal belief', which has since become the dominant point of view certainly was on the side of a denial: one should refrain from attributing physical reality to any retrodicted coordinate or momentum values.

It is of course easy to apply Heisenberg's way of reasoning to the experiments discussed in the first two sections of this chapter. The problem involved reduces simply to a question of computation: can one predict the outcome of all experiments by using the mathematical apparatus of quantum mechanics? As is well known, the answer of this question is in the affirmative, and from this Heisenberg would conclude that no further questions remain to be asked. If, to be specific, the neutron wave functions are $\psi_1(\mathbf{x}_1)$ and $\psi_2(\mathbf{x}_2)$ for points $\mathbf{x}_1$ and $\mathbf{x}_2$ lying respectively on the first and second trajectory behind the silicon slab, then the probability for observing a particle in a neutron counter aligned with the ith trajectory is

$$P_i = N_i \int_{R_i} d\mathbf{x}_i |\psi_i(\mathbf{x}_i)|^2, \quad i = 1, 2,$$

where N_1 and N_2 are normalization constants and R_1 and R_2 the regions occupied by the counters. In practice these regions will not overlap, so that P_1 and P_2 will represent probabilities for mutually exclusive events. If, on the other hand, a counter is placed in the region R behind the *third* slab of the neutron interferometer, the probability for neutron detection will be

$$P = N \int_R d\mathbf{x} |\psi_1(\mathbf{x}) + \psi_2(\mathbf{x})|^2,$$

with N denoting a normalization constant, since both wave functions are different from zero in R. The interference term $\psi_1^*\psi_2 + \psi_2^*\psi_1$ in this formula gives rise to oscillations of P when the relative phase of ψ_1 and ψ_2 is varied by some physical means. It is thus clear that all the observed features of neutron interferometry are correctly predicted by the mathematical formalism of quantum mechanics.

3.9. The Consciousness Interpretation

Recall that in Chapter 2 we discussed de Broglie's paradox by considering the case of an electron that is confined to a box B with perfectly reflecting walls which can be divided into two closed compartments B_1 and B_2 by a double sliding wall (Figure 2.1). One may assume that these compartments are disunited and then transported to two widely separated locations, B_1 to Paris and B_2 to Tokyo, say. The new situation is described quantum mechanically by two wave functions, $\psi_1(x, y, z, t)$ defined in B_1 and $\psi_2(x, y, z, t)$ defined in B_2. The probability that one will find an electron on opening any one of the two boxes can be calculated as a volume integral of the squared modulus of the corresponding wave function. Let P_1 (for B_1) and P_2 (for B_2) be these two probabilities, with $P_1 + P_2 = 1$.

Suppose next that the box in Paris (i.e., B_1) is opened and an electron found to be present. This allows one to conclude that the electron cannot be in Tokyo, which means that P_2 is known at that time to be zero, even before B_2 is opened. There is thus a sudden jump of P_2 to the value zero from the nonzero value it had before B_1 was opened. This reasoning would not be essentially different if it were carried out for the same problem in classical physics. The only novelty is that probabilities in quantum theory are calculated from the squared moduli of wave functions. However, this linkage of wave function with probability allows the further conclusion that the observation made in Paris changes the wave function ψ_2 in Tokyo, reducing *it* to zero. Therefore, the wave functions, too, undergo sudden jumps when observations are made. This dramatic type of quantum behavior is not limited to the situation encountered in the de Broglie's paradox; it is instead a general concomitant of quantum observations. Indeed, in all acts of measurement (barring some very particular cases where the value of the observed quantity is known prior to measurement) one encounters a sudden 'collapse' of the wave function or, as it is often called, a *reduction of the wave packet.*

The fact that there is such a reduction of the wave packet was hard to accept for physicists like Einstein and de Broglie who considered the wave function as something real, existing in space and time.

The reduction is, on the other hand, perfectly acceptable to Wigner who thinks that

> . . . the wave function is only a suitable language . . . which is relevant for predicting the future behavior of the system. (Good, 1962, p. 287.)

It should then be clear, according to Wigner, that the wave packet reduction takes place because

> . . . the impression which one gains at an interaction, called *the result of an observation*, modifies the wave function of the system. The modified wave function is, furthermore, in general unpredictable before the impression gained at the interaction has entered our consciousness: it is the entering of an impression into our consciousness which alters the wave function because it modifies our appraisal of the probabilities for different impressions which we expect to receive in the future. It is at this point that the consciousness enters the theory unavoidably and unalterably. (*Ibid.*, p. 289.)

The wave function of a system is usually thought to be a description of the state of that system or, if one prefers, of its material properties. But this conception is incompatible with the idea that our consciousness can modify the wave function or, rather, it becomes compatible with Wigner's idea only if one assumes flatly the superiority of consciousness over matter.

Wigner takes the latter suggestion rather seriously, and a fuller discussion of this point can be found in his 1962 paper, entitled 'Remarks on the Mind—Body Question', from which our above quotations were taken. This paper ends with a proposal to build a 'psycho-electric cell' for checking the action of our psyche over matter and concludes that

> . . . it will remain remarkable, in whatever way our future concepts may develop, that the very study of the external world led to the conclusion that the content of the consciousness is an ultimate reality. (*Ibid.*, p. 285.)

The first to introduce the consciousness line of thought into the wave-packet reduction problem was von Neumann who stressed the importance of the human observer being endowed with consciousness:

. . . no matter how far we calculate . . . at some time we must say: and this is perceived by the observer. That is, we must divide the world into two parts, the one being the observed system, the other the observer. (von Neumann, 1955, pp. 419—420.)

It is an act of subjective perception which generates the wave-packet reduction. According to von Neumann, an explanation of this kind can never contradict human experience, since

. . . experience only makes statements of this type: an observer has made a certain (subjective) observation; and never any like this: a physical quantity has a certain value. (*Ibid*., p. 420.)

Von Neumann's ideas were accepted and developed in a still clearer way by London and Bauer who stressed

. . . the essential role played by the consciousness of the observer in this transition from the mixture to the pure case. Without his actual intervention a new ψ function would never be obtained. (London and Bauer, 1939, p. 42.)

For these authors

. . . it is not a mysterious interaction between the apparatus and the object which during the measurement produces a new ψ for the system. It is only the consciousness of an 'I' who can separate himself from the old function $\Psi(xyz)$ and, because of his observation, construct a new objectivity in henceforth attributing a new function $\psi(x) = u_k(x)$ to the object. (*Ibid.*)

Wigner's distinctive contribution consisted in developing the consciousness approach to a philosophically larger extent. Recently, Zweifel (1974) added to the discussion by introducing an 'interaction potential' between the mind of the observer and the measuring apparatus. Several other papers have broached similar ideas. Cochran, for instance, in 1971 discussed the double-slit experiment, concluding:

Each electron passes through only one hole, but it is aware of the existence and location of the other hole when it is open, and it chooses different angles of diffraction when the second hole is open — angles that will enable it to form a part of the characteristic diffraction pattern. Instead of being something that has both particle properties and wave properties, the electron in this concept is a particle that has a degree of consciousness. The consciousness of the electron is a periodic pulsation with a characteristic frequency that is determined by the energy of the electron, and it does not involve an extended wave. The electron exhibits its particle aspects in interactions in which it gains or loses energy, and it exhibits its degree of consciousness in interactions in which its energy remains constant, such as diffraction. Since an electron going through a hole can deflect its course in a great many ways, a calculation of its possible angles of diffraction involves a large number of possibilities and takes the form of the quantum mechanical wave function. The wave function describes the choices open to the electron and the relative probabilities that these choices will be realized. (Cochran, 1971.)

Apart from these extreme positions, it is not difficult to give a 'consciousness interpretation' of the experiments discussed in the first part of the present chapter. Consider first the Jánossy—Naray experiment: Here the photon is observed by means of the interference detecting apparatus IP of Figure 3.1. We can therefore have no knowledge of the path along which it propagated, and our ignorance is reflected in the simultaneous use of two superimposed photon wave functions, one describing a photon transmission by the semitransparent mirror and the other a photon reflection. In this way the interference fringes express our ignorance of the photon propagation.

Quite different is the situation obtaining in the case of the experiment with phototubes plus beamsplitter. If one of these phototubes detects a photon arrival, the entering of this knowledge into our consciousness modifies our appraisal of the probabilities for the different impressions that we expect to receive in the future. In particular, we can predict that the second counter will not detect any photon even if it is placed further from the beamsplitter than the first counter. This means that the wave function describing a photon propagation on the second trajectory goes suddenly to zero as soon as our consciousness acquires a modified knowledge of the system.

To the objection, 'How can our knowledge modify or even destroy something objectively real?' the answer would be that perhaps nothing else is real beyond our conscious activity and that, even if some other reality exists, it certainly cannot be conceived as propagating in space and time. In this way, the objective reality becomes at best irrelevant and the important features of physics are recognized to be the subjective, conscious elements.

3.10. Delayed Choices

The well-known physicist John Archibald Wheeler is today perhaps the most articulate and consistent of all Bohr's heirs, enriching the latter's ideas with noteworthy original contributions. His conception of quantum phenomena hinges around two fundamental points (Wheeler, 1980):

1. "No elementary phenomenon is a phenomenon until it is a registered (observed) phenomenon." This point of view quite obviously takes up once again Bohr's idea that the role of the observer cannot be eliminated and that we should consider as an 'atomic phenomenon' only an act of measurement, including of necessity in the concept a detailed description of the instrument used. Stated briefly, Wheeler agrees with Bohr in rejecting the existence of an atomic reality independent of man and his observations.

2. Our experimental choices may *modify the past*, not only the recent past of some tiny fractions of a second ago, but even the remote past of billions of years ago. By using what he calls a 'delayed choice' experiment, Wheeler believes that

> a choice made in the here and now has irretrievable consequences for what one has the right to say about what has *already* happened in the very earliest days of the universe, long before there was any life on Earth.

His reasoning is based on the experimental apparatus of Figure 3.8. A source S of extremely low intensity emits photons, one at a time, with a long

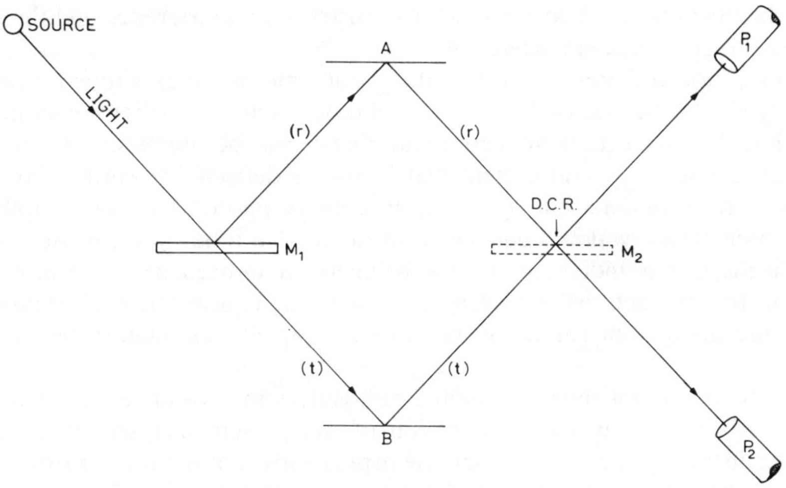

Fig. 3.8. Wheeler's delayed choice experiment. M_1 is a semitransparent mirror, *A* and *B* are fully reflecting mirrors, *DCR* is the delayed choice region where a second beamsplitter M_2 can be located. P_1 and P_2 are photomultipliers.

time interval between one photon and the next. The light is incident on a semitransparent mirror M_1 and divides into two parts, the reflected part (r) which is again reflected by the totally reflecting mirror *A* and sent towards the photomultiplier P_2, and the transmitted part (t) which is reflected by the totally reflecting mirror *B* and sent towards the photomultiplier P_1. If there are no other elements in the experimental apparatus other than those described up to now, we know what the outcome of the experiment will be: every photon will be detected either by P_1 or by P_2 and it will never happen that half a photon is detected by P_1 and half by P_2. But it is self-evident that this is the same as saying that the photon has followed the path *tBt* (if detected by P_1) or the path *rAr* (if detected by P_2). In other words, the photon has always been on one of the two trajectories, which is equivalent to saying that it has possessed localization, thus indicating a curpuscular behavior. Summarizing, if no apparatus is inserted in the region indicated by DCR in Figure 3.8, then the instrument serves overall to show up the corpuscular behavior of photons.

Let us now suppose, instead, that we do a second experiment, somewhat different from the first one, in which a second semitransparent mirror M_2 is inserted in the way indicated in Figure 3.8 in the delayed-choice region (DCR). The thickness of M_2 is calculated as a function of the wavelength of the light emitted by the source in such a way that the superposition of the transmitted part of r (r_t) and the reflected part of t (t_r) generates a wave of zero intensity. Thus P_2 will not detect the arrival of any photon and all arrivals will be detected by P_1. In other words, things may be arranged such that the beam ($t_t + r_r$) corresponds to constructive interference and the beam ($t_r + r_t$) to destructive interference.

Here, too, we know already what the outcome of the experiment is, which is very similar to that of Jánossy and Naray, discussed earlier: every photon will interfere with itself in such a way that it will be impossible for it to be deflected toward P_2 and certain that it will be deflected toward P_1. But this result can be understood by saying that every photon has always followed both trajectories, which is equivalent to saying that it has never possessed any localization, thus indicating a wave behavior. Summarizing, if we insert M_2, having the characteristics already described, in region DCR of Figure 3.8, then the instrument serves overall to show up the undulatory behavior of light.

Up to this point there is nothing that differs in substance from what has already been said in the first section of the present chapter, except for a greater insistence on the sole particle aspect shown in the former experiment and on the sole wave aspect in the latter case. But it is exactly here that the essence of Wheeler's idea resides: he excludes *a priori* and without discussion the possibility that the photon might at one and the same time be a wave *and* a particle. If one excludes this, the game becomes easy: Wheeler proposes the creation of a mechanism whereby M_2 is inserted or excluded at the last moment when the photon has already interacted with M_1 and is being propagated along the path r and/or t. If M_2 is inserted, the photon will behave as a wave, as if it had been propagated along both paths. If, on the other hand, M_2 is not inserted, the photon will behave as a particle, as if it had been propagated along just one of the possible paths. But this seems absurd because the *choice* of whether to insert M_2 or not is *delayed* (i.e., it is made later) with respect to the interaction of the photon with M_1. It seems absurd, says Wheeler, but it is not if one admits that the future choice of the position of M_2 could influence the past interaction between the photon and M_1.

We are dealing, in substance, as is quite clear, with an inversion of the cause—effect relationship, in other words with a total upset of the temporal order of phenomena. It is sufficient to shut oneself out from the ideas of Einstein and de Broglie, as Wheeler does, to find oneself in an infinite sea of troubles: here it must even be said that the reality of what existed in the past is open to modification by making choices in the present or in the future! It must not be thought that Wheeler is the only physicist to defend these retroactions in time: it will be seen in Chapter 5 that others like Rayski, Rietdijk, Costa de Beauregard and Stapp have proposed this idea of retroaction as a solution of the Einstein—Podolsky—Rosen paradox.

But let us return to Wheeler and his generalization in regard to the cosmological scale of retroaction.

Two astronomical objects, known as 0957 + 561 A and B (quasars) were, in the immediate period after their discovery, considered to be two distinct since one was separated from the other by an angle of six seconds. More recent studies have, however, shown that all the physical properties of the two objects are identical (for example, the degree of redshift of their light) and astronomers are now convinced that we are dealing with the same object seen double. This doubling is probably due to a 'gravitational lens', i.e., to the presence of a very massive galaxy between ourselves and the quasar in question which deflects light from the quasar and makes it reach the Earth along two different paths. Wheeler observes that one may apply the scheme used in Figure 3.8, where, instead of M_1, there is now a 'gravitational lens' and, instead of S, the quasi-stellar object in question and where the distances are billions of light years (See Figure 3.9). By insertion, or not, of the semitransparent mirror M_2 we can force the next photon that arrives from the quasar to have followed both the possible trajectories or to have followed only one of them, respectively. But in this case, light from the quasar passed near the gravitational lens billions of years ago. This, for Wheeler, means that we are able to influence the past even on time scales comparable to the age of the universe: the powers of human observations really do seem limitless! Wheeler comments on this:

It is wrong to think of the past as 'already existing' in all detail . . . What we have the right to say of past space-time, and past events, is decided by choices — of what measurements to carry out — made in the near past and now. The phenomena called into being by these decisions reach backward in time in their consequences . . . , back even to the earliest days of the universe. Registering equipment operating in the here and now has an undeniable part in bringing about that which appears to have happened. Useful as it is under everyday circumstances to say that the world exists 'out there' independent of us, that view can no longer be upheld. There is a strange sense in which this is a 'participatory universe'.

Wheeler develops this line of thought to its final consequences, right up to imagining that everything that exists and that has existed stems from human acts of observation:

Universe: I am a giant machine. I supply the space and time for your existence. There was no

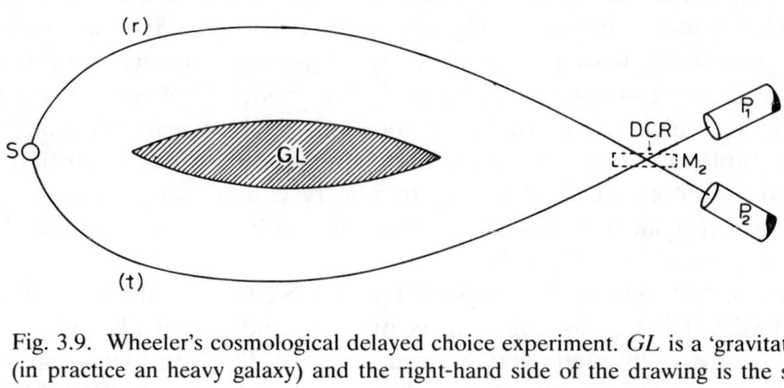

Fig. 3.9. Wheeler's cosmological delayed choice experiment. *GL* is a 'gravitational lens' (in practice an heavy galaxy) and the right-hand side of the drawing is the same as in Figure 3.8.

before before I came into existence, and there will be no after after I cease to exist. You are an unimportant bit of matter located in an unimportant galaxy.

Human being: Yes, oh universe, without you I would not have been able to come into being. Yet you, great system, are made of phenomena; and every phenomenon rests on an act of observation. You could never exist without elementary acts of observation such as mine.

Of course, not all physicists who are today sympathetically inclined toward the Copenhagen ideas are in agreement with Wheeler's positions. To quote just one example, I should like to recall the position of the British physicist Rudolf Peierls, who in one of his papers (Peierls, 1985) asks himself the explicit question:

"Does the world not exist if there is no observer to see it? Does the observer create the world around him?" My answer is that this is a non-question because the word 'existing' is undefined. In application to everyday objects we mean by their existing that we can see or touch them, etc., or that we could do so in suitable circumstances. This is not the meaning implied in the question. Of course if there is no observer to have any knowledge of a system there is no quantum mechanical description of it, since this is in terms of an observer.

Peierls' position is, however, also characterized by a rejection of the wave—particle duality, exactly as in the case of Wheeler, even if we are dealing with a different rejection because it is based on the idea that the electronic wave function cannot have any physical reality.

3.11. How to do what Complementarity Forbids

As we saw in the previous chapter, Bohr expressed his highly appreciative opinion about von Neumann's theorem and pointed out that one of his own papers, in more elementary ways, arrived at essentially the same conclusion. This connection between Bohr and von Neumann is rarely if at all discussed in the scientific literature, but is very important, especially because the paper Bohr had in mind was clearly the one he wrote three years before (Bohr, 1935). That paper contained only two references. The first one is to the Einstein, Podolsky, Rosen paper (1935) whose conclusions Bohr tries to undermine, and the second one to Bohr's Cambridge University Press book (Bohr, 1961), which is regarded even today as the best exposition of Bohr's idea of complementarity (for instance, it contains his 1927 contribution to the Como Conference). That complementarity can be seen as an impossibility argument against causal space-time reformulations of quantum theory has been shown in Section 3.6 above.

It is nowadays commonly accepted that von Neumann's theorem, in all of its proposed versions, has failed in its purpose and is actually incapable of barring the causal completions of quantum theory. The best understanding of this fact has been provided by Bohm's hidden variable models (Bohm, 1952), which did in fact exactly what von Neumann's theorem was supposed to forbid, provide a causal formulation of the existing quantum theory. As we saw in Section 2.5, further simple models exist which reproduce a limited but very meaningful part of quantum theory, the eigenvalues and the probabilities for all the spin observables for a spin-$\frac{1}{2}$ particle. Few people seem to have realized that the existence of these models does not only provide an overcoming of von Neumann's theorem, but of Bohr's complementarity as well. Today we know how to give *a causal description in space and time* of many quantum phenomena (double-slit experiment, barrier penetration, neutron interferometric results, and so on), while complementarity declared such a description *impossible*. There is even a movie (Dewdney, 1988) in which deterministic particle trajectories reproduce quantum mechanical interferences in several interesting cases, the main idea being the association of a suitable 'quantum potential' with the action of the wave on the particle of the studied dual quantum system. The result is that the wave and the particle, which together constitute a quantum system, both behave deterministically in space and time. Thus the arbitrariness of the conclusions that Bohr and other people thought to be able to deduce from the complementarity principle is brought out in a most striking and convincing manner. This is not the same, of course, as saying that one should subscribe to the validity of these models: their value is mostly methodological in showing in a detailed rational way the obsolescence of all the 'impossibility proofs', complementarity included.

The roots of complementarity can best be exposed by discussing the example of space localization (position measurement) and causality implementation (momentum measurement). Space localization can be obtained by measuring position with infinite precision ($\Delta x = 0$). After such a measurement the wave function of the observed particle becomes the delta function $\delta(x - x_0)$, if x_0 is the obtained result. But a delta function can be written as a superposition with constant weight of all possible plane waves, and this implies that absolutely nothing is known about momentum. All eventual knowledge about momentum, prior to the position measurement, is in this way completely lost. No evidence can therefore exist about momentum conservation, since no knowledge about momentum is available. A concrete realization of localization in space thus implies a necessary abandonment of the causal description. Symmetrically, in a different experiment, one could choose to implement the causal description by measuring momentum with infinite precision: the wave function would then become, as a result of the measurement, a plane wave. But this would immediately imply that nothing is known about position, resulting in a complete loss of the description of the studied phenomenon in space. A concrete implementation of the causal description thus 'forces' the physicist to abandon all description in space. The two possibilities (space-time and causality) thus appear to be mutually incompatible, to exclude each other. Bohr concludes then that in the atomic world it is in principle impossible to give a picture of quantum processes as developing causally in space-time.

The previous conclusion applies to any two noncommuting observables.

Consider, for example, a spin-$\frac{1}{2}$ particle and its spin-component observables σ_x, σ_y, σ_z. Any two of them do not commute, and this means that they cannot be measured simultaneously. Consider a system in the initial state

$$|\psi_{z+}\rangle = \begin{pmatrix} 1 \\ 0 \end{pmatrix}.$$

If the σ_x component is measured there can be only two results, ± 1. Therefore the spin state *after* the measurement becomes

$$\text{either} \quad |\psi_{x+}\rangle = \frac{1}{\sqrt{2}}\begin{pmatrix} 1 \\ 1 \end{pmatrix} \quad \text{or} \quad |\psi_{x-}\rangle = \frac{1}{\sqrt{2}}\begin{pmatrix} 1 \\ -1 \end{pmatrix}.$$

In either case, the σ_y component becomes completely unknown, as can easily be checked. Bohr would say that the implementation of the reality of σ_x has made σ_y completely undetermined. The opposite reasoning can also be made: σ_y can be measured and become known, but then it is σ_x which becomes necessarily completely unknown. One can then say, with Bohr, that σ_x and σ_y are complementary aspects of reality: either σ_x is real, or σ_y is real, but never the two of them at the same time.

As we saw in the previous chapter a model of spin-$\frac{1}{2}$ measurements exists which agrees with all the empirical predictions of quantum theory for the single spin-$\frac{1}{2}$ system. The model has the following features:

(1) It provides a causal picture in space and time of spin measurements;
(2) it describes all spin components as simultaneously predetermined by objective properties of the measured system.

Therefore the model does twice what complementarity tried to forbid. The model contains assumptions both about the measured system and about the measuring apparatus: of course, the role of the apparatus cannot be eliminated in the quantum domain, given the existence of a minimal action (Planck's constant h).

The proposed model could be concretely realized in a mechanical workshop by means of spinning spheres and with a suitably built 'measuring apparatus'. The resulting simulation of quantum spin would be perfect in all respects, except for the numerical value of h, which is however a rather trivial multiplying factor of spin eigenvalues. Naturally the model is not a really believable reconstruction of quantum spin: its existence shows merely that a door is open and that much better models should be possible. In the above model all the conceivable spin components are simultaneously predetermined by the hidden variable λ. Notice however that simultaneous measurements of different spin components are impossible because the hidden variables should come out to be redistributed according to two different and obviously incompatible distributions. It should also be stressed that Heisenberg's relations for spin components are valid as scatter relations over the statistical ensemble.

The interdictions coming from complementarity have been violated and the arbitrariness of Bohr's conclusions comes out very clearly to lie in the *assumption* that if two observables cannot be measured simultaneously, then they cannot both be real. It is all as simple as that!

References

Books

Bohr, N.: 1961, *Atomic Theory and the Description of Nature*, Cambridge University Press, Cambridge.

Born, M.: 1964, *Natural Philosophy of Cause and Chance*, Dover, New York.

Bonse, U., and Rauch, H. (eds): 1979, *Neutron Interferometry*, Clarendon Press, Oxford.

de Broglie, L.: 1956, *Une tentative d'interprétation causale et non linéare de la mécanique ondulatoire: la théorie de la double solution*, Gauthier-Villars, Paris.

de Broglie, L.: 1971, *La réinterprétation de la mécanique ondulatoire: Principes généraux*, Gauthier-Villars, Paris.

Dirac, P. A. M.: 1958, *The Principles of Quantum Mechanics* (4th edn), Clarendon Press, Oxford.

Fock, V.: 1978, *Fundamentals of Quantum Mechanics*, MIR Publishers, Moscow.

George, A. (ed.): 1953, *Louis de Broglie, physicien et penseur* (*Collection Les savants et le monde*), Albin-Michel, Paris.

Good, I. J. (ed.): 1962, *The Scientist Speculates*, Heinemann, London.

Heisenberg, W.: 1949, *The Physical Principles of Quantum Theory*, Dover, New York. (First published by The University of Chicago Press in 1930.)

Heisenberg, W.: 1959, *Physics and Philosophy*, Harper and Row, New York.

Jammer, M.: 1966, *The Conceptual Development of Quantum Mechanics*, McGraw-Hill, New York.

Landé, A.: 1965, *New Foundations of Quantum Mechanics*, Cambridge University Press, London.

London, F., and Bauer, E.: 1939, *La théorie de l'observation en mécanique quantique*, No. 775 of *Actualites Scientifiques et industrielles: Exposés de physique générales*, Hermann, Paris.

Millikan, R. A.: 1947, *Electrons* (+ *and* −), *Protons, Photons, Neutrons, Mesotrons and Cosmic Rays*, The University of Chicago Press, Chicago.

Peierls, R.: 1985, in P. Lahti and P. Mittelstaedt (eds.), *Symposium on the Foundations of Modern Physics*, World Scientific, Singapore.

Przibram, K. (ed.): 1967, *Letters on Wave Mechanics*, The Philosophical Library, New York.

Rosenfeld, L.: 1973, 'The Wave-Particle Dilemma' in J. Mehra (ed.), *The Physicist's Conception of Nature*, Reidel, Dordrecht.

Rozental, S. (ed.): 1967, *Niels Bohr*, North-Holland, Amsterdam; Wiley, New York.

Tagliagambe, S.: 1972, in L. Geymonat (ed.), *Storia del pensiero filósofico e scientifico*, Vol. VI, Garzanti, Milan.

von Neumann, J.: 1955, *Mathematical Foundations of Quantum Mechanics*, Princeton University Press, Princeton.

van der Waerden, B. L. (ed.): 1967, *Sources of Quantum Mechanics*, North-Holland, Amsterdam.

Articles

Bartell, L. S.: 1980, *Phys. Rev.* **21D**, 1968.
Beltrametti, E. G., and Cassinelli, G.: 1976, *Riv. Nuovo Cimento* **6**, 321.
Bohm, D.: 1952, *Phys. Rev.* **85**, 166 and 180.
Bohr, N.: 1913, *Phil. Mag.* **26**, 1.
Bohr, N.: 1918, *K. Dan. Vidensk. Selsk.* 8 Raekke IV, 1; translation in van der Waerden, 1967.
Bohr, N.: 1935, *Phys. Rev.* **48**, 696.
Bohr, N., Kramers, H. A., and Slater, J. C.: 1924, *Phil. Mag.* **47**, 785.
Born, M.: 1926, *Z. Phys.* **37**, 863.
Bothe, W., and Geiger, H.: 1925, *Z. Phys.* **32**, 639.
Clauser, J. F.: 1974, *Phys. Rev.* **D9**, 853.
Cochran, A. A.: 1971, *Found. Phys.* **1**, 235.
Compton, A. H.: 1923, *Phys. Rev.* **21**, 483.
Compton, A. H.: 1927, Nobel Lecture.
Compton, A. H., and Simon, A. W.: 1925a, *Phys. Rev.* **25**, 306.
Compton, A. H., and Simon, A. W.: 1925b, *Phys. Rev.* **26**, 289.
Dagenais, M., and Mandel, L.: 1978, *Phys. Rev.* **A18**, 2217.
Davisson, C. J., and Germer, L. H.: 1927, *Phys. Rev.* **30**, 705.
de Broglie, L.: 1925, *Ann. Phys.* (*Paris*) **3**, 22.
de Broglie, L.: 1929, Nobel Lecture.
de Broglie, L.: 1977, *Ann. Fond. L. de Broglie* **2**, 1.
de Broglie, L.: 1979, *Ann. Fond. L. de Broglie* **4**, 13.
Dewdney, C.: 1988, 'Particle Trajectories and the Explanation of Quantum Phenomena' in A. van der Merwe *et al.* (eds.), *Microphysical Reality and Quantum Formalism*, Kluwer Academic Publishers, Dordrecht.
Einstein, A.: 1905a, *Ann. Phys.* (*Leipzig*) **17**, 132.
Einstein, A.: 1905b, *Ann. Phys.* (*Leipzig*) **18**, 639.
Einstein, A.: 1917, *Phys. Z.* **18**, 121.
Einstein, A.: 1954, Letter from A. Einstein to L. de Broglie, quoted in de Broglie (1979).
Einstein, A., Podolsky, B., and Rosen, N.: 1935, *Phys. Rev.* **47**, 777.
Fock, V. A.: 1964, 'My Answer to Professor Niels Bohr', *Voprosy Filosofii*, No. 8.
Jánossy, L., and Naray, Zs.: 1958, *Suppl. Nuovo Cimento* **9**, 588.
Mayer, E., and Gerlach, W.: 1914, *Ann. Phys.* (*Leipzig*) **45**, 177.
Millikan, R. A.: 1914, *Phys. Rev.* **4**, 73.
Pfleegor, R. L., and Mandel, L.: 1967, *Phys. Rev.* **159**, 1084.
Rauch, H., Treimer, W., and Bonse, U.: 1974, *Phys. Lett.* **47A**, 369.
Schrödinger, E.: 1926a, *Ann. Phys.* (*Leipzig*) **79**, 361.
Schrödinger, E.: 1926b, *Ann. Phys.* (*Leipzig*) **81**, 109.
Selleri, F.: 1985, *Riv. Storia Scienza* **2**, 545.
Slater, J. C.: 1925, *Nature* **116**, 278.
Thomson, G. P.: 1928, *Proc. Roy. Soc.* **A117**, 600.
Wheeler, J. A.: 1980, *Delayed-Choice Experiments and the Bohr-Einstein Dialog*. Paper presented at the joint meeting of the American Philosophical Society and the Royal Society, London, June 5, 1980.
Wootters, W. K., and Zurek, W. H.: 1979, *Phys. Rev.* **19D**, 473.
Zweifel, P. F.: 1974, *Int. J. Theor. Phys.* **10**, 67.

Chapter 4

Properties of Quantum Waves

In recent times some clarification of the complex issue of dualism is starting to emerge. In the first place, some of its proposed formulations are either in trouble or completely ruled out when confronted with the empirical evidence. The Einstein—de Broglie formulation, on the other hand, is indicated as *possible* by neutron interferometry, atomic decay in the presence of mirrors, and laser light amplification. Some further experiments, presently existing only at the level of theoretical proposals, could significantly strengthen the case for this type of duality and for 'empty' waves.

4.1. Quantum Waves and Quantum Potential

The construction of the mathematical apparatus of quantum theory took place in the years 1924—1927 with what has sometimes been described as a 'systematic guessing, guided by the Principle of Correspondence'. It was a type of highly creative formal game in which the classical equations of motion (e.g., Hamilton's equations) were at first retained, but the mathematical entities describing the physical observables were of an entirely new type. In fact, the first quantum mechanical description of the observables position and momentum was given by Heisenberg in 1925 by means of infinitely-dimensional matrices. These matrices were later to become the linear Hermitian operators which represent still today the quantum observables. The new 'kinematical' description of the quantum world (the dynamics remained largely the same as in classical theory, as already stated, at least in the Copenhagen—Göttingen formulation) allowed for a new degree of freedom in the mathematical apparatus, the commutator of any two operators. As is well known, also this problem was solved within the general formal development of the theory, by means of the correspondence principle discovered by Bohr. This type of development did not require a clarification of the physical basis of the theory, not even of the nature of duality.

In fact, we saw in the previous chapter that almost every important contributor to the development of quantum mechanics had a different interpretation of the empirical evidence concerning the dual nature of atomic systems. In the present chapter we shall discuss in some detail the Einstein—de Broglie duality, which is based on the idea that both the particle aspect and the wave aspect are real. Obviously this view represents a paradigmatic shift with respect to the pragmatic, formal 'game' which is still the dominant one in theoretical physics. The difference involved is well emphasized by Lochak:

> Louis de Broglie is an intuitive spirit, concrete and realist, in love with simple images in three-dimensional space. He does not grant ontological value to mathematical models, in particular to geometrical representations in abstract spaces; he does not consider and does not use them other than as convenient mathematical instruments, among others, and it is not in their handling that his physical intuition is directly applied; faced with these abstract representations, he always keeps in mind the idea that all phenomena actually take place in physical space, so that these mathematical modes of reasoning have a true meaning in his eyes only insofar as he perceives at all times what physical laws they correspond to in usual space. (Lochak, 1982, p. xxi.)

By considering several performed and proposed experiments, we show in this and the following sections that the Einstein—de Broglie duality is compatible with the existing empirical evidence and that it suggests several interesting ideas. It will be shown that a number of experiments are possible, dealing with the amplification of light and with the interference of light and that of massive particles (neutron interferometry), for which the Einstein—de Broglie duality gives quantum mechanical predictions different from those of the usual ('Copenhagen') interpretation of quantum theory.

The discussions of the present section focus on an interpretation of the Schrödinger equation that was originally proposed by de Broglie (1927) and Bohm (1952), but which has been further developed in recent times and extended to include Maxwell's equations for the electromagnetic field. In the latter case it seems possible, at least in principle, to carry out experimental checks of the de Broglie—Bohm idea which can distinguish it from the more traditional Copenhagen—Göttingen interpretation of quantum mechanics.

Considering Schrödinger's equation for a particle with mass m in an external potential V,

$$i\hbar \frac{\partial \psi}{\partial t} = -\frac{\hbar^2}{2m} \nabla^2 \psi + V\psi, \tag{4.1}$$

and, setting

$$\psi = R \exp\left(\frac{iS}{\hbar}\right), \tag{4.2}$$

one can easily show that the complex Equation (4.1) is equivalent to the two following real equations

$$\frac{\partial R^2}{\partial t} + \nabla \cdot \left(R^2 \frac{\nabla S}{m} \right) = 0, \tag{4.3}$$

$$\frac{\partial S}{\partial t} + \frac{1}{2m} (\nabla S)^2 + V - \frac{\hbar^2}{2m} \frac{\nabla^2 R}{R} = 0. \tag{4.4}$$

It is interesting to note that Planck's constant does not enter at all in Equation (4.3), while in Equation (4.4) it appears only in the last term, which we henceforth designate by Q:

$$Q = - \frac{\hbar^2}{2m} \frac{\nabla^2 R}{R}.$$

The quantity Q, which from (4.4) is seen to have the dimensions of energy, is usually called the *quantum potential.*

Equation (4.3) and also Equation (4.4), apart from the term Q, are well known in classical physics. To begin with, Equation (4.3) clearly expresses the conservation of a fluid of density

$$\rho(x, y, z, t) = R^2(x, y, z, t)$$

and velocity

$$\mathbf{v}(x, y, z, t) = \left(\frac{1}{m} \right) \nabla S(x, y, z, t).$$

In fact, since $R^2 = |\psi|^2$, Equation (4.3) is precisely the well-known continuity equation which one usually derives from Schrödinger's equation and commonly interprets in terms of conservation of total probability, with R^2 representing the probability density and $\mathbf{v}$ the local velocity of the probabilistic fluid.

In the usual formulation of quantum mechanics one does not go beyond these statements, but under de Broglie's conception of duality one can think in terms of particles embedded in a wave field. For a given function ψ, we can then imagine a statistical ensemble of particles randomly distributed within their identical wave functions with a density (over the statistical ensemble) of value $|\psi|^2 = R^2$ and with a local velocity of the probabilistic fluid that is defined by $\mathbf{v} = (1/m)\nabla S$. In this manner, $\mathbf{v}(x, y, z, t)$ very naturally becomes identifiable with the velocity *of the particle* occupying the point (x, y, z) at the time t, thus $m\mathbf{v} = \nabla S$ with its momentum, and, finally, the second and third terms of Equation (4.4) with its kinetic and potential energy, respectively. But then it is apparent that Equation (4.4) has very nearly the structure of the classical Hamilton—Jacobi equation,

$$H\left(q, \frac{\partial S}{\partial q}\right) + \frac{\partial S}{\partial t} = 0,$$

for a Hamilton's principal function $S(x, y, z, t)$, the only novelty being the appearance of the quantum potential Q as an extra term.

The interpretation of the quantum potential is, according to de Broglie, the following: If the particle is embedded in a plane wave

$$\psi_0(x, y, z, t) = R_0 \exp[2\pi i(\mathbf{k}\cdot\mathbf{x} - \nu t)],$$

with R_0 constant, the potential Q obviously vanishes everywhere and with it the quantum force $-\nabla Q$ acting on the particle. The latter will thus propagate along a straight line in obeying the principle of inertia. The situation changes, however, if the plane wave falls on a screen with a slit: the well-known phenomenon of diffraction then takes place, and beyond the slit one has to deal with a more or less conical wave. The amplitude of such a wave is not constant over all of space, which means that Q and therefore $-\nabla Q$ are now in general not everywhere zero. This implies that the wave exerts a force on the particle and thus bends its path in those cases where the particle passes through the slit. But this is a welcome feature, since energy (associated with the particles) is found on a second screen to have a diffraction pattern-like distribution, with maxima and minima, which in all cases occupies a surface larger than that belonging to the slit in the first screen.

In this manner one can infer that the quantum potential Q represents in all circumstances the action of the wave on the particle. It can be shown that the diffraction pattern-like distribution of the particles is indeed generated by Q even for a uniform distribution of particles arriving at the slit. The following is a qualitative way to understand this fact. If R vanishes at a certain point P, while $\nabla^2 R$ does not, then the quantum potential, from its definition, tends here to $-\infty$. Such a behavior of Q corresponds to an infinite repulsion, and the particle will thus never be found at P. For reasons of continuity, it is clear that the particle will also rarely be found in the regions where $R = |\psi|$ is small; that is, in the regions where distructive interference generates minima of $|\psi|^2$ (Bohm, 1952).

On the basis of this kind of reasoning, we arrive at a double but self-consistent interpretation of R: it gives the particle density for a statistical ensemble and, at the same time, the amplitude of the wave. It allows us to reinterpret in physical terms all the well-known quantum phenomena for the single particle. Consider, for instance, the phenomenon of barrier penetration. The wave function for such a situation is a time-dependent wave packet. As a result, the quantum potential Q fluctuates in such a way that occasionally it becomes negative enough to cancel the positive barrier potential V

and thus to allow the passage of a particle through the barrier before Q has the opportunity to return to a significantly less negative value (Bohm, 1952).

The de Broglie—Bohm approach to quantum physics has recently been reconsidered by Philippidis, Dewdney, and Hiley (PDH, 1979), who showed in particular how a detailed calculation of the quantum potential for the usual two-slit experiment can give rise to interference without the need to abandon the notion of a well-defined particle trajectory. Their result is that every single particle follows a well-defined trajectory and that, in spite of this, if the initial particle distribution over the two slits is quantum-mechanical, the presence of the quantum potential in (4.4) ensures that the particle probability density remains equal to $|\psi|^2$ for all subsequent times. The numerical calculations of these authors were performed using data based on the experiment performed by Jönsson (1961). Thus the electron's energy was 45 keV, the separation between the slits 10^{-4} cm, and the slit width 10^{-5} cm. In Figure 4.1 we reproduce the particle trajectories calculated by PDH. Initially the trajectories from each slit fan out in a manner consistent with the diffraction pattern for a single slit. The subsequent 'kinks' in the trajectories coincide with the troughs in the quantum potential. They arise because a particle entering the region of a trough experiences a force, in a direction perpendicular to the symmetry line of the slits, which accelerates the particle rapidly through the trough into a plateau region where the forces are again weak. In consequence, most of the trajectories run along the plateau regions, giving rise to the bright fringes, while the troughs coincide with the dark fringes.

An interesting feature of this result is that *no trajectory crosses the symmetry plane*: this means that all particles observed on the lower (upper)

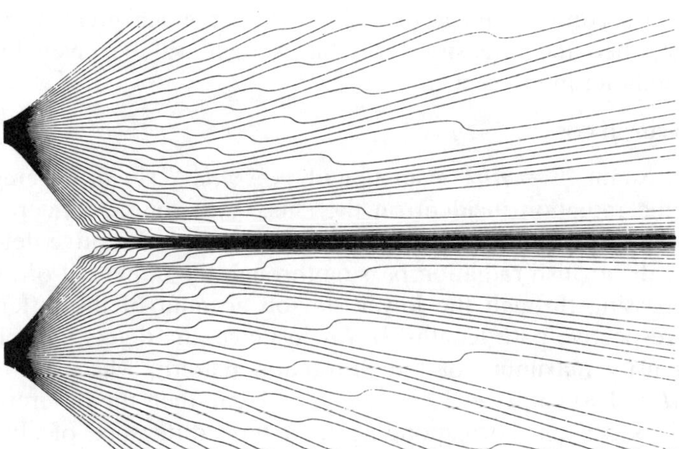

Fig. 4.1. Particle trajectories for the double-slit experiment as calculated by Philippidis, Dewdney, and Hiley. (From Philippidis *et al.*, 1979.)

half of the second screen must come from the lower (upper) slit — a result charged with philosophical implications for those who believed that the double-slit experiment did not allow any simple interpretation.

The PDH picture of the double-slit experiment is very simple:

> The approach through the quantum potential retains a pointlike particle, and each particle in the original ensemble follows a well defined trajectory which passes through one or the other of the slits. This ensemble produces the required interference pattern and, at the same time, shows that the final position of the particle on the screen allows us to deduce through which slit it actually passes. Thus it is possible to retain the trajectory concept and, at the same time, account for the interference. There is no longer a mystery as to how a single particle passing through one slit 'knows' the other slit is open. The information is carried by the quantum potential, so that we no longer have a conceptual difficulty in understanding the results obtained in very low intensity interference experiments. (*Ibid.*)

An interesting feature of the results found by PDH is that no particle trajectory crosses the axis of symmetry between the slits. Exactly the same result had been obtained by Richard Prosser (1976) in his study of the double-slit experiment using photons. He observed that lines of constant phase of the electromagnetic wave always intersect the axis of symmetry between the slits at right angles, and hence the flow lines, which are orthogonal to the phase lines, cannot cross it. From this he concluded that there is no photon trajectory which can cross the symmetry axis and that consequently photons illuminating the right-hand part of the screen must always come from the right-hand slit, no photons issuing into this region via the left-hand slit.

From this conclusion, Prosser arrives at the possibility of carrying out a highly interesting experiment on the nature of duality. Consider a plane screen with two parallel slits of width b whose centers are a distance d apart (Figure 4.2). If a second screen is placed a distance much greater than d behind the first, then the intensity I on this screen will be given by the formula for Fraunhofer interference,

$$I = A\beta^{-2} \sin^2 \beta \cos^2 \gamma, \tag{4.5}$$

where A is a constant, $\beta = \pi b \lambda^{-1} \sin \theta$, and $\gamma = \pi d \lambda^{-1} \sin \theta$, λ being the wavelength of the radiation incident on the slits. As a source a microwave generator with a sharply pulsed output is used, and a time-sensitive detector placed at P can distinguish radiation passing through the upper slit of Figure 4.2 from that passing through the lower slit, on account of the difference $d \sin \theta$ in the respective path lengths. If, for instance, the detector is placed at the first intensity maximum of the diffraction pattern, where $\sin \theta \simeq 3\lambda/2b$, and if $d = 100$ cm, $b = 3$ cm, and $\lambda = 1$ cm, then the difference in the path length is 50 cm, corresponding to a time difference of about 2 nanoseconds. Time intervals of this magnitude can now be detected, so that Prosser's experiment appears to be feasible. A similar experiment could also be performed at optical wavelengths using a sharply pulsed laser as a source and a suitably fast photodetector.

The physical interest of Prosser's proposed experiment is the following: If some of the photons arriving at *P* had traveled through one slit and some through the other, one would expect the detector to register two types of pulses, separated by time intervals corresponding to the different times of flight for the two cateogries of photons involved. On the other hand, if the photons had traveled through one slit only, then only a single type of pulse would be detected. One is therefore able to distinguish between these two alternatives by observing the nature of the detector's response (Figure 4.2). Such an experiment would be of no interest as far as a decision between the Copenhagen and the Einstein—de Broglie descriptions of duality is concerned. They can in fact be shown to lead to the same prediction for Prosser's experiment: interference should disappear as soon as the two times of flight become observably different. From the Copenhagen point of view this is so because the knowledge of the path followed by the photon implies a 'reduction' of the wave function, and then a disappearance of the interference effects. From the Einstein—de Broglie point of view the same is true because the two wave packets coming from the two slits no longer overlap if one has to be able to distinguish the times of flight of the particle in the two possible trajectories.

Prosser's experiment is nevertheless interesting because it can at least lead to a falsification of the idea, still held by many persons, that photons are classical particles, but that a 'field-like' deformation of space is produced between the screens by the presence of the two slits. Such a 'deformation' would exist independently of the particles crossing it, but would act on them when they pass, by bending their trajectories in such a way as to build the interference-like statistical distribution on the second screen.

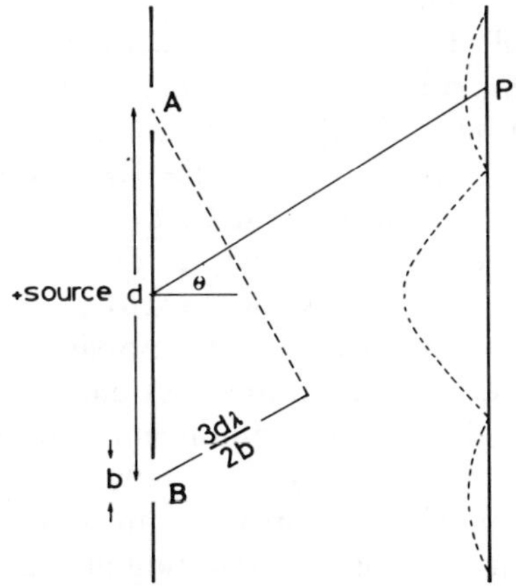

Fig. 4.2. Time-of-flight measurements proposed by Prosser for the double-slit experiment could falsify an unusual interpretation of quantum theory. (Redrawn from Prosser, 1976.)

4.2. Experiments on the Nature of Duality

The idea that the empty waves introduced in quantum theory by Einstein, de Broglie, Slater, and Bohr could give rise to stimulated emissions, in spite of the fact that they do not carry any energy-momentum, has been expressed several times. Thus de Broglie wrote:

> If an arbitrary cause has generated the emission of a quantum of light . . . its wave . . . passing on nearby atoms will stimulate other emissions . . . . (de Broglie, 1923 (1982).)

Similarly Bohr asserted about the action of his 'virtual radiation field':

> . . . an induced transition in an atom is not directly caused by a transition in a distant atom for which the energy difference between the initial and the final stationary state is the same. On the contrary, an atom which has contributed to *the induction of a certain transition in a distant atom through the virtual radiation field* [the italics are ours] . . . may nevertheless itself ultimately perform another of these transitions. (Bohr *et al.*, 1924; reprinted in van der Waerden, 1967, p. 166.)

Another interesting statement from the same source concerns the interference between the incident virtual plane wave and the outgoing spherical virtual waves emitted by 'illuminated' atoms:

> . . . these atoms will act as secondary sources of virtual wave radiation which interferes with the incident radiation. If the frequency of the incident waves coincides closely with the frequency of one of the virtual harmonic oscillators corresponding to the various possible transitions, the amplitudes of the secondary waves will be especially large, and these waves will possess such phase relations with the incident waves that they will diminish or augment the intensity of the virtual radiation field, *and thereby weaken or strengthen its power of inducing transitions in other atoms* [our italics]. (*Ibid.*, p. 167.)

The idea is thus very simple: it is the wave and not a localized particle-like structure which induces stimulated transitions in atoms.

In this and in the next section we shall discuss some indirect evidence in favor of the foregoing picture and present some possible experiments which could in principle allow this question to be settled. Our reasoning will refer only to energy-momentum conserving processes, since the BKS idea of nonconservation must be regarded as untenable in view of the fact that a multitude of atomic and subatomic processes has shown a complete consistence with the relevant conservation laws. We shall therefore refer to stimulated *emission*, where energy-momentum conservation is possible also if no external contribution is added to the energy of the emitting system.

Blake and Scarl (1979) have investigated laser light amplification in a laser gain tube. A schematic representation of their apparatus is shown in Figure 4.3: a beam of light, generated by a He—Ne laser at a wavelength of 6328 Å, traverses a He—Ne laser gain tube which is capable of emitting at the same wavelength. The *gain*, defined as the ratio of the intensity of the laser beam, with the discharge on, divided by the intensity with the discharge off (after

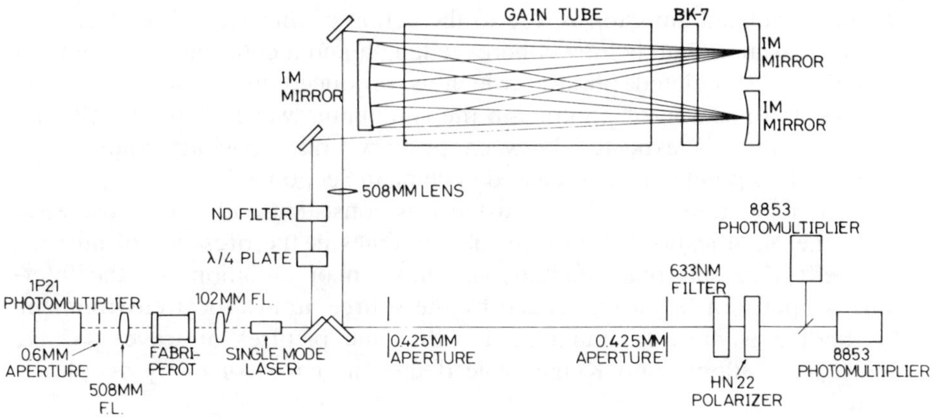

Fig. 4.3. Set up of the Blake–Scarl experiment. (Redrawn from Blake and Scarl, 1979.)

the amplifier-tube spontaneous emission had been subtracted) was 1.30 ± 0.02. At such low amplifications the gain is known to be linear for the intensity levels used in this experiment. Physically one can say that it is possible, in a first approximation, to neglect the further amplification of the light generated in the gain tube by the laser beam. In these circumstances one might expect the stimulating photon and the stimulated photon to be related in time, *if the process of stimulated emission is generated by the energy-momentum of the electromagentic field*, since these physical entities are always detected as localized in space and time. The amplified laser light along with the unavoidable spontaneous emission from the gain tube traverse an angular-spatial filter consisting of two 0.425 mm pinholes and an interference filter which lets through only light of wavelength (6328 ± 5) Å. A beamsplitter and two photomultipliers serve as the two-photon detection system.

Measuring the *coincidence rate* (number per unit time of photon detections by the two photomultipliers having a time difference less than 1 ns), Blake and Scarl established that, after background subtraction, the coincidence probability was the same *for all intensities of the laser light*. This result was the outcome of four experimental runs, in which the ratios of amplified laser light to amplifier tube chaotic light in the same mode were 0, 0.5, 1, and 2, respectively. Blake and Scarl accordingly concluded that "no correlations between incident and induced photons seem to be necessary in order to explain the results".

Surprised by their finding, they went on to state that, "Stimulated emission illustrates one of the apparent difficulties of current photon descriptions of the electromagnetic field."

A natural explanation of Blake and Scarl's discovery can be given in terms of the empty wave. Suppose a photon is a localized entity carrying energy-momentum and that it is accompanied by an extended empty wave. Suppose that the stimulated emission is due to the action of this wave. For laser light the wave has an infinitely long coherence length and a constant probability of generating a stimulated photon, which is produced independently of the presence and position of photons in the stimulating wave. Consequently, no time correlation is expected between primary and secondary photons, as observed. This point will be discussed in detail in Section 4.4.

Another phenomenon whose existence is consistent with the empty-wave idea is the modification of fluorescence lifetimes in the presence of mirrors, which effect reveals that spontaneous emission of radiation has the interesting property of being influenced by the source atom's local environment. This finding is hard to understand with some pictures of wave—particle duality, and Milonni and Knight voiced the misgivings of many people by writing:

> *The experimental observation that fluorescence lifetimes depend upon the atom's surroundings was apparently met with some surprise: how could the spontaneous emission of a photon be affected by the local environment, since the atom can only 'see' its surroundings by emitting a photon in the first place?* (Milonni and Knight, 1973.)

In their subsequent discussion, Milonni and Knight conclude that the radiated field (or the photon 'wave function') contains all the information needed for changing the atomic emission probability, provided only a time $t = 2d/c$ has elapsed since the atom entered the region, at a distance d from it, where the mirror is placed (Figure 4.4).

The marked dependence on the distance from a mirror of the decay time

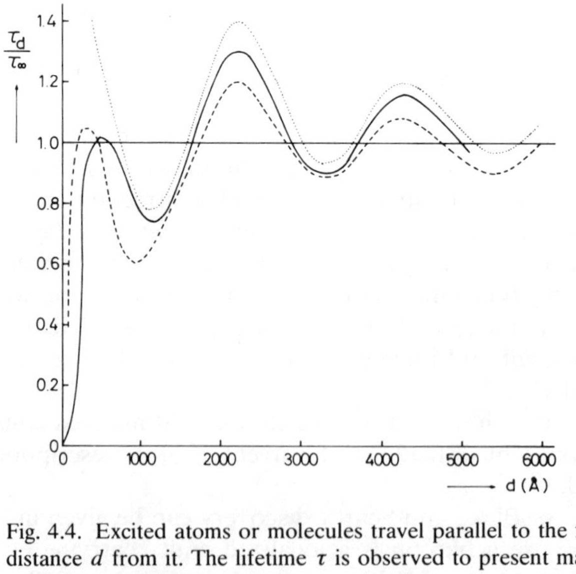

Fig. 4.4. Excited atoms or molecules travel parallel to the flat surface of a mirror, at a distance d from it. The lifetime τ is observed to present marked oscillations when d is varied and to approach the free lifetime τ_0 when d becomes very large.

for the luminescence of a molecule has been discussed by Kuhn (1970) who considered experimental data collected by Drexhage. One can, in particular, quantitatively explain the observed oscillations of the decay time with distance from the mirror by invoking Hertz's classical equations for dipole radiation. The radiated field arrives at the molecule, after being reflected by the mirror, with a retardation time $2d/c$. The decay time of the luminescence depends on the phase shift produced by this retardation and thus on the ratio of the distance of the oscillator from the mirror and the wavelength of the emitted light.

In quantum mechanical terms, the luminescence lifetime phenomenon can be attributed to stimulation or inhibition of the emission *of the light quantum*, the stimulating field being the radiation field of the atom capable of emitting the quantum itself. These results are extremely interesting since it appears virtually impossible to give any rational explanation of their existence in terms of a purely undulatory (or a purely corpuscular) description of radiation.

A number of experiments on very-low-intensity photon beam interference after splitting and recombination have been performed in Pisa by Adriano Gozzini (private communication). A narrow beam of circularly polarized monochromatic light from a sodium source is allowed to fall on a calcite crystal, from which two spatially separated beams, designated as 'ordinary' (O) and 'extraordinary' (E), emerge in linearly polarized states at right angles to one another (Figure 4.5a). The intensity of the source is reduced to such a low level that each of the two photomultipliers intercepting the O and E beams registers about 3000 pulses s^{-1}. Assuming a 20% efficiency of the photomultipliers, this reading is seen to correspond to about 30 000 photons passing through the apparatus every second. For a 30-cm long apparatus and photon wave packets of 100 cm coherence length, one can easily show that the probabilty for more than one photon being in the apparatus is about 10^{-4}, i.e., very small indeed. In fact, no coincidence above the normal background is observed between the O and E photomultipliers. Everything happens here as if the energy-momentum of every photon elected, after entering the calcite crystal, to travel along either the O beam or the E beam, similar to what we found in the cases of the half-silvered mirror and of the neutron interferometer discussed in a previous chapter. The three following experimental situations were investigated:

1. Two identical calcite crystals A and B are placed on the light beam. The separation of the beam into O and E components when traversing A would lead to an enlarged separation in B, since the ordinary beam would cross A and B with no directional change, while the extraordinary beam would be deviated twice in the same direction. A recomposition of O and E is however possible if a $\frac{1}{2}\lambda$ plate is put in the path of each of the two beams in such a way as to exchange their linear polarizations, transforming the ordinary beam into the extraordinary one, and *vice versa*. Under these

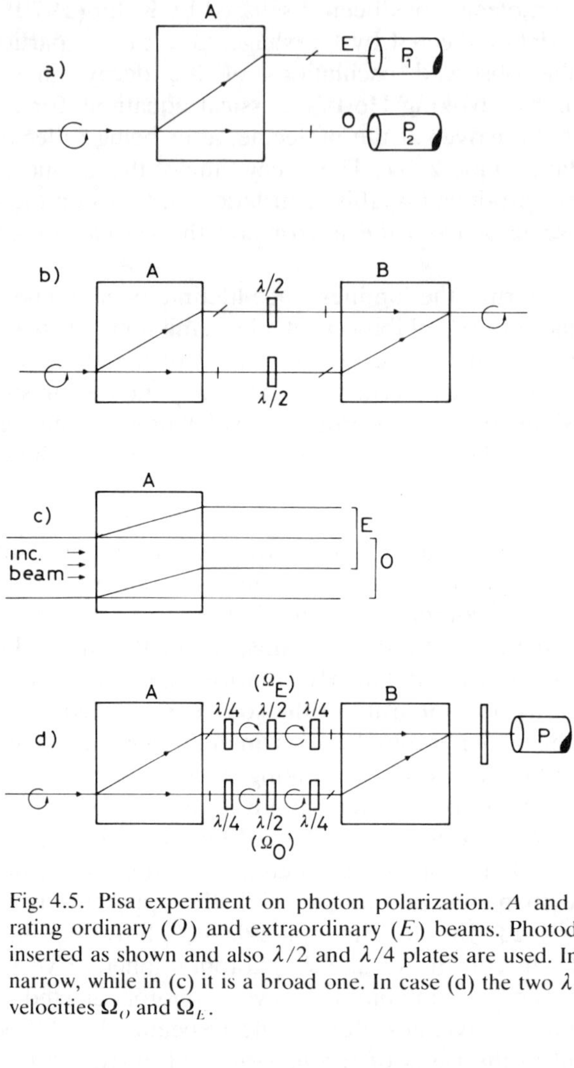

Fig. 4.5. Pisa experiment on photon polarization. A and B are calcite crystals separating ordinary (O) and extraordinary (E) beams. Photodetectors (P_1, P_2 and P) are inserted as shown and also $\lambda/2$ and $\lambda/4$ plates are used. In (a), (b) and (d) the beam is narrow, while in (c) it is a broad one. In case (d) the two $\lambda/2$ plates rotate with angular velocities Ω_O and Ω_E.

conditions (Figure 4.5b) only one beam emerges from B, and Gozzini found that it had exactly the same (circular) polarization as the incoming beam, even at the lowest intensities ('one photon at a time'). The only possible interpretation in spacetime terms of this result is that the $\frac{1}{2}\lambda$ plates act in the same way on energized and on empty packets. Polarization appears more as a property of the wave than of the particle.

2. On using a single crystal and increasing the cross section of the incident beam, the exit beam is composed of three regions. In the first of these only the O beam is present, in the second only the E beam, while in the third (intermediate) region both beams are present (Figure 4.5c). In the

last region, Gozzini observed, even at very low intensity, an interference pattern identical with that observed at high intensity: The E—O splitting with opposite polarizations is therefore a process affecting every single incoming photon, in agreement with the finding of the first experiment (Fig. 4.5b).

3. Using again two identical crystals A and B, one inserts $\frac{1}{4}\lambda$ plates in the paths of the O and E beams, changing their linear polarization states into ones of circular and opposite polarizations. Behind these, two rotating $\frac{1}{2}\lambda$ plates are placed which intercept the two beams in such a way as to reverse the circular polarization and change the frequency. If Ω_O and Ω_E are the angular velocities of the rotating plates inserted in the O and E beam, respectively, one expects the photon frequencies to vary by the corresponding amounts $\pm 2\Omega_O/2\pi$ and $\pm 2\Omega_E/2\pi$. (The positive sign pertains to a spin of the plate parallel to that of the photon crossing it.) A second $\frac{1}{4}\lambda$ plate is inserted in both beams (Figure 4.5d) in such a way that they generate linear polarizations suitable for recombination in the second crystal B. The recombined beam is then made to traverse a polaroid and feed a photomultiplier whose current output is measured by means of a low-frequency spectrum analyzer. Beats are observed at the frequencies $2(\Omega_O + \Omega_E)/2\pi$ or $2(\Omega_O - \Omega_E)/2\pi$, depending on whether the spins of the two plates are opposite or parallel to one another; all this, once more, at very low photon intensities. The implication is clear: a rotating plate is able to change the frequency of a photon wave both for energized and for empty packets. Since the relation $E = h\nu$ is expected to hold without exception, in the case of an energized packet some energy must be supplied to the photon in such a way that it will compensate exactly for the change of frequency. No energy should, on the other hand, be donated to the empty packet. We conclude that these experiments are not inconsistent with the empty-wave approach to duality.

A central problem concerning the nature of quantum waves, and their ability to generate stimulated emissions, is to decide whether these waves have an energy content or not. As is well known, de Broglie favored a positive answer to this question, while Einstein and Bohr were rather convinced of the opposite view. An appropriate *empirical* attack, if it can be found, is obviously the best way to settle the controversy and to decide where the truth lies.

We believe that a feasible experiment could be the following. The energy of the 14.4 keV γ-ray emitted by a nucleus of ^{51}Fe has been measured with a precision of something like 10^{-15} (fractional error) by using the Mössbauer effect. This was done in particular by Pound and Rebka (1960) who could measure the 'apparent weight of photons' in the earth's gravitational field. An energy of 14.4 keV corresponds to a wavelength of about 1 Å. Detailed study of radiation quanta in the ångstrom range of wavelengths was accordingly initiated by Bonse and Hart (1966), using powerful x-ray interferometers. As a result, we now know with good precision the coherence length of the 14.4 keV wave packets, their transverse dimensions, and so on. Using this knowledge, it should therefore be easy to conceive and build a device which necessarily absorbs part of the wave of a γ-ray traversing it. Measurements with the aid of the Mössbauer effect on γ-rays that have passed through such a partial absorber could therefore reveal a measurable energy loss, if it should exist.

The proposed test appears interesting also because of the following argument. If quantum waves carried some energy-momentum, it would be natural to think of this as applying also for massive particles. In the case of a nonrelativistic neutron one would therefore expect a Hamiltonian operator of the type

$$H = -\frac{\hbar^2}{2m}\nabla^2 + V + F(|\psi|), \tag{4.6}$$

where the first two terms on the right are the familiar kinetic and potential energy operators, respectively, for the particle, while the third term, $F(|\psi|)$, represents the energy associated with the wave modulus $|\psi|$. Such a Hamiltonian would then lead to the generalized (nonlinear) Schrödinger equation

$$i\hbar\frac{\partial\psi}{\partial t} = -\frac{\hbar^2}{2m}\nabla^2\psi + V\psi + F(|\psi|)\psi. \tag{4.7}$$

Bialynicki-Birula and Mycielski (1976) studied this equation and showed it to have many interesting properties, irrespective of the exact form of the functional $F(|\psi|)$. One particularly appealing form of F is however

$$F(|\psi|) = -b\ln(a|\psi|^2), \tag{4.8}$$

where b has dimensions of energy. Shimony (1979) proposed to investigate Equation (4.7) with F given by (4.8), using low-intensity neutron beams and neutron interferometers. The best upper limit on b was reported by Zeilinger at the Perugia conference (1982) and is

$$b \lesssim 10^{-15}\,\text{eV} \simeq 1.6\times10^{-27}\,\text{erg}.$$

Since the logarithmic factor is expected to be of the order one, this value is also a measure of an upper limit on the energy associated with the wave:

$$|F(|\psi|)| \lesssim 1.6\times10^{-27}\,\text{erg}.$$

Now, the neutrons under investigation have a de Broglie wavelength of about 1 Å, which corresponds to a kinetic energy

$$T \simeq 1.3\times10^{-13}\,\text{erg}.$$

It follows then that the upper limit for $|F|$ is about 10^{-14} T, *if* one assumes that F has the form (4.8).

In view of the theoretical uncertainties and of the fact that the Mössbauer effect allows one to measure fractional energy losses smaller than 10^{-15}, it seems therefore a matter of no small interest to carry out the proposed experiment on the partial absorption of photonic wave packets.

4.3. Stimulated Emission

When Einstein wrote in 1949 his 'Reply to Criticisms' in *Albert Einstein: Philosopher—Scientist* (Schilpp, 1970), he stated that the problem of wave—particle duality was 'probably the most interesting subject' to discuss:

> I now come to what is probably the most interesting subject which absolutely must be discussed in connection with the detailed arguments of my highly esteemed colleagues Born, Pauli, Heitler, Bohr, and Margenau. They are all firmly convinced that the riddle of the double nature of all corpuscles (corpuscular and undulatory character) has in essence found its final solution in the statistical quantum theory . . . In what follows I wish to adduce reasons which keep me from falling in line . . . . (*Ibid.*, p. 666.)

In the same book, Bohr refers to Einstein's picture of the electromagnetic field by recalling some discussions they had on such problems:

> The discussions, to which I have often reverted in my thoughts, added to all my admiration for Einstein a deep impression of his detached attitude. Certainly, his favored use of such picturesque phrases as 'ghost waves (*Gespensterfelder*) guiding the photons' implied no tendency to mysticism, but illuminated rather a profound humour behind his piercing remarks. (*Ibid.*, p. 206.)

The idea of the empty wave was however not only Einstein's. As has been shown previously, it was entertained in different forms also by de Broglie and Bohr, and it can even be said to belong to the line of thought of the Copenhagen school, even if Heisenberg preferred a 'withdrawal into the mathematical scheme' to the ambiguous formulation of duality contained in the complementarity principle. In the previous section we saw that some important experiments on the nature of duality can naturally be interpreted in terms of an empty wave accompanying the propagation of energized and localized photons. The same is true for the Jánossy—Naray experiment and for the neutron interferometric experiments discussed in the previous chapter. In neutron interferometry one can interpret the existing evidence by assuming that the neutron wave is resolved into two different parts following macroscopically separated trajectories and that a neutron 'particle' carrying energy, momentum, and mass remains associated only with one of these parts.

How can one ever hope to reveal the presence of a wave which does not carry energy or momentum? This problem can have an answer if it is noticed that one does not only measure energy changing processes but probabilities as well: the wave could therefore reveal its presence by modifying decay probabilities for an unstable system.

Such an effect, if it occurs, could probably be seen with existing technology using the following apparatus.

An atomic source emits a well-defined spectral line of wavelength λ selected with an optical filter. The source is designed to randomly emit the photons, one by one, in such a way that there is never more than one photon at a time in the whole apparatus. Each photon is made to fall on a semitransparent mirror M, which 'splits' it into a transmitted part and a reflected one. As is well known, no coincidences above the casual background are observed if a phototube is placed on each of the transmission and reflection paths. This means that the property enabling a photon to discharge a photomultiplier is not split by M but is made to choose between the two paths. More accurate analyses reveal that the entire energy $E = hc/\lambda$ is always found on either one or the other of the two paths. Thus, whatever the carrier of energy is, we see then that *it* is not subdivided by the mirror. Something *is* however, as shown by the low-intensity permanence of the interference pattern.

We wish to study the possibility that what is split is just Einstein's *Gespensterfeld* (or Bohr's virtual wave). It is therefore assumed that the mirror diverts the entire energy in one direction even as it splits the wave into transmitted and reflected components. Suppose we interpose a phototube P_1 (Figure 4.6) in the reflection path and concentrate our attention on the cases where this counter is triggered. According to our postulated picture, one must then conclude that only a virtual wave is present in the transmitted beam.

The transmitted beam is made to pass through a laser gain tube (LGT), where the associated wave packet, though devoid of energy and momentum, has a chance to reveal its existence by generating a zero-energy-transfer stimulated emission. Indeed, one may assume that the molecules filling the

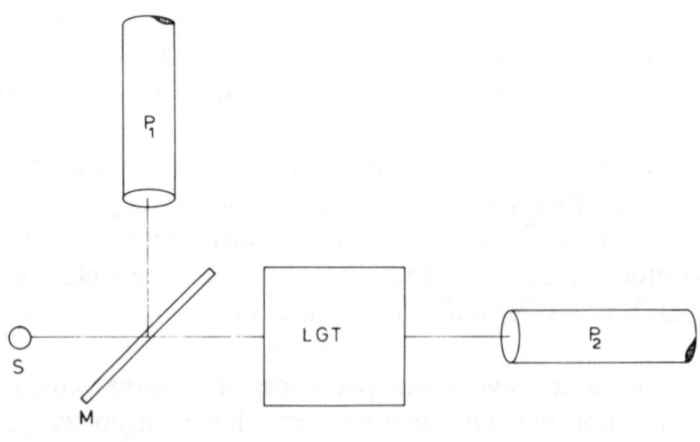

Fig. 4.6. Experiment for the detection of empty waves. S is the source, M is a semitransparent mirror, LGT a laser gain tube (amplifier) and P_1 and P_2 are photodetectors.

LGT are maintained in an excited energy band which on de-excitation, gives rise to radiation that includes the wavelength λ of the incoming wave. The emitted photon could then be revealed by the phototube P_2 that follows the LGT. In this way, P_1P_2 coincidences above the casual background would reveal the existence of a zero-energy undulatory phenomenon transmitted by M. The spacetime propagation of this entity could be demonstrated by checking that P_1P_2 coincidences disappear whenever an obstacle is inserted before LGT in the transmitted beam.

A positive outcome of this experiment would definitely indicate that something propagates in space and time which does not carry energy-momentum but which can, all the same, induce transitions. This would almost amount to the discovery of a new level of reality and would tend to put on a firmer basis the old controversy concerning the nature of the wave—particle duality.

We shall in what follows review some earlier and recent proposals for the detection of empty waves.

The problem of the empirical meaning of the idea that empty waves exist in space and time, independently of our observations, was discussed in two papers (Selleri, 1969, 1982) where the general solution was indicated in the following way: the wave can manifest its reality, independently of the localized energy and momentum which sometimes is associated with it, if it gives rise to changes in the transition probabilities of the systems with which it interacts.

It is well known that all the elementary entities (photons, electrons, neutrons, etc.) have the fundamental dualistic nature. Therefore, for each of them there is an associated wave, and a first blind guess could be that all *waves* are of the same nature, even though they are described by very different mathematical structures within quantum mechanics. Therefore, one could try to discover whether the waves associated with a beam of neutrinos (which are very weakly interacting particles and can therefore be thought to penetrate a piece of matter without any transfer of energy-momentum) are by any chance capable of modifying transition probabilities of unstable systems. Specifically, let a beam of neutrinos traverse a piece of matter containing unstable entities — nuclei or excited atoms or molecules. The lifetime of the unstable entities is measured under these conditions and compared with the lifetime for the same systems in the absence of a neutrino beam. If a difference should be observed, its only logical explanation would be that it was due to the action of the waves accompanying the neutrinos. Naturally one would try to arrange for the transition frequency of the unstable systems to equal one of the frequencies available in the neutrino beam.

Another possible way to approach the problem is to consider a source emitting photons that can reach, with certain probability density, some region R of a screen where they are absorbed. Imagine that within R one places an unstable (e.g., fluorescent) substance σ which has a suitably long lifetime and a transition frequency equal to that of the incoming photons. The latter will arrive randomly as chunks of energy-momentum at different points of R, but the waves associated with them will always impinge on all of R, including therefore the unstable substance σ. It is conceivable that the wave, when it interacts with the atoms of σ, will perceptibly change their lifetime.

A possible proposal to study stimulated emissions due to an energy-momentumless wave is one that considers the double-slit experiment. The incoming photon flux is assumed to be so low that, if a counter is placed immediately behind each of the slits and one of them counts a photon at time t_0, the probability that the other one will detect a photon in the interval $(t_0 - \Delta t, t_0 + \Delta t)$ is negligible. The proposed experiment should then have a counter C behind the first slit and a fluorescent substance σ behind the second slit. Fluorescent emissions by σ are monitored only in time intervals $2\Delta t$ surrounding the instants at which C has counted an incoming photon, in the hope that the detection rate in such intervals (during which one must imagine that σ is continuously washed by empty waves) will be different from the normal one.

An apparatus with interesting properties was proposed by Szczepański (1976). It consists of a light source S which emits monochromatic photons split by a semitransparent mirror BS (Figure 4.7). One part of the beam is detected by D_1, the second part by D_2. In front of D_2 a two-level atom A is placed with an excitation energy equal to the energy of the photons emitted by S. One then measures the delayed coincidences of the recordings of D_1 and D_2 whose respective distances from BS, d_1 and d_2, are variable.

Szczepanski viewed the main interest of this experiment as residing in the possibility of distinguishing the classical description of photons, resurrected by Jaynes (1973), from their quantum description. He stressed that:

From both the classical and the quantum points of view, the wave packet (photon) emitted by S is split by BS, and the situation is similar in both theories as long as the detector D_1 does not record the photon. After it does, the quantum and the classical descriptions differ

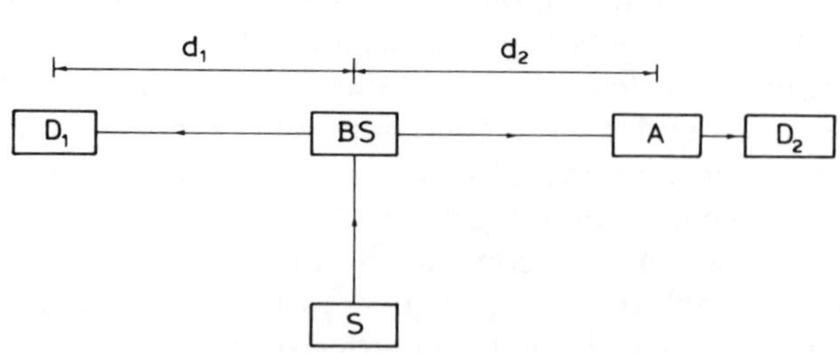

Fig. 4.7. Szczepański's experiment. S is the source and BS a beamsplitter. D_1 and D_2 are photodetectors and A is an amplifier. The $D_1 - D_2$ coincidence rate is expected to be different depending on the relative values of the A—BS distance (d_2) and D_1—BS distance (d_1).

dramatically. Classically, the detection of the wave packet by D_1 does not affect in any way the second part of the wave packet, i.e., the radiation propagating in the direction of D_2. (Szczepański, 1976, pp. 431 and 432.)

Quantum mechanically, the recording of the photon by D_1 causes, according to Szczepański,

An immediate vanishing of the inducing field propagating toward D_2 because of the reduction of the photon's [wave packet]. Thus, the [packet] emitted by S is able to induce an emission from A only until it has not been detected by D_1. The dependence of [the (delayed) coincidence rate] upon the distances d_1 and d_2 is then the main information the experiment is supposed to yield. (*Ibid.*, p. 432.)

We note that the classical description was already excluded by several experiments (Adam *et al.*, 1955; Davidson and Mandel, 1967; Clauser, 1974) and that the reduction of the wave packet is only one of several parallel formulations of quantum measurement theory. In the so-called statistical interpretation of quantum mechanics (see, e.g., Ballentine, 1970), the reduction of the wave packet does not take place. In all cases Szczepański's experiment can indeed allow one to check whether the reduction of the wave packet occurs: if it does, as many people seem to believe, then the $D_1 D_2$ coincidences should disappear as soon as $d_2 > d_1$.

The importance of a photon amplifier ('laser gain tube') for the verification of an active role of the empty waves in photon interactions was stressed by Selleri (1982).

A proposal for the detection of empty waves advanced by Selleri and Vigier (1983) considers the use of superfluorescence (*SF*); that is, of the coherent decay of many atoms which was predicted by Dicke (1954) and studied experimentally by different groups (see, e.g., Vrehen, 1979). Since *SF* involves the coherent decay of something like 10^8 atoms, triggered by a single spontaneous emission due to quantum fluctuations of the electromagnetic field in vacuum, the natural idea would be to see whether an empty wave could give rise to *SF* pulses. The basic apparatus would be different from the one shown in Figure 4.6 only because the *LGT* would be replaced by superfluorescent atomic cesium pencil.

Garuccio and Vigier, in collaboration with A. Rapisarda (1982), proposed the use of an apparatus which was conceived for observing interferences of the Mandel—Pfleegor type between two light beams emitted from two independent lasers, under conditions in which very likely no more than one single photon is present in the experimental region at any given time. The main innovation was an additional device capable of identifying which laser emits any particular photon that appears in the interference pattern. Their setup, which is shown in Figure 4.8, uses a very low intensity *incoherent* source of monochromatic photons. These photons are made to cross a semitransparent mirror M_1, and the transmitted beam is amplified by a laser gain tube *LGT* that is inserted into its path. Behind the *LGT* a second

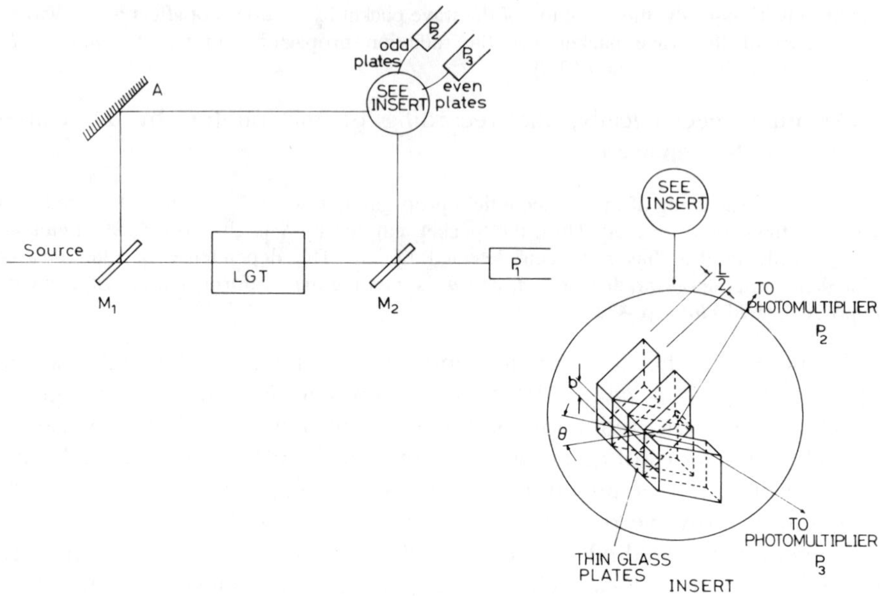

Fig. 4.8. Experiment proposed by Garuccio, Rapisarda and Vigier. *LGT* is the laser gain tube (amplifier), P_1, P_2, and P_3 are photodetectors, M_1 and M_2 are semitransparent mirrors and *A* is a totally reflecting mirror.

semitransparent mirror M_2 splits the beam again in two parts, and one considers only those cases in which a phototube P_1 placed in the M_2-reflected part counts a photon.

The proposal of Garuccio *et al.* depends on an idea opposite to the one discussed in an earlier-cited reference (Selleri, 1982), inasmuch as it assumes that the experiment there proposed has been performed and has failed (at least in the sense that a reasonably low upper limit has been set on the photon production rate for incoming empty waves).

If this assumption were true, one could conclude that, when a photon has been detected by P_1, no energetic photon is propagating along the M_1-reflected path. The photon detected by P_1 is therefore necessarily connected to a situation where the single photon emitted by the source has been transmitted by M_1 and has traversed the *LGT*. If the latter has generated an amplification, it will happen rather often (in 33% of the cases if *two* photons come out of *LGT*) that the photon revealed by P_1 is accompanied by a twin photon that is transmitted by M_2 and travels, following a certain optical trajectory, to an interference-detecting system (*IR*), where it is superimposed on the M_1-reflected beam. Interferences should show up in *IR* if an empty wave has propagated in the M_1-reflected beam and could be detected by using in *IR* an instrument of the type devised by Pfleegor and Mandel (1967) and built with a stack of thin glass plates, each of which has a thickness corresponding to a half-fringe width. The plates are cut and arranged so that any photon falling on the odd plates are 'fed' to one photomultiplier, P_2, while photons incident on the even plates are fed to another, P_3. Detection of any difference in the coincidence rates

$$P_1 \wedge P_2 \text{ versus } P_1 \wedge P_3$$

would constitute evidence, under the stated conditions, that something not carrying energy-momentum is propagating along the M_1-reflected trajectory.

A necessary consequence of this idea is that an obstacle put on the latter trajectory should destroy the interference.

In order that the effect be observable, it is obviously necessary that every single act of amplification does not give rise to a random change of the phase of the wave packet crossing *LGT*. This condition is known to be satisfied in those cases where the amplifying gas volume has dimensions small compared with a radiation wavelength (see, e.g., Dicke, 1954). A conclusion in favor of phase stability after amplification was obtained by Friberg and Mandel (1983). Therefore, it should be possible to observe interference effects even if an amplifier is inserted in the experimental apparatus. One can obtain the same conclusion by considering in some detail the process of amplification. It has already been observed that the wave emitted by an atom is able to increase or to decrease the time of decay if a nearby mirror reflects this wave back on the emitting atom. This can only be understood by assuming that the wave can either stimulate or inhibit atomic emissions, depending on the phase relation between the e.m. undulatory oscillation and the internal oscillation of the excited atom (the beat frequency between two electronic waves). On the other hand, we have already observed that the latter idea is capable of quantitatively explaining the maxima and minima in the lifetime curve of Drexhage (Kuhn, 1970).

A natural question arises at this point: How is it possible that a photonic wave, passing through the region containing several excited atoms, has a net effect in the direction of *stimulating emissions* if the wave is equally capable of stimulation and inhibition, depending only on phase differences? One could naively expect its net effect to be zero, since atoms of arbitrary relative phase will be encountered by the propagating wave. The mystery disappears if one remembers that atoms should be treated as nonlinear oscillators, so that the applied field induces the correct phase automatically in the excited atoms (Sargent *et al.*, 1974).

It is therefore the very phase-locking that guarantees atomic *stimulation* which also give rise to an emitted e.m. wave more intense than, but still in phase with, the incoming wave! It has been calculated that about 300 to 400 wavelengths must interact with the excited atom before the phase-locking can take place. This number is negligible for an ordinary wave packet, which can contains about a million wavelengths, in the case of optical photons, but it is very large for atomic emissions in the presence of a mirror, where the auto-interaction of the atom takes place over distances of a few wavelengths. These observations could explain the difference of the two effects and, at the same time, provide a further argument in favor of the feasibility of such an experiment.

One of the possible realizations of the empty-wave experiments (Gozzini, 1982) is based on the Doppler-free n-photon transition theory developed by Vasilenko, Chebotayev, and Shishaev (1970), which was discussed by Cagnac *et al.* (1973) and verified experimentally by Grynberg *et al.* (1976). Another one has been proposed by Giovanelli (1989).

The proposal by Garuccio *et al.* depends on the idea that the traditional approach to photon stimulation is correct and therefore that the stimulation is due to the action of the energetic quanta. There is however a rather strong argument against this, as will be shown in detail in the following section. It will nevertheless be clear that the detection of an interference, under conditions in which one knows in every single case the trajectory followed by the energetic quantum, is still possible, thanks to a proposal made by Tarozzi (1985).

4.4. Quantitative Empty Wave Amplification

The coincidence rate for the experiment of Figure 4.6 will now be calculated under the assumption that *waves* give rise to stimulated emissions (Selleri, 1987).

Coherent light pulses of finite and known duration are emitted from a laser and are known to impinge on the semitransparent mirror M during the time intervals

$$(t_0, t_0 + \Delta t), (2t_0, 2t_0 + \Delta t), \ldots,$$

with $\Delta t \ll t_0$. The times at which these pulses enter the detectors P_1 and P_2 are calculated simply by adding to the previous times the ratio between the detector—mirror distance and the speed of light.

Coherent pulses being characterized by a Poisson distribution, the probability $p(m)$ for finding m photons in a pulse is

$$p(m) = (m!)^{-1} \, e^{-\langle m \rangle} \langle m \rangle^m \tag{4.11}$$

if $\langle m \rangle$ signifies the average photon number.

The mirror M will split each pulse into a transmitted part (containing n photons) and a reflected part (with $m - n$ photons). The probability $\pi(m, n)$ that n photons will be transmitted through M, if m arrive at its surface, is assumed to be

$$\pi(m, n) = \left(\frac{1}{2}\right)^m \binom{m}{n}. \tag{4.12}$$

Denote by $D_1(m - n)$ the probability that the photomultiplier P_1 will detect at least one of the $m - n$ reflected photons and by $D_2(n + r)$ the probability that the photomultiplier P_2 will detect at least one of the $n + r$ photons emerging from the *LGT*. If η_1 and η_2 represent the quantum efficiencies of P_1 and P_2, respectively, one can then write

$$\begin{aligned} D_1(m - n) &= 1 - (1 - \eta_1)^{m-n}, \\ D_2(n + r) &= 1 - (1 - \eta_2)^{n+r}. \end{aligned} \tag{4.13}$$

In fact, photons are here treated as independent systems which can be detected (or remain undetected) completely independently of the destiny of other photons. This is, of course, the usual treatment of photon detection in textbooks. One cannot avoid observing that such a treatment is strongly suggestive of a corpuscular behavior of photons. The existing scientific literature avoids such a description and prefers to oscillate pragmatically between the corpuscular and undulatory consideration of photons, depending on one's immediate needs. Anyway: η_1 is the probability that P_1 will detect a given photon; $1 - \eta_1$ is the probability that it will not; $(1 - \eta_1)^{m-n}$ is the probability for P_1 not detecting any one of the $m - n$ incoming photons; $1 - (1 - \eta_1)^{m-n}$ is the probability for P_1 detecting something. The argument for $D_2(n + r)$ is similar.

It is useful to introduce two further probabilities. $q_{n+r}(x)$ is the probability that $n + r$ photons will be found at a depth x inside the *LGT* if n enter from the left (see Figure 4.9). Only photon emission is assumed possible, so that $r > 0$. $C_{12}(m)$ is the probability of a $P_1 P_2$ coincident count if m photons are impinging on M. Coincident counts are defined as taking place within appropriate time intervals not exceeding Δt. One can obviously write

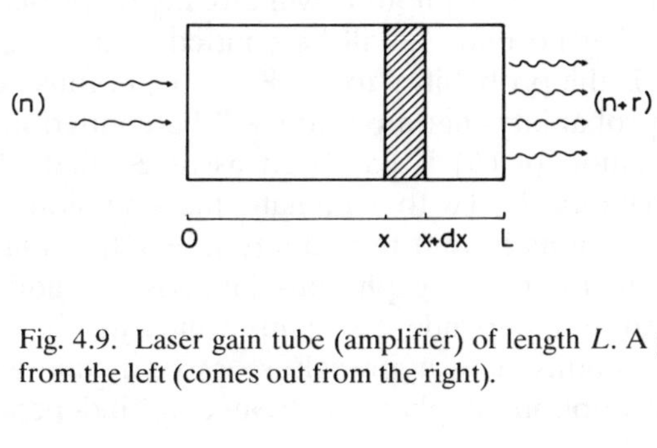

Fig. 4.9. Laser gain tube (amplifier) of length L. A pulse with n ($n + r$) photons enters from the left (comes out from the right).

$$C_{12}(m) = \sum_{n=0}^{m} \pi(m, n) D_1(m-n) \sum_{r=0}^{\infty} q_{n+r}(L) D_2(n+r). \tag{4.14}$$

The inclusion of the term with $m = n$ is possible because $D_1(0) = 0$, consistent with (4.13). Note also that $C_{12}(0) = 0$. In a statistical interpretation of quantum theory, the so-called 'coherent pulse' is a statistical ensemble of pulses, with average photon number $\langle m \rangle$, in which every given photon number m is present with a frequency given by (4.11). The probability of a $P_1 P_2$ coincident count for an incoming 'coherent pulse' is therefore given by

$$C_{12} = \sum_{m=0}^{\infty} p(m) C_{12}(m). \tag{4.15}$$

Equation (4.14) and (4.15) allow concrete empirical predictions to be deduced from the two models for stimulated emission which will be introduced next.

(i) *Corpuscular Model*

Consider the amplifier of Figure 4.9 and suppose that a pulse containing n photons enters from the left. What is the probability $q_{n+r}(x + \mathrm{d}x)$ of finding $n + r$ photons at depth $x + \mathrm{d}x$ inside the amplifier?

One can obviously write

$$\begin{aligned} q_{n+r}(x + \mathrm{d}x) = {} & q_{n+r}(x)\{1 - [(n+r)P + S]\,\mathrm{d}x\} + \\ & + q_{n+r-1}(x)[(n+r-1)P + S]\,\mathrm{d}x, \end{aligned} \tag{4.16}$$

where $P\ \mathrm{d}x$ is the probability that a photon stimulates the emission of another photon in the interval $x - x + \mathrm{d}x$ ($P > 0$); and $S\ \mathrm{d}x$ is the probability that a photon traveling in the same direction of the incoming photons will be spontaneously emitted in the interval $x - x + \mathrm{d}x$ ($S > 0$).

Note that (4.16) can be interpreted in the following way. The probability of finding $n + r$ photons at depth $x + \mathrm{d}x$ is given by the sum of two terms. The first term is the probability that $n + r$ photons will already be present at depth x, times the probability that no photon will be emitted in the interval $x - x + \mathrm{d}x$. The second term is the probability for $n + r - 1$ photons being present at depth x, times the probability that a photon will be emitted in the interval $x - x + \mathrm{d}x$. In Equation (4.16) it has been assumed that, if an incoming photon has a probability $P\ \mathrm{d}x$ to stimulate the emission of a photon, then $n + r$ incoming photons have a probability $n + r$ times larger to stimulate the same emission. In this way photons give rise to additive effects, that is, they are treated as independent systems which incoherently add their emission-stimulating actions. This approach is obviously consistent with a purely corpuscular description of photons, treated as independent particles propagating in the lasing medium, but cannot be reconciled with an undulatory description, since in the latter case one would expect some coherence of the generated effect. From Equation (4.16), one gets

$$q'_{n+r}(x) = -[(n+r)P + S]q_{n+r}(x) + \\ + [(n+r-1)P + S]q_{n+r-1}(x). \qquad (4.17)$$

This is a set of coupled differential equations of the first order which can be solved by standard methods. The solution consistent with the initial condition that exactly n photons be present for $x = 0$ is

$$q_{n+r}(x) = (n+\gamma)(n+1+\gamma)\ldots \\ \ldots(n+r-1+\gamma)\frac{1}{r!}(1-\mathrm{e}^{-Px})^r\,\mathrm{e}^{-(n+\gamma)Px}, \qquad (4.18)$$

where $\gamma = S/P$. From this result it is not difficult to verify that

$$\sum_{r=0}^{\infty} q_{n+r}(x) = 1, \qquad (4.19)$$

showing that probabilities are correctly normalized, and

$$\langle n+r\rangle = \sum_{r=0}^{\infty} (n+r)q_{n+r}(x) = n + (n+\gamma)(\mathrm{e}^{Px}-1) \qquad (4.20)$$

The foregoing calculation refers to the case of an exact number (n) of incoming photons. It is the basis for the study of the amplification of an incoming 'coherent state', characterized by the frequency (4.11) with which n photons are present in the incoming pulse of average photon number $\langle n\rangle$. One can thus say that, in the present corpuscular approach, the amplification of a coherent state leads from an average *initial* photon number $\langle n\rangle$ to an average photon number at depth x given by

$$\langle n(x)\rangle = \sum_{n=0}^{\infty} p(n)[n + (n+\gamma)(\mathrm{e}^{Px}-1)] \\ = \langle n\rangle + (\langle n\rangle + \gamma)(\mathrm{e}^{Px}-1). \qquad (4.21)$$

This last result is obviously independent of the precise form of $p(n)$ and holds also for different photon distributions, e.g., it holds for chaotic light.

The previous considerations are much more relevant to the study of photon amplification treatments usually found in textbooks than one would at first suspect. In fact, the time-dependent changes of the probabilities q_{n+r} are usually obtained from 'rate equations', which can be deduced by studying a cavity filled with radiation of a fixed frequency and considering the transitions between atomic levels with rates proportional to the number of photons in the electromagnetic (e.m.) field. This approach essentially repeats the considerations of the previous page with the trivial replacement of time by the x coordinate. The equations thus found in textbooks and in scientific papers are viewed as the standard 'quantum mechanical' equations for the photon distribution, in spite of the simple phenomenological approach used in their derivation. Other, more sophisticated arguments are sometimes employed, but they cannot be considered as truly quantum mechanical because:

(a) they contain a large number of approximations and cannot therefore be considered deductions from the axioms of the theory
(b) they lead to the same result deduced here from a corpuscular treatment. It is clearly not possible to deduce the same result with a truly quantum-mechanical approach, since the corpuscular nature of the photon distribution is evident in the way the probabilities depend on the photon numbers.

This practical and conceptual equivalence exists, for instance, in the case of the results obtained by Loudon (1978), Scully (1969), and Haken and Weidlich (1969). It exists also for the more recent results obtained by Friberg and Mandel (1983). Our conclusion is thus that the photon amplification treatments found in textbooks and in scientific papers deal essentially with corpuscular amplification processes and neglect the possibility that photonic *waves* instead may be responsible for the stimulated emissions.

(ii) *Undulatory Model*

It will next be assumed that all stimulated emissions are due to the action of the e.m. wave, independently of the presence of the energetic quanta. We shall furthermore assume that it makes sense to talk about the 'intensity' of the wave, which will be written as $I(x)$ at a depth x inside the amplifier. The exact definition of $I(x)$ is not of much interest here, although it can be expected to be given by the squared amplitude of the undulatory phenomenon. A further, necessary, feature of $I(x)$ must be some connection with the energy contained in in the e.m. field; that is, with the average number of quanta present inside the wave packets of intensity $I(x)$.

That the *average* photon number is the relevant parameter can be understood rather easily. In the present approach, atoms are assumed to emit continuously and deterministically a wave in which they release an energetic quantum $h\nu$ at random times. Also, the action of optical devices used for the attenuation of the incoming e.m. field has a deterministic effect on the wave but a random effect on the quanta. For instance, a semitransparent mirror reduces the intensity of the transmitted wave exactly by a factor 1/2, but it transmits or reflects the quantum with a 50—50 probability in a random way. Therefore, beyond a semitransparent mirror one will find in some cases an *empty wave* with no energetic quantum and, in other cases, a wave reduced in intensity associated with a full quantum. In such a picture of the e.m. field it is therefore not possible to keep a strict proportionality between wave intensity and energy content. It will be shown later that the present approach guarantees instead a proportionality between wave intensity $I(x)$ and *average* number of quanta at a depth x inside the amplifier. This is enough for all practical purposes.

It is a very simple matter to show that $I(x)$ must satisfy the equation

$$I'(x) = P_0 I(x) + S_0 \tag{4.22}$$

if P_0 and S_0 are positive parameters describing the stimulated and spontaneous emission processes, respectively, and if the stimulated increase of $I(x)$ is assumed proportional to $I(x)$ itself. The solution of the latter equation satisfying the initial condition $I(0) = I_0$ is

$$I(x) = (I_0 + \gamma_0)(e^{P_0 x} - 1) + I_0, \tag{4.23}$$

where

$$\gamma_0 = \frac{S_0}{P_0}. \tag{4.24}$$

Attributing the stimulated emission of quanta to $I(x)$, the equation replacing (4.16) is (see also Selleri, 1989)

$$\begin{aligned} q_{n+r}(x + \mathrm{d}x) = q_{n+r}(x)\,[1 - cI(x) - S]\,\mathrm{d}x + \\ + q_{n+r-1}(x)\,[cI(x) + S]\,\mathrm{d}x, \end{aligned} \tag{4.25}$$

where c is a positive constant to be fixed later.

From (4.23) and (4.25) it is easy to obtain

$$q'_{n+r}(x) = -q_{n+r}(x)\,[a\,e^{P_0 x} + b] + q_{n+r-1}(x)\,[a\,e^{P_0 x} + b], \tag{4.26}$$

where

$$\begin{aligned} a &= c(I_0 + \gamma_0), \\ b &= -c\gamma_0 + S. \end{aligned} \tag{4.27}$$

Equation (4.26) is equivalent to a set of coupled differential equations of the first order. It can be solved with standard techniques, and the solution consistent with the initial condition that exactly n photons be present in the beam for $x = 0$, is

$$\begin{aligned} q_{n+r}(x) = \frac{1}{r!}\left[\frac{a}{P_0}(e^{P_0 x} - 1) + bx\right]^r \times \\ \times \exp\left\{\frac{a}{P_0}(1 - e^{P_0 x}) - bx\right\}. \end{aligned} \tag{4.28}$$

From these results it is not difficult to show that Equation (4.19) still holds and that

$$\langle n+r\rangle = \sum_{r=0}^{\infty} (n+r)q_{n+r}(x) = n + \frac{a}{P_0}(e^{P_0 x} - 1) + bx. \qquad (4.29)$$

The previous calculation refers to the case of an exact number (n) of photons. It can be used for the study of the amplification of an incoming 'coherent state', characterized by the frequency $p(n)$ of Equation (4.11). One can easily show that such an amplification leads from an average initial photon number $\langle n\rangle$ to an average final photon number at depth x given by

$$\langle n(x)\rangle = \langle n\rangle + \frac{a}{P_0}(e^{P_0 x} - 1) + bx. \qquad (4.30)$$

By comparing (4.30) with (4.21), one sees that the possibility exists of keeping the wave intensity and the average photon number constantly proportional to one another. In fact, the condition

$$I(x) = k\langle n(x)\rangle \qquad (4.31)$$

($k > 0$) can easily be shown to imply

$$b = 0 \qquad (4.32)$$

and

$$\begin{aligned} I_0 + \gamma_0 &= k\,\frac{a}{P_0}, \\ I_0 &= k\langle n\rangle. \end{aligned} \qquad (4.33)$$

By recalling (4.27), the first Equation (4.33) leads to

$$c = \frac{P_0}{k}, \qquad (4.34)$$

while the second Equation (4.33) is an initial condition already included in (4.31) for $x = 0$.

Two important results have thus been obtained. The first one, Equation (4.32), implies that the important relations (4.28), (4.29), and (4.30) should all be considered with $b = 0$. This will be done systematically in the following. The second result, Equation (4.34), taken together with the vanishing of b in the second Equation (4.33), leads to

$$S_0 = kS, \qquad (4.35)$$

relating the coefficients of spontaneous emission of the wave (S_0) and of the particles (S).

In conclusion, Equation (4.30) can be written

$$\langle n(x)\rangle = \langle n\rangle + \left(\langle n\rangle + \frac{S}{P_0}\right)(e^{P_0 x} - 1). \tag{4.36}$$

Comparison with Equation (4.21) (case of corpuscular stimulation) shows that the two results are identical provided that

$$P_0 = P. \tag{4.37}$$

This is a very important result. Particle-generated stimulation and wave-generated stimulation give rise to the same *average* photon number and cannot be discriminated between by experiments measuring only the (average) energy of the outgoing e.m. field. Therefore, the ordinary empirical evidence cannot rule out the present proposal. Only more sophisticated experiments of photon counting in *amplified* laser light can verify or falsify the present undulatory model.

(iii) *Experimental Discrimination Between the Two Models*

One can easily discriminated experimentally between the two models for stimulated emission discussed above. In the case of exactly n photons entering the amplifier from the left (see Figure 4.9) different probabilities apply for the creation of r additional photons in the amplifier which propagate within the incoming light pulse.

Corpuscular model. Stimulated emission is due exclusively to the action of the energy quanta $h\nu$ and has nothing to do with the wave. We have

$$q^{(1)}_{n+r}(x) = \frac{(n+r)!}{n!r!}(1 - e^{-Px})^r e^{-(n+1)Px}, \tag{4.38}$$

which is identical to (4.18) if $\gamma = 1$.

Undulatory model. Stimulated emission is only due to the action of the wave, and the presence or absence of the quanta $h\nu$ is totally irrelevant. One has

$$q^{(2)}_{n+r}(x) = \frac{1}{r!}[(\langle n\rangle + 1)(e^{Px} - 1)]^r \exp\{(\langle n\rangle + 1)(1 - e^{Px})\}, \tag{4.39}$$

which agrees with (4.28) if $b = 0$, $P_0 = P$, and $a = P_0(\langle n\rangle + 1)$.

The two previous predictions for photon distribution can be discriminated between experimentally with photon counting experiments. For instance, they give rise to very different mean-square deviations in photon number. Another, even simpler way to choose between these two possibilities is to perform the experiment of Figure 4.10, which will be discussed next.

Equations (4.14) and (4.15) allow concrete empirical predictions to be deduced from the two models for stimulated emission given by (4.38) and

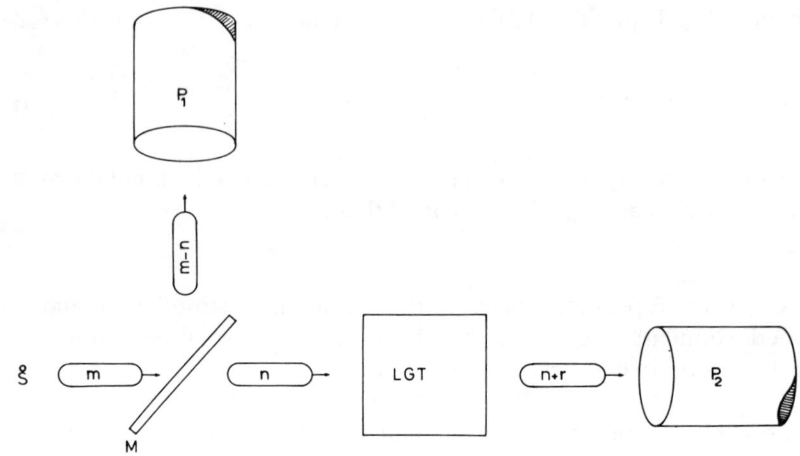

Fig. 4.10. Experiment for the detection of empty waves. Notation is same as in Figure 4.6. The number of photons in a given light pulse is denoted by: m for incoming, n for transmitted, $m - n$ for reflected, while r is the number of photons created in the amplifier.

(4.39). An intermediate step for a concrete calculation of $C_{12}(m)$ is the evaluation of the last sum of (4.14), which will be called $\Sigma(n)$ in the following. In the case of particle-generated stimulation, one gets

$$\begin{aligned}\Sigma^{(1)}(n) &= \sum_{r=0}^{\infty} q_{n+r}^{(1)}(L) D_2(n+r) \\ &= e^{-(n+1)PL} \sum_{r=0}^{\infty} \binom{n+r}{n} (1 - e^{-PL})^r [1 - (1 - \eta_2)^{n+r}] \\ &= e^{-(n+1)PL} \{ (e^{-PL})^{-n-1} - \\ &\quad - (1 - \eta_2)^n [1 - (1 - \eta_2)(1 - e^{-PL})]^{-n-1} \},\end{aligned} \tag{4.40}$$

where use has been made of the result

$$\sum_{r=0}^{\infty} \binom{n+r}{r} B^r = (1 - B)^{-n-1} \tag{4.41}$$

valid for all $B < 1$. In conclusion, (4.40) can be written

$$\Sigma^{(1)}(n) = 1 - (1 - \eta_2)^n A_0^{n+1}, \tag{4.42}$$

where

$$A_0 = [1 - \eta_2 + \eta_2 \, e^{PL}]^{-1}. \tag{4.43}$$

This result can now be inserted in (4.14) together with (4.12) and (4.13) to yield

$$C_{12}^{(1)}(m) = \sum_{n=0}^{\infty} \left(\frac{1}{2}\right)^m \binom{m}{n} [1 - (1 - \eta_1)^{m-n}] \times$$

$$\times [1 - (1 - \eta_2)^n A_0^{n+1}], \tag{4.44}$$

which can easily be transformed into

$$C_{12}^{(1)}(m) = 1 - \left(1 - \frac{\eta_1}{2}\right)^m - A_0 \left[\frac{1 + (1 - \eta_2)A_0}{2}\right]^m +$$

$$+ A_0 \left[\frac{1 - \eta_1 + (1 - \eta_2)A_0}{2}\right]^m. \tag{4.45}$$

Equation (4.45) has been deduced by using the well-known binomial formula

$$(\alpha + \beta)^m = \sum_{n=0}^{\infty} \binom{m}{n} \alpha^n \beta^{m-n}.$$

Finally, if the result (4.45) is inserted in (4.15), it is not difficult to show that

$$C_{12}^{(1)} = \left[1 - \exp\left\{-\frac{\langle m \rangle \eta_1}{2}\right\}\right] \times$$

$$\times \left[1 - A_0 \exp\left\{\langle m \rangle \frac{(1 - \eta_2)A_0 - 1}{2}\right\}\right]. \tag{4.46}$$

In order to deduce (4.46), one needs only to remember the exponential series

$$\exp(B) = \sum_{m=0}^{\infty} \frac{1}{m!} B^m,$$

which is convergent for all real values of B.

Coming now to the case of the wave-generated stimulation, one has

$$\Sigma^{(2)}(n) = \sum_{r=0}^{\infty} q^{(2)}_{n+r}(L) D_2(n+r)$$

$$= \exp\{(\langle n \rangle + 1)(1 - e^{PL})\} \sum_{r=0}^{\infty} \frac{1}{r!} \times$$

$$\times [(\langle n \rangle + 1)(e^{PL} - 1)]^r [1 - (1 - \eta_2)^{n+r}] \tag{4.47}$$

if $\langle n \rangle$ is the average photon number, per pulse, entering the amplifier. Note that one can take

$$\langle n \rangle = \frac{\langle m \rangle}{2},$$

where $\langle m \rangle$ is the average photon number, per pulse, impinging on the semitransparent mirror M, since M is assumed to generate a 50—50 transmission-reflection probability. On setting

$$\Omega = \left(\frac{\langle m \rangle}{2} + 1 \right)(e^{PL} - 1), \tag{4.48}$$

Equation (4.47) can be written

$$\Sigma^{(2)}(n) = e^{-\Omega} \sum_{r=0}^{\infty} \frac{1}{r!} \Omega^r [1 - (1 - \eta_2)^{n+r}]$$

$$= 1 - (1 - \eta_2)^n \exp\{-\eta_2 \Omega\}. \tag{4.49}$$

This result one can now insert in (4.14), together with (4.12) and (4.13), to obtain

$$C^{(2)}_{12}(m) = \sum_{n=0}^{m} \left(\frac{1}{2}\right)^m \binom{m}{n} [1 - (1 - \eta_1)^{m-n}][1 - (1 - \eta_2)^n e^{-\eta_2 \Omega}]$$

$$= 1 - \left(1 - \frac{\eta_1}{2}\right)^m - \left(1 - \frac{\eta_2}{2}\right)^m e^{-\eta_2 \Omega} +$$

$$+ \left(1 - \frac{\eta_1 + \eta_2}{2}\right)^m e^{-\eta_2 \Omega}. \tag{4.50}$$

Therefore

$$C_{12}^{(2)} = \sum_{m=0}^{\infty} p(m) C_{12}^{(2)}(m)$$

$$= \{1 - e^{-\langle m \rangle \eta_1/2}\}\{1 - e^{-\langle m \rangle \eta_2/2}\, e^{-\eta_2 \Omega}\}.$$

This is the predicted coincidence probability for the case of wave-generated stimulations, which should be compared with the analogous prediction for the case of particle-generated stimulation; that is, with Equation (4.46).

It is a very simple matter to show that the probability of a coincidence for the experiment of Figure 4.10 *with the amplifier removed*, all other conditions remaining the same, is

$$E_{12} = \left[1 - \exp\left(-\frac{\langle m \rangle \eta_1}{2}\right)\right] \cdot \left[1 - \exp\left(-\frac{\langle m \rangle \eta_2}{2}\right)\right]. \tag{4.52}$$

This can easily be shown to be the common limit of (4.46) and (4.51) when the amplifier is removed, mathematically obtainable for $P = 0$, which in turn implies $A_0 = 1$ and $\Omega = 0$, as is obvious from Equations (4.43) and (4.48), respectively.

Dividing (4.46) and (4.51) by E_{12}, as given by (4.52), one obtains

$$\Gamma^{(1)} = \frac{C_{12}^{(1)}}{E_{12}} = \frac{1 - A_0 \exp\{\langle m \rangle [(1 - \eta_2) A_0 - 1]/2\}}{1 - \exp\{-\langle m \rangle \eta_2/2\}} \tag{4.53}$$

and

$$\Gamma^{(2)} = \frac{C_{12}^{(2)}}{E_{12}}$$

$$= \frac{1 - \exp\{-\langle m \rangle \eta_2/2\} \exp\{(\langle m \rangle/2 + 1)\,(e^{PL} - 1)\}}{1 - \exp\{-\langle m \rangle \eta_2/2\}}. \tag{4.54}$$

These results should be easy to analyze experimentally.

Numerical comparisons of these two last equations have been made systematically for different values of the experimental parameters $\langle m \rangle$, η_2, and e^{PL}, the latter being usually called 'the amplification'. As an example, Figure 4.11 shows $\Gamma^{(1)}$ and $\Gamma^{(2)}$ for $\langle m \rangle = 1$ and $\eta_2 = 0.1$ as a function of the amplification e^{PL}. As one can see, a difference of about 20% is obtained for $e^{PL} = 5$. It has been checked that the percentage difference between $\Gamma^{(1)}$ and $\Gamma^{(2)}$ grows to a maximum of about 30% for $e^{PL} \simeq 12$ and decreases thereafter.

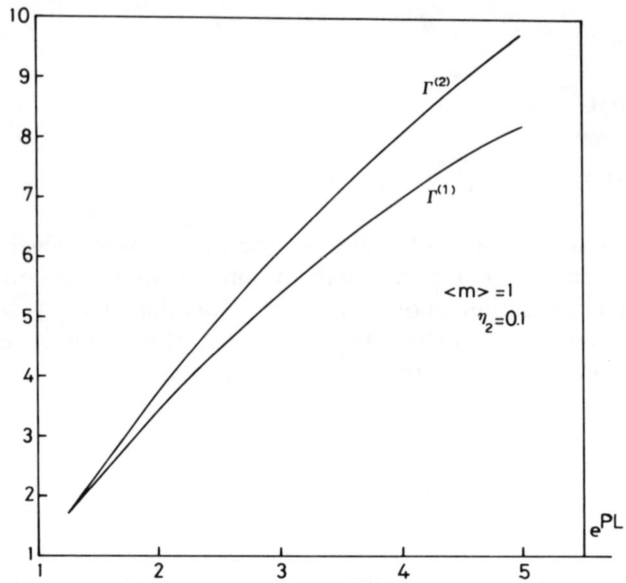

Fig. 4.11. Predicted coincidence rates for the empty-wave detection experiment of Figure 4.10. Upper and lower curves show the predictions of the undulatory and corpuscular stimulation models, respectively. Experimental parameters have been taken to be $\langle m \rangle = 1$ and $\eta_2 = 0.1$.

4.5. Two Further Experimental Proposals

The present section is devoted to the discussion of two recent proposals for the detection of effects generated by an empty wave. In both cases the empty wave is obtained with the help of a beamsplitter that sends the corpuscular component of a quantum system to a particle detector. The first proposal, due to Tarozzi (1985), is based on the study of optical photons, while the second one (Croca, 1987) uses neutron interferometry.

A simplified version of Tarozzi's experiment is shown in Figure 4.12. We will present it in the following in a highly schematized way, assuming the incidence of single photons at well-defined times and the absence of spontaneous emissions from the laser gain tube. The discussion of the experiment in more physical terms is certainly possible by using the approach of the previous section. The observable effect will probably descend from the qualitative to the quantitative level, but there can be little doubt that an observable phenomenon should remain. Single atomic emissions, containing a well-defined energy $h\nu$, are emitted by a source during time intervals separated by comparatively large times, during which no emission takes place. These wave packets impinge on a semitransparent mirror M_1. We are interested only in the cases in which the energy quantum is *reflected*, the

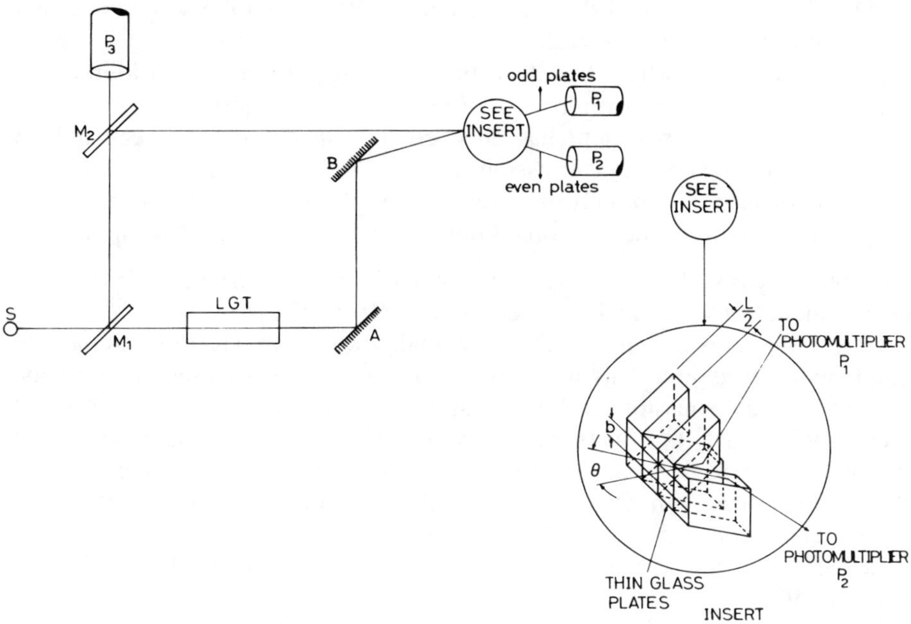

Fig. 4.12. Experiment proposed by Tarozzi. S is the source, LGT a laser gain tube, M_1 and M_2 are semitransparent mirrors and A and B are totally reflecting mirrors. IR is the interference detecting apparatus and P_1, P_2, and P_3 are photodetectors.

reflection being ascertained with a (future) detection in the phototube P_3. The M_1-reflected part of the wave packet strikes a second semitransparent mirror M_2, which splits it into two parts: only events in which the energy quantum is detected by P_3 will be considered. In such cases it is clear that only an empty wave will be reflected by M_2. The aim of the experiment is to detect an effect generated by this empty wave. To understand the idea, consider now the part of the atomic emission transmitted by M_1. It will not contain the energy quantum, reflected by M_1, and will therefore also consist of an empty wave. On its way, this wave will encounter a laser gain tube, and, as already discussed in the previous sections, it will give rise to stimulated emissions. Let us say that the amplifier is such that one or more photons are produced in the LGT and inserted into the previously empty wave packet. After being reflected in the mirror M, these photons arrive in an interference-detecting device IR, where they are superimposed on the M_2-reflected empty wave. The IR device could be of the Pfleegor—Mandel type discussed before. It is connected with two photomultipliers P_1 and P_2, and a counting difference between these two instruments is a sign of an interference taking place between the A-reflected photons and the M_2-reflected empty wave.

Suppose that:

(1) it has been checked that photons are emitted from the source one at a time at large time intervals;
(2) only events in which P_3 detects the incoming photon are considered;
(3) spontaneous emission from the *LGT* can be neglected;
(4) a previous experiment has checked that an empty wave can indeed give rise to stimulated emissions;
(5) an obstacle eventually inserted on the M_2-reflected beam has been shown to make the counting difference between P_1 and P_2 disappear.

Then the only possible interpretation of the observed interference is that it is a consequence of the arrival of the empty wave from M_2.

Of all these assumptions the only really weak one is the absence of spontaneous emission. This is, however, not a really big problem, since the amount of this emission is known and can eventually be remeasured if necessary, so that its effects can be subtracted from the final result. That empty waves can give rise to stimulated emissions has been shown to be a likely conclusion with the analysis of the previous sections, so that this experiment seems really feasible. Moreover, single photon sources are now available, after the discovery of the 'antibunching effect' (Dagenais and Mandel, 1978).

It is obviously important to carry out also for this experiment the quantitative analysis of the previous section.

There is a second way to detect the existence of empty waves, as shown by Croca. The starting point is an idea published by J. and M. Andrade e Silva (1980), according to which one should superimpose three wave packets: the first two should be the interfering results of the action of a beamsplitter on a single incoming quantum system emitted by the source S_1; the third one should be an empty wave-packet emitted by a second source S_2. The two sources are incoherent, so that the interferences of the third wave with the first two, though physically present in every single act of superposition of such three wave packets, is different from case to case and vanishes in the statistical ensemble of many repetitions of the phenomenon.

If one writes for the overall interference figure:

$$I = I_1 + I_2 + I_3 + 2(I_1 I_2)^{1/2} \cos \delta_{12} + \\ + 2(I_1 I_3)^{1/2} \cos \delta_{13} + 2(I_2 I_3)^{1/2} \cos \delta_{23}, \quad (4.55)$$

where I is the final intensity, I_1, I_2, and I_3 are the intensities of the three beams, and δ_{12}, δ_{13}, and δ_{23} are the phase-differences of the respective waves, it is clear that the incoherent nature of the quantum systems emitted by S_1 and S_2 gives rise to randomly distributed values of δ_{13} and δ_{23} and therefore leads *on the average* to the relations

$$\cos \delta_{13} = 0, \cos \delta_{23} = 0.$$

Inserting this in the previous relation (4.55), one obtains

$$I = I_1 + I_2 + I_3 + 2(I_1 I_2)^{1/2} \cos \delta_{12}. \tag{4.56}$$

The effect of the third wave has not disappeared, since the term I_3 is still present in (4.56). This affects the so-called fringe visibility parameter V, defined by

$$V = \frac{2(I_1 I_2)^{1/2}}{(I_1 + I_2 + I_3)}. \tag{4.57}$$

The mathematical explanation of the presence of I_3 in the r.h.s. of (4.56) has been given above and is obvious. The physical reason for the presence of this term is the following: All wave packets coming from the source S_2 have been assumed to be empty, so that by the intensity I_3 is meant only the squared modulus of the third wave function. The third wave is however physically present in the interference region and has the effect of forcing the particle contained in one of the other two packets sometimes to behave as if it had traveled along the third path. This explanation is of course very similar to that of de Broglie for the Pfleegor—Mandel experiment, which was discussed in the previous chapter.

This beautiful idea of J. and M. Andrade e Silva was however marred by the fact that no convincing description of a concretely feasible experiment was provided. Only recently has Croca found a way to realize the experiment. Consider the apparatus of Figure 4.13, consisting of two parts:

1. A normal *neutron interferometer* wherein neutrons, coming from a reactor one at a time with constant velocity (the technique for achieving this is well known and often used) give rise to the usual interference phenomenon in the intensities of the two outgoing beams: One can assume, for instance, that the interference gives rise to $I_1 \gg I_2$.

2. A *source of empty waves* consisting of a beamsplitter *BS* made of a single slab — a silicon crystal with atomic planes perpendicular to the face of the crystal. When a neutron strikes this slab at an angle θ with the normal to the surface of magnitude 20—30°, Laue scattering takes place and the neutron wave is split into two parts. The first part propagates in a direction parallel to the one of incidence, while the second one is reflected by the atomic planes and thus is diffracted by the crystal at an angle θ. If a neutron detector is put on the diffracted trajectory and if the neutron is actually detected, it is safe to assume that only an empty neutronic wave propagates along the first path. An electronic circuit opens a counting gate for the detectors placed on the trajectories coming out of the interferometer. The size of this temporal 'gate' can be of about 1 ns, consistent with the time of passage of a neutronic empty wave having a group velocity of 10^5 cm s^{-1} and a longitudinal length of 10^{-4} cm. Fast neutron detectors with a time resolu-

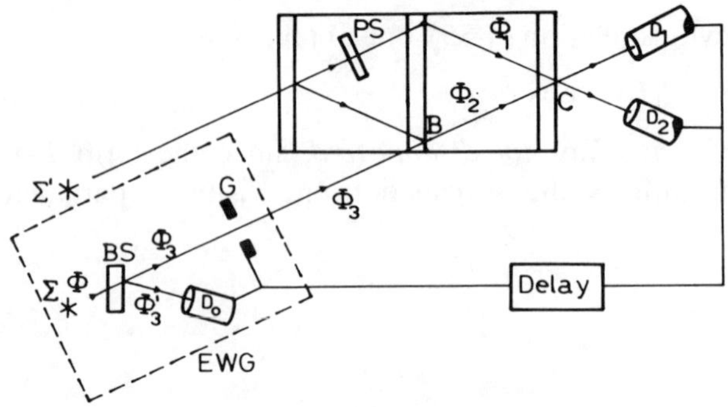

Fig. 4.13. Experiment proposed by Croca. *EWG* is the empty-wave generator, *BS* a beamsplitter, *G* an electronically controlled gate, D_0, D_1, D_2 are neutron detectors. In the neutron interferometer *PS* is the phase shifter.

tion of 10^{-9} s are needed for this experiment, but this seems to be not too difficult to obtain with 1.5 MeV alpha particles emitted in the collision of reactor neutrons with boron atoms.

As shown in Figure 4.13, the empty wave packets issuing from the empty-wave generator are perfectly aligned with one of the internal beams of the interferometer. In this way the geometrical superposition of the three wave packets on the third slab of the interferometer is achieved. The experiment should detect a shift of neutrons from I_1 to I_2.

If the predicted effect is discovered, one could say, from the Copenhagen point of view, that a beam of nothing, coming from the EWG of Figure 4.13 gives rise to a modified interference of the two interferometric beams. *The Einstein—de Broglie conclusion would be that neutronic empty waves have been discovered*!

The technical difficulties of this experiment include:

1. The need to obtain two neutron beams of the same speed.
2. Detectors with a resolving time of 1 ns.
3. Perfect alignment of one of the beams in the interferometer with the beam coming out of the EWG.
4. Neutronic wave packets with passing time of about 10^{-9} s.

These points have been considered in detail by Croca with the conclusion that the proposed experiment should be feasible.

4.6. Triple-Slit Experiments

As we saw according to de Broglie and Einstein a quantum object is composed of a small particle which is constantly localized in space, and of an objectively real wave $\phi(x, y, z, t)$ which is a physical process propagating in space and time satisfying a well-defined equation of motion, for example the nonrelativistic Schrödinger equation. The usual quantum mechanical wave function ψ is proportional to the physical wave ϕ

$$\psi(x, y, z, t) = c\phi(x, y, z, t), \tag{4.58}$$

where c is a constant that must be chosen in such a way that the integral over the whole space of $|\psi(x, y, z, t)|^2$ be equal to unity. This is required by the probability interpretation of ψ, while ϕ, being a real physical field, cannot of course be normalized at will. Like in classical physics an observation modifies the predicted probabilities and this implies that in some regions of space ψ (and thus c, but not ϕ) has to go to zero: The reduction of the wave packet takes place for ψ but not for ϕ!

De Broglie seems to have overlooked the fact that a nonzero value of ϕ in regions where $\psi = 0$ leads to the prediction of phenomena which are absent in the usual ('Copenhagen') interpretation of quantum theory. The present section reviews some of these phenomena, following a basic idea due to Croca (1987) [See also: Croca, Garuccio and Selleri (1988) and Selleri (1988)].

In the Philippidis—Dewdney—Hiley discussion of the double-slit experiment presented in our Section 4.1, one has to do with a double meaning of R (modulus of the wave function): it is the amplitude of the physical wave and its square gives the particle density in the statistical ensemble of many repetitions of the same physical situation.

Initially the particle is embedded in a plane wave with a constant amplitude R_0: Therefore the quantum potential Q vanishes and with it the quantum force and the particle propagates along a straight line. The situation is different beyond the first screen, since a double diffraction takes place and two cylindrical waves

$$\phi_1 = R_1 \exp\left(\frac{iS_1}{\hbar}\right) \quad \text{and} \quad \phi_2 = R_2 \exp\left(\frac{iS_2}{\hbar}\right) \tag{4.59}$$

emerge from the two slits with variable amplitudes R_1 and R_2. These waves overlap on the second screen and give rise to a total undulatory phenomenon

$$\phi = \phi_1 + \phi_2 = R \exp\left(\frac{iS}{\hbar}\right), \tag{4.60}$$

where

$$R = \left[R_1^2 + R_2^2 + 2R_1R_2 \cos\left(\frac{S_1 - S_2}{\hbar}\right)\right]^{1/2} \tag{4.61}$$

and

$$S = \hbar \tan^{-1}\left\{\frac{R_1 \sin(S_1/\hbar) + R_2 \sin(S_2/\hbar)}{R_1 \cos(S_1/\hbar) + R_2 \cos(S_2/\hbar)}\right\}. \tag{4.62}$$

From these equations one can calculate the quantum potential, the particle velocity and, therefore, the trajectory for any given initial position in the first or second slit. Notice that the double-slit experiment has actually been carried out with neutrons by Zeilinger (1986).

Of course in de Broglie's picture the wave itself cannot be detected directly because it carries very little energy, or no energy at all. Only particles are directly observed and every one of them passes either through the first or through the second slit. Nevertheless every particle feels the action of both waves ϕ_1 and ϕ_2 (or, rather, of the total wave ϕ) and modifies its trajectory accordingly. The interaction of the particle with the waves is thus very similar to that of a classical electron interacting with several electric fields $\mathbf{E}_1$, $\mathbf{E}_2$, . . . , $\mathbf{E}_n$ generated by different sources. The interaction takes place with the sum of the fields:

$$\mathbf{E} = \mathbf{E}_1 + \mathbf{E}_2 + \cdots + \mathbf{E}_n$$

Generalizing (4.60) it will be assumed that a quantum particle propagating in a region of space where several waves ϕ_1, ϕ_2, . . . , ϕ_n are present interacts with the total wave given by

$$\phi = \phi_1 + \phi_2 + \cdots + \phi_n. \tag{4.63}$$

The previous assumption will be made also in the case when the waves arise from different sources. In the frame of de Broglie's interpretation of quantum phenomena there exists direct empirical evidence that this extension is correct, at least for photons. In fact, interference effects produced by the superposition of light beams from two independent single-mode lasers have been investigated experimentally by Pfleegor and Mandel (1967). They found that interference takes place even under conditions in which, with high probability, one photon is absorbed well before the next one is emitted by one or the other laser (see Figure 3.5). The causal interpretation of this experiment was given by de Broglie and Andrade e Silva (1968) who wrote: "A photon coming from one laser or the other and arriving in the interference zone is guided — and this seems to us physically certain — by the superposition of the waves emitted by the two lasers . . .".

Essentially the same words were used by de Broglie in a later paper (de Broglie, 1977). This idea represented for de Broglie an important modification of previous statements according to which ". . . the wave of a particle can interfere only with other segments of itself and not with a wave of a different particle" (de Broglie, 1960). Naturally this modification was absolutely necessary if all empirical phenomena, including the Pfleegor—Mandel result, had to be explained. Anyway we consider the 1968 position more physical and more natural from a general realistic point of view, than the 1960 idea which is very similar to the Copenhagen conception that two waves of two different quantum objects cannot interact with one another because they are represented by a unique wave field $\psi(x_1, x_2)$ in configuration space.

We therefore assume the following physically obvious generalization of de Broglie's 1968 position:

A particle propagating in a region of space where several wave packets $\phi_1, \phi_2, \ldots, \phi_n$ of the same nature and with the same frequency are simultaneously present interacts locally with their algebraic sum ϕ through the quantum potential, even when the waves come from different sources. This interaction has the effect, over the statistical ensemble, that the particle density becomes proportional to $|\phi|^2 = R^2$.

In order to obtain some of the startling consequences of this assumption consider the triple-slit experiment of Figure 4.14. Two *incoherent* sources Σ_1 and Σ_2 emit quantum objects of the same nature (photons, electrons, neutrons, . . .) and of the same energy, so that the quantum waves emitted by the two sources shall have the same frequency. The sources are assumed to have a very low intensity of emission, i.e., quantum objects are emitted one at a time at widely separated time intervals. Emissions from the upper source Σ_1 can hit on the two upper slits only: from these slits two cylindrical waves ϕ_1 and ϕ_2 will be emitted to the right, sometimes together with a particle which has crossed either the first or the second slit. Only cases in which this particle is observed by detector D_1 (set on the second screen) will interest us.

Emissions from the lower source Σ_2 can hit on the two lower slits; one of them is however constantly closed by particle detector D_2 and the slits that can be crossed by quantum waves in Figure 4.14 are only three. Only cases in which D_2 does detect a particle interest us, and in such cases an empty quantum wave ϕ_3 can be assumed to be emitted from the third slit. The

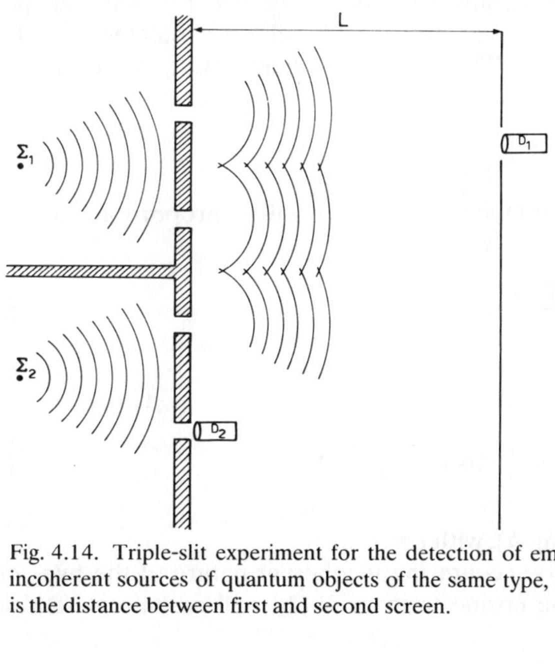

Fig. 4.14. Triple-slit experiment for the detection of empty waves. Σ_1 and Σ_2 are two incoherent sources of quantum objects of the same type, D_1 and D_2 are detectors and L is the distance between first and second screen.

complication due to the simultaneous emission of two particles from Σ_2 will be dealt with later.

The proposed experiment is based on coincident detections from D_1 and D_2, with the coincidence time chosen in a suitable way. Assume that the two sources emit wave packets with coherence length ℓ and that they do not spread during propagation. This assumption is valid for optical photons and for x-rays, since phase and group velocity both equal the velocity of light, but does not hold for electrons or neutrons. The complications due to the spreading of the wave packet can be dealt with but the following discussion neglects them and therefore it applies, strictly speaking, to photons only.

We assume ℓ to be known from previous measurements, as well as the group velocity u. In such conditions every wave packet will take a time

$$\Delta t = \frac{\ell}{u} \tag{4.64}$$

for crossing any given point. Precisely this Δt defines the coincidence time. If D_2 fires at time t_2 then D_1 must fire at time

$$t_1 = t_2 + L/u \pm \Delta t/2 \tag{4.65}$$

if the double detection is to be counted as a true coincidence from electronics (L is the distance between the two screens). *This definition of 'coincidence' guarantees that the waves accompanying the two particles detected by* D_1 and D_2 *overlap in the region just before the second screen.*

By applying our fundamental assumption we can say that the particle coming from the first or the second slit will feel the simultaneous action of three waves: ϕ_1, ϕ_2, and ϕ_3. The undulatory field acting on the particle is given by

$$\phi = \phi_1 + \phi_2 + \phi_3 \tag{4.66}$$

and its probability density on the second screen will be proportional to

$$|\phi|^2 = \sum_{i=1}^{3} |\phi_i|^2 + \sum_{i<j} I_{ij}, \tag{4.67}$$

where

$$I_{ij} = 2\,\mathrm{Re}\,\phi_i^*\phi_j = 2R_iR_j \cos\left(\frac{S_i - S_j}{\hbar}\right) \tag{4.68}$$

if one writes $\phi_i = R_i \exp(iS_i/\hbar)$, with $i = 1, 2, 3$.

We must now take into account the incoherent nature of the two sources Σ_1 and Σ_2. Obviously the cosine term in (4.68) will assume at random all possible values between +1 and −1 in different events, so that the average over the statistical ensemble of the interference terms between ϕ_1 and ϕ_2 and between ϕ_2 and ϕ_3 will certainly vanish:

$$I_{13} = I_{23} = 0. \tag{4.69}$$

Therefore the interference figure on the second screen reduces to

$$|\phi|^2 = \sum_{i=1}^{3} R_i^2 + 2R_1R_2 \cos\left(\frac{S_1 - S_2}{\hbar}\right). \tag{4.70}$$

This prediction of the particle density on the second screen is different from that expected on the basis of the Copenhagen interpretation of quantum theory. In the latter approach particle detection in D_2 leads to the immediate disappearance of the wave emerging from the third slit (ϕ_3) which is not a physical wave, but only a mathematical tool for the calculation of probabilities. The disappearance of ϕ_3 is of course nothing but the well-known 'reduction of the wave function'.

In order to get an idea of the quantitative difference between the two predictions consider a region on the second screen where the three waves have roughly the same amplitude and write here:

$$R_1^2 = R_2^2 = R_3^2 = I_0.$$

Equation (4.70) becomes then

$$|\phi|^2 = 2I_0\left[\frac{3}{2} + \cos\left(\frac{S_1 - S_2}{\hbar}\right)\right]. \tag{4.71}$$

In the Copenhagen interpretation the term 3/2 has to be substituted with 1. Therefore the difference between the two predictions amounts to 25% in the worst possible case!

The two sources must satisfy several conditions for a meaningful experiment. First of all they must be equal as far as the velocity of the emitted systems is concerned, so that the particle crossing one of the first two slits can be carried also from the wave emitted by the third slit. Of course a perfect identity of the emitted velocities is not needed: only the frequency spectra of the two wave packets should have a good overlap, so that a particle can feel at home (in de Broglie's sense) on all waves. The coherence length ℓ of the wave packets emitted from the two sources should also be the same, but this is practically certain if the previous condition is satisfied. The intensities of the two sources must be so low that it is very rare to find more than one particle in the space region between the two screens.

There is always a nonzero probability that two particles emitted by Σ_2 give rise to $D_1 - D_2$ coincidence. In order to make sure that the particles detected by D_1 come almost always from Σ_1 *it is convenient that* Σ_1 *be more intense* *than Σ_2, let us say by a factor of a hundred.* This difference must be obtained without any change of the wave accompanying a particle emitted by Σ_2: a suitable toothed wheel can do the job, since it absorbs completely some quantum objects while it lets through unchanged some other ones.

A different experiment of the same type is shown in Figure 4.15. Here the source Σ_1 emits only toward the first slit, while Σ_2 emits toward the second slit, the third slit and the slit closed by detector D_2. Everything else is the same as in the previous experiment: in particular Σ_1 emits with a 100 times larger intensity than Σ_2. Therefore the detection of $D_1 - D_2$ coincidences is almost always due to the activation of detector D_1 by a particle which passed *from the first slit.* In the region between the two screens such a particle is still acted upon by the total wave given by (4.66), but this time ϕ_1 gives vanishing (average) interferences with the other two waves. Therefore one has in (4.67)

$$I_{12} = I_{13} = 0 \tag{4.72}$$

and the particle density on the second screen becomes proportional to

$$|\phi|^2 = 2I_0 \left[\frac{3}{2} + \cos\left(\frac{S_2 - S_3}{\hbar} \right) \right]. \tag{4.73}$$

This result is only formally similar to (4.70) because *now the interference term results from space-time superposition of two empty waves.* Such oscillations, if actually present, would constitute a truly extraordinary phenomenon

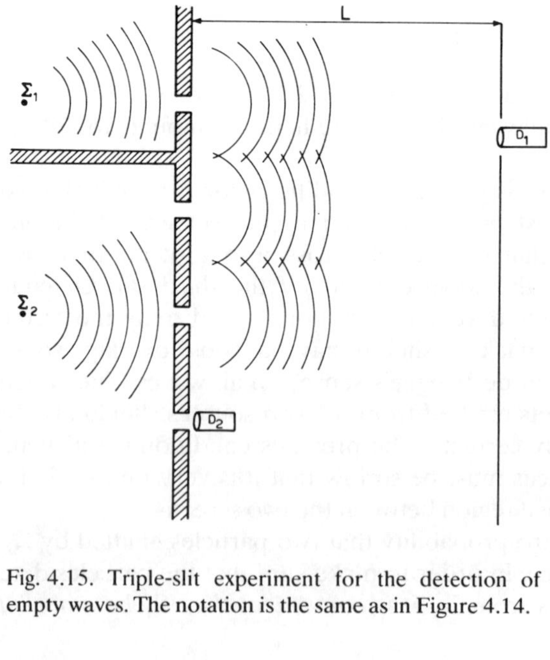

Fig. 4.15. Triple-slit experiment for the detection of an interference between two empty waves. The notation is the same as in Figure 4.14.

since they arise from waves that in the Copenhagen interpretation should both be set equal to zero after the discovery by D_2 of the particle with which they were born. In other words the Copenhagen prediction for the experiment in Figure 4.15 is a distribution proportional to I_0, and therefore devoid of those oscillations which characterize interference. The difference of the two predictions is really striking!

Lastly we will answer an objection which is often advanced against the basic idea of the present chapter. If empty waves exist, it is objected, there should be many of them in the apparatus and the observation of self-interference should be impossible. The answer is that absorbers of particles also shield very efficiently from empty waves. In fact, in the realistic interpretation of the double-slit experiment quantum waves must be assumed to arise only from the slits and not from other parts of the screen in which the slits are pierced. More generally, a constant care of the experimentalists is to protect their apparata from unwanted particles, which could generate a sizable noise. The same protection must then work against the presence of empty waves: in fact empty waves can be produced only in carefully selected *subsets* of propagations and if the whole flux is taken into account there is every reason for believing that the particle probability is in all cases proportional to the squared amplitude of the wave. Since particles in Figures 4.14 and 4.15 are experimentally certain to arise only from the three slits, there cannot exist a different source of empty waves.

4.7. The Bohm–Aharonov Effect

As is well known, Maxwell's equations assume a particularly elegant form when written in terms of the 4-vector potential $A_\mu(x) = (\mathbf{A}(x), i\phi(x))$ at every spacetime point $x = (\mathbf{x}, ict)$. In fact, they can be written

$$A_\mu(x) = -\left(\frac{4\pi}{c}\right) j_\mu(x), \tag{4.74}$$

where $j_\mu(x) \equiv (\mathbf{j}(x), ic\rho(x))$ is the 4-current and A_μ satisfies the Lorentz condition

$$\frac{\partial A_\mu}{\partial x_\mu} = 0. \tag{4.75}$$

The 4-vector potential A_μ can undergo a socalled *gauge transformation of the second kind,*

$$A_\mu \to A'_\mu = A_\mu + \frac{\partial \Lambda}{\partial x_\mu}, \tag{4.76}$$

where Λ is a scalar function of x that satisfies the homogeneous D'Alembertian equation

$$\Box\Lambda = 0, \tag{4.77}$$

by virtue of which neither (4.74) nor (4.75) undergoes any change when subjected to (4.76). Considering the electromagnetic tensor $F_{\mu\nu}$, which is antisymmetric in μ, ν and composed of six independent quantities, *viz.*, the components of the electric field **E** and of the magnetic field **B**, one sees that its well-known expression in terms of the 4-potential,

$$F_{\mu\nu} = \frac{\partial A_\nu}{\partial x_\mu} - \frac{\partial A_\mu}{\partial x_\nu}, \tag{4.78}$$

is also left unchanged by the transformation (4.76) obeying (4.77). From the physical point of view, the 4-potential A_μ has long been considered meaningful not in itself but only through its connection (4.78) with the fields. The argument apparently was: the fields generate forces which accelerate the charges, thereby modifying their basic physical properties, energy and momentum; without fields there are no forces and therefore no physics at all.

This line of reasoning, which can hardly be questioned in the classical domain, was shown to be generally incorrect by Aharonov and Bohm (1959) who discovered a physically observable quantum mechanical action on charged particles that is ascribable to A_μ under conditions such that $F_{\mu\nu} = 0$ (zero-field condition). Now the fields can be deduced from (4.78) and turn out to be

$$\begin{aligned} \mathbf{E} &= -\nabla\phi - \frac{1}{c}\frac{\partial \mathbf{A}}{\partial t}, \\ \mathbf{B} &= \nabla \times \mathbf{A}. \end{aligned} \tag{4.79}$$

Obviously, then, in order to have $\mathbf{E} = \mathbf{B} = 0$ it is enough that: (i) the gradient of ϕ be zero; (ii) **A** be time-independent; (iii) the curl of **A** be zero. But these conditions require neither **A** nor ϕ to vanish. Two physical examples in which either ϕ or **A** is different from zero, but where one still has $\mathbf{E} = \mathbf{B} = 0$, are the following:

1. *Charged Tubes.* Suppose there is a particle inside a 'Faraday cage' which is connected to an external generator, causing the potential of the cage to alternate in time. The effect of the latter is to add to the Hamiltonian of the particle a potential-energy term $e\phi$, which within the cage is a function of time alone. We therefore write

$$H = H_0 + e\phi(t),$$

where H_0 is the Hamiltonian when the generator is not working, e is the particle's charge, and ϕ is the potential function. If $\psi_0(\mathbf{x}, t)$ is the solution of Schrödinger's equation for the unperturbed Hamiltonian H_0, then the solution for H will be

$$\psi(\mathbf{x}, t) = \psi_0(\mathbf{x}, t)\, \mathrm{e}^{-iS/\hbar}, \tag{4.80}$$

with

$$S(t) = e \int_o^t \phi(t')\, \mathrm{d}t', \tag{4.81}$$

as one may easily check by taking the time derivative of $\psi(\mathbf{x}, t)$.

Now consider a more complex experiment in which a low-intensity electron beam is split into two parts, each of which is then allowed to enter a cylindrical metal tube. After the beams have passed through their respective tubes, they are combined to interfere coherently in a region F. The beams are assumed to be composed of wave packets that are long compared with the de Broglie wavelength but short in relation to the length of the tubes. Moreover, the potential in each tube is arranged to be different from zero only when a wave packet is well inside the tube. The full wave function reaching F is then given by

$$\psi = \psi_1^0\, \mathrm{e}^{-iS_1/\hbar} + \psi_2^0\, \mathrm{e}^{-iS_2/\hbar},$$

where

$$S_1 = e \int_o^t \phi_1(t')\, \mathrm{d}t' \quad \text{and} \quad S_2 = e \int_o^t \phi_2(t')\, \mathrm{d}t'$$

are the phase shifts due to the first and second tube, respectively, the integrals extending over the times during which the potentials are acting. It is evident that the interference of the two parts composing ψ will depend on $(S_1 - S_2)$. Thus there is a physical effect due to the potentials, even though no force is ever actually exerted on the electron. And this effect is evidently essentially quantum mechanical, since it influences the phenomenon of interference.

2. *Magnetized Solenoid.* By means of a current flowing through a very closely-wound cylindrical solenoid (say, of radius r and axis in the z direction), one is able to create a magnetic field $\mathbf{B}$ that is essentially confined within the solenoid. However, the vector potential $\mathbf{A}$ cannot be zero everywhere outside the solenoid, because the total flux Φ of $\mathbf{B}$ through every closed circuit containing the symmetry center of the solenoid can be written as

$$\Phi = \int \mathbf{B} \cdot \mathrm{d}\mathbf{S} = \oint \mathbf{A} \cdot \mathrm{d}\boldsymbol{\ell}.$$

Thus *outside the solenoid* we have here a situation for which $\mathbf{B} = 0$ but $\mathbf{A} \neq 0$. Obviously one also has $\mathbf{E} = 0$ and $\phi = 0$.

As before, a low-intensity beam of electrons is split into two parts, which travel on opposite sides of the solenoid, totally avoiding it. As in the earlier example, the beams are brought together at *F*. It is not difficult to show that the wave function appropriate to this problem is

$$\psi = \psi_1^0 \, e^{-iS_1'/\hbar} + \psi_2^0 \, e^{-iS_2'/\hbar}, \tag{4.82}$$

with

$$S_1' = -\left(\frac{e}{c}\right)\int_{ABP} \mathbf{A}(\mathbf{x}')\cdot d\boldsymbol{\ell}' \tag{4.83}$$

and

$$S_2' = -\left(\frac{e}{c}\right)\int_{ACP} \mathbf{A}(\mathbf{x}')\cdot d\boldsymbol{\ell}', \tag{4.84}$$

where $P(x)$ is the point within the region F where ψ is calculated and the integrals are extended over the trajectories (ABP and ACP for S_1' and S_2', respectively) that the first and second wave packets actually describe.

We sketch the proof of the result (4.82)—(4.84) for the first trajectory: the wave function ψ_1^0 for an inactive solenoid is a solution of the Schrödinger equation

$$i\hbar \frac{\partial \psi_1^0}{\partial t} = \frac{1}{2m}(-i\hbar\nabla)^2\psi_1^0 + V\psi_1^0.$$

Since it is easy to show that

$$-i\hbar\nabla \, e^{-iS_1'/\hbar} = (e/c)\mathbf{A}(\mathbf{x}) \, e^{-iS_1'/\hbar},$$

it follows, for

$$\psi_1 = \psi_1^0 \, e^{-iS_1'/\hbar},$$

that

$$-i\hbar\nabla\psi_1 = \left(-i\hbar\nabla\psi_1^0 + \frac{e}{c}\mathbf{A}(\mathbf{x})\psi_1^0\right) e^{-iS_1'/\hbar},$$

whence

$$\left(-i\hbar\nabla - \frac{e}{c}\mathbf{A}(\mathbf{x})\right)\psi_1(x) = (-i\hbar\nabla\psi_1^0(x))\, e^{-iS_1'/\hbar}.$$

Similarly one can show that

$$\left(-i\hbar\nabla - \frac{e}{c}\mathbf{A}(\mathbf{x})\right)^2\psi_1(x) = (-\hbar^2\nabla^2\psi_1^0(x))\, e^{-iS_1'/\hbar}.$$

Using this result, it follows that from

$$\left(-\frac{\hbar^2}{2m}\nabla^2\psi_1^0 + V\psi_1^0 - i\hbar\frac{\partial\psi_1^0}{\partial t}\right)\mathrm{e}^{-iS_1'/\hbar} = 0,$$

since S_1' is independent of time, one can deduce

$$\left(\frac{1}{2m}\right)\left(-i\hbar\nabla - \frac{e}{c}\mathbf{A}(\mathbf{x})\right)^2\psi_1(x) + V\psi_1(x) - i\hbar\frac{\partial\psi_1(x)}{\partial t} = 0,$$

which is indeed Schrödinger's equation for our particle in an external potential that, besides V, involves also of the vector potential **A**. In the present case, the interference at F will depend on $S_1' - S_2'$ and thus directly on the vector potential **A**, even though the magnetic and electric fields are zero on the trajectories that are actually described by the wave packets.

Experimental investigations of the Bohm—Aharonov effect carried out in the early sixties (Chambers, 1960; Fowler *et al.*, 1961; Boersch *et al.*, 1961; Möllenstedt and Bayth, 1962) fully confirmed the existence of this phenomenon. The validity of these experiments has been questioned on the ground that electrons could have been affected by the inevitable leakage of magnetic field from the whiskers and solenoids used (Bocchieri *et al.*, 1980), but this objection can be considered practically dispelled after the paper by Tonomura *et al.* (1982). Their experiment employed electron holography and was based on the use of tiny toroidal magnets (a few microns in diameter) prepared with a technique that combines electron beam lithography and the evaporation of permalloy thin films (80% N and 20% F). This approach allowed the authors to avoid leakage effects. They furthermore employed a holographic interference microscopy technique, both to obtain contour maps of the electron interference and to *detect* quantitatively any eventual leakage that might have come from the magnets. Measurements confirmed the absence of any magnetic field leakage, up to a flux larger than h/e. A flux of this size would have generated a phase change much smaller than that due to the Bohm—Aharonov effect. The full agreement of the experimental results with the theoretical predictions confirmed the presence of a vector potential **A** even in the absence of a magnetic field.

A direct visualization of the Bohm—Aharonov effect has been obtained by Philippidis *et al.* (1982), who used the quantum-potential approach of Philippidis *et al.* (1979) that we discussed in the first section of the present chapter.

4.8. Further Ideas about Wave—Particle Duality

Some further theoretical developments relevant to the interpretation of the wave—particle duality are reviewed in the present section. We shall discuss in particular some research results that have done much to fill the gap between the particle and wave aspects of atomic systems, in contrast with Bohr and Fock's formulations of complementarity. Bartell (1982) has shown that one can study intermediate wave—particle behavior and not only pure-wave or pure-particle behavior. Furthermore, Noyes has conceived of an apparatus which can exhibit at the same time, although in disjoint sets of events, the wavelike interference pattern and the single-slit choices typical of localized (particle-like) structures. Recent developments have also shown that several lines of investigation which many people previously believed to be possible representations of atomic reality are actually in contradiction with existing experiments. Most notable among the impugned ideas are the following: (i) the Landé formulation, which is excluded by neutron interferometry; (ii) the semiclassical interpretation of the electromagnetic field, which is excluded by the absence of coincidences on two opposing sides of a beamsplitter; (iii) the stochastic interpretation of quantum mechanics, which faces several difficulties, among them its inability to explain interference patterns.

The possibility of building an apparatus that can reveal, through minor modifications (shifting of a screen), either the wave aspect or the particle aspect of light has been discussed by Wheeler (1978) and generalized by Bartell (1980). In Wheeler's proposal, the experimenter is free to choose *either* to observe which slit has been traversed *or* to accumulate interference fringes. Bartell was able to modify the proposed apparatus in such a way as to detect *intermediate* particle—wave behavior. A similar proposal has been put forward also by Wootters and Zurek (1979).

The interference screen is placed at the focal plane of the lens, at a distance L from the slits 1 and 2, and on it the Fraunhofer interference pattern can be observed. If, alternatively, it is desired to observe the trajectories of the photons, the screen is taken away and the photons are observed by the counters C_2 and C_1, which, because of the particular optical geometry of the apparatus, can only detect light from the lower and upper slit, respectively. The apparatus incorporates Gaussian transmission filters, which eliminate the secondary diffraction maxima characteristic of slits with abrupt edges.

If the movable screen is put a distance $\ell > L$ from the slits, one observes in general intermediate particle—wave behavior. The intensity $I(x)$ at a position x on the screen (the origin of x is the symmetry point) is predicted by elementary diffraction theory to have the form

$$I(x) = k[F(x)]^2 \cosh(ax)\,[1 + V(x)\cos bx], \tag{4.85}$$

where k, a, and b are constants dependent only on the geometry of the apparatus and, in the case of k, on the intensity of the incoming light; the function

$$F(x) = \exp\left(-\frac{x^2}{2\sigma^2}\right)$$

is the slit form factor, σ being a constant; and $V(x)$ is the important 'fringe-visibility' function, which is related to the slit transmission probabilities $P_1(x)$ and $P_2(x)$, for radiation arriving at the location x, via

$$V(x) = 2[P_1(x)P_2(x)]^{1/2}. \tag{4.86}$$

If the slit discrimination ratio $P_1(x)/P_2(x)$ is selected to be 4:1, for example, the resultant fringe visibility is only 20% lower than that of an ideal two-slit pattern. In other words, the interference figure is very clearly visible in points where the probabilities for an incoming photon to pass through slit 1 and slit 2, respectively, are 80% and 20%. Even if these probabilities are taken to be 95% and 5%, the visibility parameter — as one can check from (4.86) — reduces to about 0.45, a little more than a factor 2 below its optimal value 1. We thus see that, in an experiment of this type, undulatory behavior, as reflected in the interference fringes, goes together with an almost-particle behavior, as reflected in the near-unit value of $P_1(x)$. This result is obtained uniquely from standard quantum mechanics without any particular interpretation. The natural suggestion seems to be that particles and waves co-exist objectively.

Another indication of the fact that the division between particle and wave aspects is less sharp than expected from the usual complementarity interpretation of quantum mechanics has been uncovered by Pierre Noyes (1980). In his operational analysis of the double-slit experiment, Noyes introduces a peculiar device, a kind of semitransparent counter, which is imagined to be capable of detecting any particle traversing it with a certain probability p and thus of not detecting the same event with probability $1-p$. In the latter case, one assumes naturally that no random phase change of the wave function takes place. Noyes' counters are supposed to be placed in both slits in such a way that a particle passing through either slit must necessarily enter a counter. Noyes was thus able to predict that three different statistical distributions will be detected by an array of (small) particle detectors placed at a sufficient distance behind the two slits and playing the role of a 'second screen' for the usual double-slit apparatus. In this way, a statistical distribution, showing an interference pattern, arises which is composed of events in which neither of the slit detectors fires, while two other distributions arise (without interference) out of the events in which one or the other of the slit detectors fires in addition to one of the detectors in the final array.

Noyes' ingenious apparatus contradicts a well-known formulation of the complementarity principle, Fock's idea of relativity with respect to the observational means, which was discussed in the previous chapter. Complementarity asserts that an experiment allowing one to determine through which slit a particle passes is complementary to (i.e., *excludes*) an experiment yielding the double-slit interference pattern. By contrast, Noyes shows that *the same apparatus* can produce complementary outcomes, such as (a) identifying which slit the electron passed through (particle behavior) *and* (b) revealing an interference pattern (wave behavior).

Alfred Landé developed a particular form of quantum theory which, in his opinion, demonstrates

> that wavelike looking atomic phenomena, such as diffraction of particles through crystals, are due to pure mechanical action of the diffraction under the rules of quantum mechanics. (Landé, 1973, p. 4.)

According to Landé there does not exist any undulatory aspect of matter:

> As thousands of experiments have shown, matter consists of discrete particles, real, concrete, countable, with definite rest masses and charges, condensed in *small spaces*. (*Ibid.*, p. 117.)

Correspondingly, Landé's electromagnetic field is a pure wave and does not have any particle-like aspect. One can then conclude that Landé accepts a classical *picture* both of material particles and of the electromagnetic field. This mode of thinking is however supplemented with quantum rules which determine, for instance, the behavior of a particle interacting with a slit, with a double slit, with the regular distribution of atoms in a crystal lattice, and so on.

Landé associates quantum rules with energy E, angular momentum p_ϕ, and linear momentum p. The first rule controls the mutual exchange of energy between two bodies which are in resonance at the frequency ν or period $\tau = 1/\nu$; it reads

$$E = \frac{h}{\tau} = h\nu \quad \text{(Planck).}$$

A similar rule holds for the angular momentum,

$$\Delta p_\phi = \frac{h}{\phi} \quad \text{(Sommerfled),}$$

where ϕ is the angular period. Every object has an angular period 2π, so that one always has the rule $\Delta p_\phi = h/2\pi$.

The third selection rule states that a body with a period ℓ in space changes its momentum parallel to ℓ only in amounts h/ℓ:

$$\Delta p = \frac{h}{\ell} \quad \text{(Duane).}$$

This so-called Duane's selection rule is the real novelty of Landé's approach, the first two selection rules, named after Planck and Sommerfeld, being well known in standard quantum theory. The chief application of Duane's rule is to a crystal with a series of lattice planes of spacing L and hence of period $\ell = L$: the crystal can therefore change its momentum perpendicular to the planes only in quanta $\Delta p = nh/L$, causing the particles interacting with it to emerge along well-defined quantized directions. Even if one ignores the difficulties Landé's theory faces in any attempt to explain Laue scattering in neutron interferometry — where, for some reason not explicable in the present approach (but very well understood in standard wave theory), the neutrons emerge only along *two* directions — the Landé scheme fails badly because of its inability to explain the oscillating intensities of the two final neutron beams if a phase-shifting material of variable thickness is inserted into one of them.

The proof of the latter statement is straightforward: with reference to Figure 3.2, let I_L and I_R be the intensities of the left and right *internal* beams, respectively, and let I'_L and I'_R be the corresponding intensities of the two *external* beams. If single particles were propagating along one or the other of the two internal trajectories, one should always have

$$\begin{aligned} I'_L &= I^D_L + I^T_R, \\ I'_R &= I^T_L + I^D_R \end{aligned} \tag{4.87}$$

where T and D denote the transmitted and diffracted parts of the intensity involved. But I^D_L and I^T_L can be measured by putting an obstacle in the R beam. Also I^D_R and I^T_R are similarly measurable. All the quantities entering into (4.87) can thus be measured independently, and *one finds that (4.87) is often badly violated* by existing experimental data for neutron interferometry.

It should therefore be evident that Landé's formulation of dualism is incapable of explaining some crucial empirical facts and that it therefore ought to be rejected as scientifically untenable.

There are some formulations of dualism in which the only truly stable entity is the wave, whereas particles enjoy only a fleeting existence: they can suddenly materialize out of field energy at some points of space or vanish at others. Such was basically the idea behind Bopp's theory (Bopp, 1947 and 1953) and, more recently, the proposal put forward by Sapogin (1980). The latter author proposes to take as fundamental the group-velocity formula

$$v = u + k\,\frac{\mathrm{d}u}{\mathrm{d}k},$$

where v is the wave packet's group velocity, u the phase velocity, and k the wave number. Assuming v constant and k independent, one gets, on integration,

$$k = \frac{c}{v - u},$$

with c constant; whence, using $u = \omega/k$,

$$k = \frac{\omega + c}{v}. \tag{4.88}$$

Now, let $f(t)$ be a (narrow) real function that represents the particle passing through a given fixed point as a function of time t. If

$$A(\omega) = \left(\frac{1}{2\pi}\right)\int_{-\infty}^{\infty} f(t')\,e^{i\omega t'}\,dt' \tag{4.89}$$

is a Fourier component of f with frequency ω, then the full wave with frequency ω in space and time is assumed to be

$$A(\omega)\,e^{i(kx-\omega t)}.$$

Summing over all ω's, the wave packet is represented by

$$\phi(x, t) = \mathrm{Re}\int_{-\infty}^{\infty} A(\omega)\,e^{i(kx-\omega t)}\,d\omega. \tag{4.90}$$

Substitution of (4.88) and (4.89) into (4.90) leads easily to the result

$$\phi(x, t) = \cos\left(\frac{c}{v}\,x\right) f\left(t - \frac{x}{v}\right), \tag{4.91}$$

whence Sapogin extracts the idea that the particle disappears at points defined by $cx/v = n\pi/2$, n being an integer. The de Broglie wave he views as the envelope of the periodical appearances and disappearances of the particle. The world accordingly consists of a tremendous number of particles moving with different velocities. Partial waves of disappearing particles can add up locally by pure chance and suddenly cause a 'random particle' to appear. A quantum wave is then a field in which particles can abruptly appear and disappear, and this process is assumed to be responsible for the inherently probabilistic nature of quantum laws. This approach encounters no problems with interference experiments, which it can interpret naturally, given the stable nature of its waves. But, on the other hand, it seems unable to explain the lack of coincidences between any two phototubes placed on the outgoing pair of trajectories associated with low-intensity light passing through a half-silvered mirror.

A further popular approach to electromagnetism has been the so-called *neoclassical theory* proposed by Jaynes and other authors. Typical of their anticorpuscular attitude with respect to the radiation field is the following statement, taken from the text on laser physics by Sargent, Scully, and Lamb (1974):

Photons are quanta of a single (monochromatic) mode of the radiation field and are not localized at any particular position and time within the cavity-like fuzzy balls; rather, they are spread out over the entire cavity. In fact, no satisfactory quantum theory of photons as particles has ever been given. On the other hand, the quantum theory of radiation seems to offer amazingly satisfactory accounts of a very wide range of radiative problems, and there is no real need to have a corpuscular theory of photons.

In this approach, the electromagnetic field is treated in purely classical terms, but quantum mechanics is used for the description of atoms and molecules. Nevertheless, several phenomena were thought to be explainable without the photon hypothesis. Such was the case for the photoelectric effect, the Hanbury-Brown and Twiss effect (Hanbury-Brown and Twiss, 1956), the so-called photon-counting experiments, photon bunching, and, finally, the random registration of photoelectric pulses generated by laser light. A careful conceptual and experimental analysis of these problems has, however, led Mandel (1976) to the conclusion:

Although semiclassical theories have had considerable success in accounting for many observed effects, . . . they fail completely in other cases, and no evidence exists that should cause us to think of giving up Q.E.D. in favor of a semiclassical theory.

One of the fundamental weaknessses of the semiclassical approach is its inability to account for the short time delays in photoemission if energy conservation has to be true without exception.

For a recent attempt to treat the electromagnetic field without particles ('stochastic optics') see Marshall (1987).

References

Books

de Broglie, L.: 1960, *Non-Linear Wave Mechanics: A Causal Interpretation*, Elsevier, Amsterdam.

Diner, S., Fargue, D., Lochak, G., and Selleri, F. (eds): 1984, *The Wave—Particle Dualism: A Tribute to Louis de Broglie on his 90th Birthday*, D. Reidel, Dordrecht.

Greenberger, D. M.: 1986, *New Techniques and Ideas in Quantum Measurement Theory*, New York Academy of Sciences, New York.

Haken, H., and Weidlich, W.: 1969, in R. J. Glauber (ed.), *Quantum Optics*, Academic Press, New York.

Honig, W. M., Kraft, D. W., and Panarella, E.: 1987, *Quantum Uncertainties*, Plenum, New York and London.

Jaynes, E. T.: 1973, in L. Mandel and E. Wolf (eds), *Proceedings of the Third Rochester Conference on Coherence in Quantum Optics*, Plenum, New York, p. 35.

Kuhn, T. S., Heilbron, J. L., Forman, P. L., and Allen, L.: 1967, *Sources for History of Quantum Physics*, The American Philosophical Society, Philadelphia.

Landé, A.: 1973, *Quantum Mechanics in a New Key*, Exposition Press, New York.

Lochak, G.: 1982, Preface to L. de Broglie, *Les incertitudes d'Heisenberg et l'interpretation probabiliste de la mécanique ondulatoire*, Gauthier-Villars, Paris.
Loudon, R.: 1978, *The Quantum Theory of Light*, Clarendon Press, Oxford.
Mandel, L.: 1976, in E. Wolf (ed.), *Progress in Optics XIII*, North-Holland, Amsterdam.
Sargent, M. III, Scully, M. O., and Lamb, W. E., Jr: 1974, *Laser Physics*, Addison-Wesley, Reading, Massachusetts.
Schommers, W. (ed.): 1989, *Quantum Theory and Pictures of Reality*, Springer-Verlag, Berlin.
Scully, M.: 1969, in R. J. Glauber (ed.), *Quantum Optics*, Academic Press, New York.
Selleri, F.: 1971, in B. d'Espagnat (ed.), *Fondamenti di Meccanica Quantistica* (Italian Physical Society, Proceedings of the International School of Physics 'Enrico Fermi', Course IL), Academic Press, New York.
Selleri, F., and Vigier, J.-P.: 1983, 'Induced Superfluorescence and the Nature of the Wave—Particle Duality' in Alwyn van der Merwe (ed.), *Old and New Questions in Physics, Cosmology, Philosophy, and Theoretical Biology: Essays in Honor of Wolfgang Yourgrau*, Plenum, New York.
van der Waerden, B. L. (ed.): 1967, *Sources of Quantum Mechanics*, North-Holland, Amsterdam.
van der Merwe, A., Selleri, F., and Tarozzi, G. (eds): 1988, *Microphysical Reality and Quantum Formalism*, Kluwer Academic Publishers, Dordrecht.
Vrehen, H. F.: 1979, in H. Walther and K. W. Rothe (eds), *Proceedings of the Fourth International Conference Rottach-Egern* (June 1979), Springer, Heidelberg.
Wheeler, J. A.: 1978, in A. R. Marlow (ed.), *Proceedings of the New Orleans Conference on the Mathematical Foundations of Quantum Theory*, Academic Press, New York.

Articles

Adam, A., Jánossy, L., and Varga, P.: 1955, *Ann. Phys.* (*Leipzig*) **16**, 408.
Aharonov, Y., and Bohm, D.: 1959, *Phys. Rev.* **115**, 485.
Alefeld, B., Badurek, G., and Rauch, H.: 1981, *Z. Phys. B* **41**, 231.
Andrade e Silva, J., and Andrade e Silva, M.: 1980, *C.R. Acad. Sci.* **290**, 501.
Arecchi, F. T.: 1965, *Phys. Rev. Lett.* **15**, 902.
Aspect, A., Roger, G., Reynaud, S., Dalibard, J., and Cohen-Tannoudji, C.: 1980, *Phys. Rev. Lett.* **45**, 617.
Ballentine, L. E.: 1970, *Rev. Mod. Phys.* **42**, 358.
Bartell, L. S.: 1980, *Phys. Rev.* **D21**, 1698.
Bialynicki-Birula, I., and Mycielski, J.: 1976, *Ann. Phys.* **100**, 62.
Blake, G. D., and Scarl, D.: 1979, *Phys. Rev.* **A19**, 1948.
Bocchieri, P., Loinger, A., and Siragusa, G.: 1980, *Nuovo Cimento* **A56**, 55.
Boersch, H., *et al.*: 1961, *Z. Phys.* **165**, 79.
Bohm, D.: 1952, *Phys. Rev.* **85**, 166, and 180.
Bohr, N., Kramers, H. A., and Slater, J. C.: 1924, *Phil. Mag.* **47**, 785.
Bonse, U., and Hart, M.: 1966, *Z. Phys.* **194**, 1.
Bopp, F.: 1947, 1952, 1953, *Z. Naturforsch.* **2a**, 202; **7a**, 82; **8a**, 6.
Cagnac, B., Grynberg, G., and Biraben, F.: 1973, *J. Phys.* (*Paris*) **34**, 845.
Cantelaube, Y.: 1984, *Phys. Lett.* **101A**, 7.
Chambers, R. G.: 1960, *Phys. Rev. Lett.* **5**, 3.
Chubarov, M. S., and Nikolayev, E. P.: 1985, *Phys. Lett.* **110A**, 199.
Clauser, J. F.: 1972, *Phys. Rev.* **A6**, 49.
Clauser, J. F.: 1974, *Phys. Rev.* **D9**, 853.
Costa de Beauregard, O.: 1982, *Phys. Lett.* **89A**, 171.
Croca, J. R.: 1987, *Found. Phys.* **17**, 971.
Croca, J. R., Garuccio, A., and Selleri, F.: 1988, *Found. Phys. Lett.* **1**, 101.

Dagenais, M., and Mandel, L.: 1978, *Phys. Rev.* **A18**, 2217.
Davidson, F., and Mandel, L.: 1967, *Phys. Lett.* **25A**, 700.
de Broglie, L.: 1923, *C.R. Acad. Sci.* **177**, 548; reprinted in *Ann. Fond. L. de Broglie* **7**, 101 (1982).
de Broglie, L.: 1927, *J. Phys.* (*Paris*) **6**, 225.
de Broglie, L.: 1974, *Found. Phys.* **4**, 321.
de Broglie, L.: 1977, *Ann. Fond. L. de Broglie* **2**, 1 (1977).
de Broglie, L., and Andrade e Silva, J.: 1968, *Phys. Rev.* **172**, 1284.
Dewdney, C., *et al.*: 1984, *Phys. Lett.* **102A**, 291.
Dicke, R. H.: 1954, *Phys. Rev.* **93**, 99.
Elbaz, C.: 1985, *Phys. Lett.* **109A**, 7.
Engelke, C. E., and Engelke, C. W.: 1983, *Found. Phys.* **16**, 905; **16**, 917.
Fowler, H. A., *et al.*: 1961, *J. Appl. Phys.* **32**, 1153.
Friberg, S., and Mandel, L.: 1983, *Opt. Commun.* **46**, 141.
Friberg, S., Hong, C. K., and Mandel, L.: *Phys. Rev. Lett.* **54**, 2011. (1985)
Ganesh Vaidyanathan, A., Spencer, W. P., and Kleppner, D.: 1981, *Phys. Rev. Lett.* **47**, 1592.
Garuccio, A., Guéret, Ph., and Selleri, F.: 1988, *Found. Phys. Lett.* **1**, 139.
Garuccio, A., Rapisarda, V., and Vigier, J.-P.: 1982, *Phys. Lett.* **90A**, 17.
Giovanelli, R.: 1989, *Found. Phys. Lett.* **2**, 27.
Gordon, P.: 186, *Phys. Lett.* **117A**, 447.
Gozzini, A.: 1982, communication at the Symposium on Wave—Particle Dualism, Perugia, April 1982; for the Symposium Proceedings, see Diner *et al.*, 1984.
Grangier, P., Aspect, A., and Vigue, J.: 1985, *Phys. Rev. Lett.* **54**, 418.
Greenberger, D. M., and Overhauser, A. W.: 1980, *Sci. Am.* **242** (5), 66.
Grynberg, G., Biraben, F., Bassini, M., and Cagnac, B.: 1976, *Phys. Rev. Lett.* **37**, 283.
Hanbury-Brown, R., and Twiss, T. Q.: 1956, *Nature* **177**, 27 and **178**, 1046.
Heidmann, A., and Reynaud, S.: 1984, *J. Phys.* (*Paris*) **45** (5), 873.
Herbert, N.: 1982, *Found. Phys.* **12**, 1171.
Hulet, R. G., Hilfer, E. S., and Kleppner, D.: 1985, *Phys. Rev. Lett.* **55**, 2137.
Jönsson, C.: 1961, *Z. Phys.* **161**, 454.
Kaiser, H., Werner, S. A., and George, E. A.: 1983, *Phys. Rev. Lett.* **50**, 560.
Klein, A. G., Opat, G. I., and Hamilton, W. A.: 1983, *Phys. Rev. Lett.* **50**, 563.
Kleppner, D.: 1981, *Phys. Rev. Lett.* **47**, 233.
Kostro, L.: 1985, *Phys. Lett.* **107A**, 429 and **112A**, 283.
Kuhn, H.: 1970, *J. Chem. Phys.* **53**, 101.
Ley, M., and Loudon, R.: 1985, *Opt. Commun.* **54**, 317.
Ley, M., and Loudon, R.: 1986, *Opt. Acta* **33**, 371.
Liebowitz, B.: 1970, *Phys. Rev.* **D1**, 720.
Loudon, R.: 1983, *Opt. Commun.* **45**, 361.
Mandel, L.: 1984, *Phys. Lett.* **103A**, 416.
Marshall, T. W.: 1987, in Honig *et al.* (1987).
Matteucci, G., and Pozzi, G.: 1978, *Am. J. Phys.* **46**, 619.
Matteucci, G., and Pozzi, G.: 1985, *Phys. Rev. Lett.* **54**, 2469.
Milonni, P. W., and Knight, P. L.: 1973, *Opt. Commun.* **9**, 119.
Milonni, P. W., and Hardies, M. L.: 1982, *Phys. Lett.* **92A**, 321.
Möllenstedt, G., and Bayth, W.: 1962, *Phys. Blätt.* **18**, 299.
Namiki, M., and Otake, Y.: 1985, in van der Merwe *et al.* (eds); see book list.
Noyes, P.: 1980, in P. Suppes (ed.), *Studies in the Foundations of Quantum Mechanics*, Philosophy of Science Association, East Lansing, Michigan.
Pfleegor, R. L., and Mandel, L.: 1967, *Phys. Rev.* **159**, 1084.
Philippidis, C., Dewdney, C., and Hiley, B. J.: 1979, *Nuovo Cimento* **52B**, 15.
Philippidis, C., Bohm, D., and Kaye, R. D.: 1982, *Nuovo Cimento* **71B**, 75.

Pound, R. V., and Rebka, G. A., Jr: 1960, *Phys. Rev. Lett.* **4**, 337.
Prosser, R. D.: 1976, *Int. J. Theor. Phys.* **15**, 181.
Rauch, H., *et al.*: 1975, *Phys. Lett.* **54A**, 425.
Rauch, H., *et al.*: 1978, *Z. Phys.* **B29**, 281.
Rauch, H.: 1984, in S. Diner *et al.* (eds); see book list.
Scarl, D. B., and Smith S. R.: 1974, *Phys. Rev.* **A10**, 709.
Sapogin, L. G.: 1980, *Ann. Fond. L. de Broglie* **5**, 285.
Scheer, J., and Schmidt, M.: 1986, *Phys. Lett.* **115A**, 322.
Selleri, F.: 1969, *Lett. Nuovo Cimento* **1**, 908.
Selleri, F.: 1982, *Found. Phys.* **12**, 1087.
Selleri, F.: 1984, in S. Diner *et al.* (eds); see book list.
Selleri, F.: 1987, *Found. Phys.* **2**, 739.
Selleri, F.: 1988, *Il Nouvo Saggiatore* (*Boll. Soc. It. Fisica*) **4**, 22.
Selleri, F.: 1989, in Schommers (1989).
Shimony, A.: 1979, *Phys. Rev.* **A20**, 394.
Szczepański, A.: 1976, *Found. Phys.* **6**, 427.
Tarozzi, G.: 1985, *Lett. Nuovo Cimento* **42**, 438.
Tersoff, J., and Bayer, D.: 1983, *Phys. Rev. Lett.* **50**, 553.
Tonomura, A., *et al.*: 1982, *Rev. Lett.* **48**, 1443; 1986, *Phys. Rev. Lett.* **56**, 792.
Umberger, D. K., Milonni, P. W., and Lieber, M.: 1984, *Nuovo Cimento* **83B**, 23.
Vasilenko, L. S., Chebotayev, V. P., and Shishaev, A. V.: 1970, *JETP Lett.* **12**, 113.
Vrehen, Q. H. F., Schuurmans, M. F. H., and Polder, D.: 1980, *Nature* **285**, 70.
Werner, S. A.: 1980, *Physics Today*, December issue.
Wootters, W. K., and Zurek, W. H.: 1979, *Phys. Rev.* **D19**, 473.
Yuen, H. P.: 1986, *Phys. Lett.* **113A**, 401 and 405.
Zeilinger, A.: 1976, *Z. Physik* **B25**, 97.
Zeilinger, A.: 1984, in S. Diner *et al.* (eds); see book list.
Zeilinger, A.: 1986, in Greenberger (1986).

Chapter 5

The Einstein—Podolsky—Rosen Paradox

Given two small objects, the first located on earth, the second a distance of one light year away, consider for a moment your reaction to the following assertion: "No large modification of the properties of the object on earth can take place as a consequence of the distant object interacting with a third body." This statement claims in fact that a measurable physical quantity should have a value more than 40% different from its value predicted by quantum mechanics. The conclusion is paradoxical, if we define a 'paradox' to consist of an argument deriving contradictory conclusions by valid deduction from acceptable premises.

5.1. The Original Formulation

"*If, without in any way disturbing a system, we can predict with certainty* (*i.e., with probability equal to unity*) *the value of a physical quantity, then there exists an element of physical reality corresponding to this physical quantity.*"

Thus reads the famous 'criterion of physical reality' proposed by Einstein, Podolsky, and Rosen (EPR) in 1935 (Einstein, Podolsky, and Rosen, 1935). It represents a rather weak criterion for detecting a physical reality. The 'element of physical reality' is to be understood simply as an attribute, a property of the studied physical system. The above formulation is very careful and rather general. It nevertheless appeared to its originators (and still does today) as 'far from exhausting all possible ways of recognizing a physical reality'.

This criterion has many applications to the macroscopic domain, where it appears as trivially true, to such an extent that we never think of the fact that we are often applying it.

For instance, if one can predict with certainty that the length of a table is 2 m (within an error of, say, $\frac{1}{2}$ cm) then it can be said that there exists an element of physical reality corresponding to the length of the table. Notice that EPR would not claim that the length itself is real. For one thing, the unit of length is conventional and thus a subjective aspect. Furthermore, what appears to us immediately as an extension in space might result from physical properties which actually lie outside our field of perception but give rise to a final sensation that is only a topological distortion of some true real properties of the table. Even in this complicated case, though, the EPR reality criterion would apply, since it assumes only the existence of an element of physical reality, without specifying its nature and its relationship to the measured value of the chosen physical quantity. Similar considerations could easily be advanced in many other macroscopic situations. Elements of physical reality should therefore exist, according to the EPR criterion, that correspond to the speed of a falling stone, the height of a tower, the weight of a body, and so on.

Einstein, Podolsky, and Rosen regarded their criterion as being in agreement with classical as well as quantum mechanical ideas of reality. To discuss the quantum mechanical case, let us consider Heisenberg's relations. Statements on which all schools of thought concerning the nature of wave–particle duality should agree, are the following:

The macroscopic manifestations of the interactions of an atomic object with measuring instruments are the basic elements from which our understanding of the properties of that object arises. In quantum physics the existence of the quantum of action h implies a finite interaction between the atomic object and any measuring apparatus. This circumstance puts limits, expressed by the Heisenberg relation $\Delta x \; \Delta p \geqslant \hbar$, on the precision with which the position x and the momentum p of the object can be measured simultaneously. (For simplicity, only one spatial dimension is here considered.) It is accordingly possible to construct instruments capable of achieving perfect localization of a particle in space ($\Delta x = 0$), but then nothing can be said at the same time about its momentum, because Heisenberg's relation can be satisfied with $\Delta x = 0$ only if Δp is infinite. Conversely, it is possible to design instruments capable of furnishing an infinitely sharp definition of a particle's momentum ($\Delta p = 0$), but only at the price of thwarting any attempt to simultaneously localize it in space (Δx becomes infinite).

The previous conclusions not only follow directly from Heisenberg's relations; they are also rigorously incorporated in the mathematical apparatus of the theory, by way of the fact that no wave function exists which assigns well-defined values to both x and p. By extreme contrast, what *do* exist are the following two types of wave functions:

(i) those describing a particle with fixed momentum p_0 but with position completely unknown (plane waves),

$$\psi = u_{p_0}(x); \tag{5.1}$$

(ii) those describing a particle with fixed position x_0 but with momentum completely unknown (delta functions),

$$\psi = v_{x_0}(x). \tag{5.2}$$

The foregoing notation is the usual one. The reader will notice that it is not a symmetrical one, since the space variable x enters both (5.1) and (5.2). This fact does not, however, in any observable sense destroy the x—p symmetry between the two instances, because spatial localization is in quantum theory characterized by the squared modulus of ψ, which quantity in the case (5.1) can be shown to be independent of x, meaning that all positions are equally probable; in the case (5.2), by contrast, $|\psi|^2$, like ψ itself, vanishes everywhere except for $x = x_0$, so that the position x_0 is certain. Of course, quantum mechanics also admits a multitude of wave functions different from (5.1) and (5.2) in which neither x nor p have well-defined values. To these functions one cannot in general apply the EPR criterion of reality, because neither the particle's position nor its momentum can be predicted with certainty. The criterion can however be applied to both the wave function (5.1), which predicts that a measurement of momentum will give the value p_0 with certainty, and the wave function (5.2), which guarantees that the measurement of position will produce the value x_0 with certainty. Consequently, one can say:

(1) an element of physical reality corresponds to the value p_0 of momentum identified in (5.1);

(2) another element of physical reality corresponds to the value x_0 of position identified in (5.2).

Notice, again, that we refrain from saying that the momentum itself or the position itself is real. Since it is not known how much is subjective in our very conception of the notions of momentum and position when applied to an atomic system, it is wise to be as cautious as possible and to state, if a physical quantity can be predicted with certainty, only that there must be *something* real determining the *a priori* known value of this quantity. In this way no commitment is made concerning the precise nature of the element of physical reality.

The above statements (1) and (2), applying respectively to the wave functions (5.1) and (5.2), are the least that can be said about the physical reality underlying the corresponding two physical situations.

Still, one may ask whether there is not perhaps something more to the physical reality of the particle which is not contained in this minimum assertion. This query is legitimate, because it is possible to view the EPR criterion of physical reality merely as a sufficient, rather than a necessary, condition of reality. Limiting our considerations to the concepts of position and momentum, we can obviously conclude that, if some further elements of reality exist in the physical systems described by (5.1) and (5.2), different from momentum, in the first case, and from position, in the second, then quantum mechanical theory must necessarily be considered incomplete. We characterize as *complete* a theory satisfying the following criterion: *Every element of physical reality attributable to a certain physical system in a given state must have a counterpart in the mathematical description provided by the theory for that physical system in the state considered.* One arrives then at the following conclusion:

Either the quantum mechanical description of reality given by the wave function is not complete or, in the physical situations described by the wave functions (5.1) *and* (5.2), *position and momentum cannot have simultaneous reality.* For, if both of the latter had simultaneous reality — and thus well-defined values — these values would, according to the criterion of completeness, enter into any simultaneously postulated complete description. But, since in actuality the only available description is provided by the wave function, which entity, as we know, does not contain well-defined values of both position and momentum, the entertained hypothesis is false, and we are left with the alternatives stated above in italics.

The foregoing inference looks very mild and reasonable at first sight, but, when used in conjunction with the quantum mechanical description of two correlated objects, it leads to paradoxical conclusions, as we shall see in a moment.

Let a microscopic system (molecule, atom, etc.) be given that is capable of decaying into two new systems α and β. Hundreds of examples of such decay processes exist in different branches of physics: The molecule NO can decay from an excited state to a state of two free atoms (N and O) which propagate in opposite directions; the Λ hyperon decays into a proton and a negatively charged π meson; and so on. Quantum theory in principle provides a wave function describing the two bodies in the final state of such decay processes. This wave function will, in general, be of the type

$$\psi = \psi(x_1, x_2),$$

where x_1 and x_2 are the coordinates of α and β, respectively. The physical interpretation of ψ is the usual one put forth by Born: The quantity $|\psi(x_1, x_2)|^2$ provides the probability density for finding α at x_1 and β at x_2.

A consequence of the quantum mechanical formalism which becomes important at this point is the fact, easy to check, that the sum of the momenta and the difference of the positions of two quantum systems always commute:

$$[x_1 - x_2, p_1 + p_2] = 0.$$

It follows that a class of wave functions can exist for which the sum of the momenta $p_1 + p_2$ of α and β has a fixed value p and the difference of the positions $x_1 - x_2$ of α and β has *simultaneously* a fixed value x. One can write one of these wave functions as

$$\psi = \psi(x, p; x_1, x_2), \tag{5.3}$$

where the four arguments have rather different meanings: x and p denote fixed values of the difference of the positions and of the sum of the momenta, respectively, while x_1 and x_2 are the space variables associated with the two particles. An example of wave function of this type will be given below [see

Equation (5.4)]. Notice that (5.3) predicts with certainty the observables x and p, which can therefore be regarded, if one wishes, as two simultaneous elements of reality. By contrast, no fixed value of the momentum p_1 of α or of the momentum p_2 of β is contained in ψ, only their *sum* being given as fixed and equal to p. Similarly, no fixed values of the individual positions x_1 and x_2 are contained in ψ, only their *difference* is given as fixed and equal to x.

Recalling the criterion for the completeness of a theoretical description given above, one can therefore conclude: *Either the description of reality given by $\psi(x, p; x_1, x_2)$ is not complete or the individual positions and momenta of α and β do not possess any physical reality.*

With a different line of reasoning one can however prove that individual positions and momenta of α and β *do* possess a physical reality, reaching thus the conclusion that the quantum mechanical description of reality is not complete. The argument goes as follows:

Suppose we are given a very large set E of similar decays into $\alpha + \beta$, to all of which the wave function ψ given in (5.3) applies. Quantum mechanics then predicts that measurements of the positions of α and β, performed on individual decay processes, will give results x_1 and x_2 that *always* satisfy the relation $x_2 - x_1 = x$. Similarly, measurements of the momenta of α and β performed on (other) individual decay processes will give results p_1 and p_2 that *always* satisfy the relation $p_1 + p_2 = p$. Let us suppose this prediction has been checked experimentally and found correct: This simply means that quantum mechanics is not wrong in its predictions.

Consider now a subset E_1 of E not previously subjected to measurement and perform position measurements on the member α of the system pair (α, β) resulting from each individual decay; let x_1', x_1'', etc. denote the values obtained. We can then predict *with certainty* that subsequent measurements of the position of the other member β will give $x_1' + x$ for the first pair, $x_1'' + x$ for the second pair, and so on.

Consequently, we can invoke the criterion of physical reality written down at the very beginning of the present section and conclude that the position of β is an element of physical reality. In fact, no interaction with α has taken place, because the distance between α and β at the time of the measurement on α may be taken as macroscopic and arbitrarily large.

It is natural to conclude that this element of physical reality exists regardless of whether or not a measurement on α has been made, because, if this were not true, the only alternative (which we wish to exclude) would be to say that the measurement made on α has created at a distance and instantaneously the element of reality of β.

Therefore, the position of β is an element of reality *for all the decays of the ensemble E.*

A parallel argument can be made for the momenta. Consider a subset E_2 of E, perform momentum measurements on the member α of the system pair (α, β) emerging from each individual decay belonging to E_2, and write

p_1', p_1'', etc., for the ensuing results. Since it can be predicted with certainty that subsequent measurements of the momentum of β will give $p - p_1'$ for the first pair, $p - p_1''$ for the second pair, and so on, it can also be concluded that the momentum of β is an element of reality for all decays of the ensemble E_2. Excluding that this element of reality is created at a distance and instantaneously by the measurement made on α, one can then extend the previous conclusion to the whole of E.

Obviously the choice of the system (α or β) on which measurements are performed is arbitrary. A symmetrical reasoning thus leads to the conclusion that also the position and the momentum of particle α are simultaneous elements of reality in the whole ensemble E.

The conclusion that either the description of reality given by $\psi(x, p; x_1, x_2)$ is not complete or the individual positions and momenta of α and β do not possess any physical reality was reached earlier. But now it has been demonstrated that the positions and momenta of α and β are real (or, better, that to them correspond elements of reality). One is thus left no choice but to infer that the quantum mechanical description of reality given by wave functions is not complete.

This conclusion seemed very important to Einstein at a time when von Neumann's theorem was established and thought to imply that no physical completion of quantum theory was possible. It might well, *at first sight*, look less important today since, as we saw in Chapter 2, von Neumann's theorem is not an obstacle any longer. Furthermore, any serious thought on the nature of a physical theory is today bound to come to the conclusion that it cannot be complete. How can a human theory ever describe completely, that is, *perfectly*, any given physical reality if we are exploring distances as small as 10^{-16} cm, and it is almost certain that it still makes sense to consider distances of order of 10^{-33} cm?

The reason why the EPR paradox is of great importance even today is that the same argument proposed by Einstein, Podolsky, and Rosen can be shown to lead to experimental predictions *in contradiction* to those of quantum mechanics. A full discussion of the assumptions implicit in the EPR argument is therefore necessary; it will be given in other sections of this chapter.

The precise form of the wave function $\psi(x_1, x_2)$ used by EPR was

$$\psi(x_1, x_2) = \int_{-\infty}^{+\infty} \exp \frac{\{i(x_1 - x_2 + x)q\}}{\hbar}\, dq, \tag{5.4}$$

which can be written in two alternative ways:

$$\psi(x_1, x_2) = \int_{-\infty}^{+\infty} \exp \frac{\{ix_1 q\}}{\hbar} \exp \frac{\{-i(x_2 - x)q\}}{\hbar}\, dq, \tag{5.5}$$

$$\psi(x_1, x_2) = h\delta(x_1 - x_2 + x). \tag{5.6}$$

These equations show that, if the particle α is found to have a momentum q, then the particle β will necessarily be found to have the momentum $-q$ (making the total momentum of the two particles zero), and that the two positions, if measured, will necessarily turn out to be related by $x_2 - x_1 = x$, with x constant.

EPR therefore actually used a wave function of type (5.3) with $p = 0$. As we saw, a knowledge of the precise form of this wave function is not essential for an understanding of their reasoning; it will become important, though, for our discussion of some criticisms raised against them.

5.2. Bohr's Answer

Bohr's answer (Bohr, 1935) to the original EPR paper appeared in the *Physical Review* of the same year (1935) and was received by the editor of this journal less than four months after he received the EPR article. Basically it can be said that Bohr does not question the correctness of the EPR reasoning once all its implicit and explicit premises are accepted. But it is exactly the premises that, according to Bohr, are vulnerable. The quantum mechanical formalism, he avers, cannot be adapted to a philosophical point of view like that of Einstein; rather, one must necessarily view quantum theory in conjunction with the philosophical interpretation developed in Copenhagen and Göttingen. Already in his opening remarks, Bohr states that the EPR paradox

> discloses only an essential inadequacy of the customary viewpoint of natural philosophy for a rational account of physical phenomena of the type with which we are concerned in quantum mechanics.

Furthermore, Bohr argues in favor of

> a final renunciation of the classical idea of causality and a radical revision of our attitude towards the problem of physical reality,

and stresses that his notion of complementarity is a 'new feature of natural philosophy', which implies 'a radical revision of our attitude as regards physical reality'.

His line of reasoning is as follows. Firstly, one must notice that all measurements on atomic and subatomic systems are necessarily prepared, carried out, and, when results are obtained, analyzed in classical terms. This is so because the physicist lives in a macroscopic world, where classical laws and, even more importantly, classical conceptions (of space, time, causality, etc.) reign and have become the unavoidable tools for the expression of all human experience. The experimental physicist will therefore naturally try to express his instrumental readings in classical terms, that is, he will *try* to express the regularities of the microscopic world as causal processes in space and time. On discovering that this end is impossible to achieve, the physicist finds himself suddenly exposed to an irrational element.

The impossibility in question has its origin in the quantum of *action* (also called Planck's constant) h whose existence implies a finite mutual *interaction* between the measured object and the measuring instrument. This interaction, according to Bohr, not only cannot be made arbitrarily small, precisely because h has a finite value, but it also gives rise to changes of the observed system which cannot even be allowed for in a logical manner. In other words, the existence of h implies a perturbation of the measured object which is completely unpredictable and thus, in a way, irrational. These facts impose a limit not only on the *extent* of the information obtainable by measurements, but also on the *meaning* that can be attributed to such information. Bohr expresses himself repeatedly in favor of a reduced meaning: The purpose of our description of nature through physics, he argues, is not to disclose the real essence of the phenomena, but only to track down, as far as possible, the relations obtaining between different aspects of our experience. This is a positivistically flavored answer to the problem of the objectivity of phenomena, which has always attracted much attention in philosophical discussions. Bohr's opinion is that an independent reality cannot be ascribed to phenomena: The word 'phenomenon' should exclusively be used to refer to observations obtained under specified circumstances, including an account of the whole experimental setup. As a consequence, Bohr considers it wrong to imagine that the task of physics is to discover the way nature is; physics is instead concerned only with what can be said about nature. Accordingly, the word 'reality' is just a word, and one that we must learn to use correctly. It is therefore not surprising to learn that, for the Danish physicist, Einstein's idea of radiation quanta, as well as the waves introduced by de Broglie and Schrödinger, are *only conceptual expedients* used for constructing the laws of quantum mechanics:

> . . . the wave mechanics forms a natural counterpart to the . . . light quantum theory of Einstein. As in this theory, so also in the wave mechanics, we are not dealing with a self-contained conceptual scheme but, rather, as especially emphasized by Born, with an expedient to formulate the statistical laws which govern atomic phenomena. (Bohr, 1961, p. 111.)

The above quotations, taken mainly from Bohr's book entitled *Atomic Theory and the Description of Nature* (Bohr, 1961), which is one of the two references cited in Bohr's 1935 paper here under discussion (the other reference is, of course, the EPR paper), should already explain why and how Bohr disagreed with Einstein. Einstein conceives of dual objects (particle + wave) as existing in space and time also when human beings are not making observations on them. Bohr opposes this point of view and adopts another philosophy: only the acts of measurements are real. One may, if one wishes, think of a measurement as an interaction between a measured object and a measuring apparatus. But such a habit ought not to obscure the point that there is in actuality no separation between object and apparatus, only a profound unity brought about by the existence of the quantum of action, which does not allow the experimenter to control the effect of the interaction. Physics must accordingly deal *exclusively* with acts of observation, and any reference to unobserved elements of reality should be banished from truly scientific reasoning. Obviously, then, no paradox exists when one considers two correlated systems described by the wave function (5.3). To see this point, consider in fact two apparatuses Q_1 and P_1 (Q_2 and P_2) capable of performing respectively position and momentum measurements on the system α (β).

If one chooses to employ the setup Q_1 and Q_2, the wave function (5.3) predicts that the measurement results x_1 and x_2 will be precisely correlated: indeed, $x_1 - x_2 = x$. If, on the other hand, one elects to use the setup P_1 and P_2, the wave function (5.3) predicts that a precise correlation of the measurement data p_1 and p_2 will emerge: $p_1 + p_2 = p$. The two apparatuses Q_1 and P_1 are mutually incompatible: One can choose to employ either Q_1 or P_1 but never the two of them simultaneously; and the same can be said of Q_2 and P_2.

From this point of view, the EPR assumption about the elements of reality immediately becomes useless: One can continue to advance this assumption, but now it merely asserts that an element of reality is *associated with any concretely performed act of measurement*, for there is no other reality that one can speak of. In particular, the EPR conclusion that position and momentum correspond to two simultaneously existing elements of reality appears totally unjustified (Bohr writes that it contains 'an essential ambiguity'), because one can never perform simultaneous measurements of position and momentum; if there is no concrete measurement, there is also nothing real to which an element of reality can be attributed.

Einstein, Podolsky, and Rosen anticipated the possibility of such a refutation, but they viewed it as unacceptable:

> One could object to this conclusion on the grounds that our criterion of reality is not sufficiently restrictive. Indeed, one would not arrive at our conclusion if one insisted that two or more physical quantities can be regarded as simultaneous elements of reality *only when they can be simultaneously measured or predicted.* On this point of view, since either one or the other, but not both simultaneously, of the quantities P and Q can be predicted, they are not simultaneously real. This makes the reality of P and Q depend upon the process of measurement carried out on the first system, which does not disturb the second system in any way. No reasonable definition of reality could be expected to permit this. (Einstein, Podolsky, and Rosen, 1935, next to last paragraph.)

It is therefore rather natural that Einstein declared himself unconvinced by Bohr's answer. The essence of their disagreement centered around their divergent conceptions of physical reality and of the nature of physics. For Einstein, physics was only a refinement of common sense that aspires to describe reality in space and time; Bohr espoused a less ambitious definition of physics — as a discipline whose task is not to discover the way nature *is* but only to impose some order on our sense impressions.

One of the first replies to the EPR paper was that of Kemble (1935) who rejected the entire argument essentially because he was convinced that the wave function is merely a subjective computational tool and not in any sense a description of objective reality. Another answer was given by Ruark (1935) who objected that elements of reality for β come into being only if an actual measurement on α is carried out. Furry (1936) carried out a formal investigation of the general quantum mechanical problem discussed by EPR and showed that their idea of an isolated quantum mechanical system with independent real properties cannot be reconciled with the formalism of quantum theory. Also Wolfe (1936) rejected the EPR argument, stressing that the wave function ψ should only be viewed as representing our knowledge of the systems that are being investigated. Another response was published by Margenau (1936) who put the blame on the quantum theory of measurement, claiming that if the wave function was taken to describe nothing but a statistical ensemble (rather than the individuals composing it) then no paradox would arise. (It will later be shown that the latter claim is totally unjustified and that the EPR paradox exists for an arbitrary interpretation of quantum theory, depending only on its empirical predictions.)

All these 'answers' were published in 1935—1936 and heralded the start of a debate that would continue to the present day. Max Jammer, in his book *The Philosophy of Quantum Mechanics* (Jammer, 1974), skillfully reconstructs this debate and recalls that

> what amused Einstein was that, while all scientists were quite positive that the argument was wrong, they all gave different reasons for their belief! (*Ibid*., p. 187.)

While it would be impossible here to go into all the details of this debate, there are some interesting contributions which we deem relevant enough to recall.

Not all the reactions to the EPR paper were negative. Schrödinger (1935) not only reaffirmed the result of EPR, but generalized it and stated that it should be recognized as a serious deficiency of quantum mechanics. He stressed that assigning a wave function *to the pair* $\alpha + \beta$ often implied that no wave function could be attributed to the individual components α and β of the pair: More precisely, all the wave functions for $\alpha + \beta$ that could not be resolved into a product of two wave functions, one for α and one for β, had this ugly defect. This meant that the systems α and β could not be treated on the assumption that they separately occupied well-defined states and that any one of them could suddenly acquire a new state when a measurement was performed on the other. This was called a 'sinister' consequence of quantum mechanics by Schrödinger.

Epstein (Jammer, 1974) criticized the EPR argument on the grounds that the wave function (5.4) is time-independent, in conflict with what is required by the time-dependent Schrödinger equation. One can observe that the whole EPR reasoning can be carried out at precisely the instant that the wave function has the expression (5.4); furthermore, the paradox exists also for spin states, as will be shown later, and in this case no time dependence is necessary.

A careful historical reconstruction of the birth of EPR's paper and of the ideas underlying it has recently been given by Jammer (1985). His conclusion is:

> The EPR paper, as composed by Podolsky, is essentially valid and, if judged from a realistic point of view, also sound. But it is, strictly speaking, just an amalgamation of two approaches, one based on what Einstein had in mind . . . , and the other based on the trend of ideas underlying the 'EPR thought experiment'. This fusion, however, was quite natural, for Einstein's road to his final conception of the issue had been paved by those thought-experiments which Rosen was able to mathematize.

A detailed reconstruction of the conceptual differences dividing Einstein from Bohr has been made by Carazza (1988), who points out that the main point is the confrontation of two philosophies. On one side is Einstein who defends realism, on the other, Bohr with his insistence that a direct operational meaning be given to all ideas for which the term 'real' is going to be applied. One witnesses thus a gigantic clash between the opposed philosophies of realism and positivism.

5.3. Two Types of State Vectors

In the present section the general quantum mechanical formalism for the description of pairs of correlated systems will be reivewed by paying particular attention to the physical implications of different formal choices. A useful tool for such a study is the Dirac formalism describing quantum states with 'bra' and 'ket' vectors.

According to the usual quantum theory, if there are two quantum systems, α in the state $|\Psi\rangle$ and β in the state $|\Phi\rangle$, then the global system $\varepsilon = (\alpha, \beta)$ is described by the state vector $|\Psi\rangle|\Phi\rangle$:

$$|\Psi\rangle|\Phi\rangle \rightarrow \varepsilon = (\alpha, \beta). \tag{5.7}$$

The opposite problem is also of great interest. Given a general state vector for ε, can one always write it as a direct product of two states separately describing the systems composing ε?

That the answer is negative can be seen in the following way. Let $\{|\Psi_i\rangle\}$ and $\{|\Phi_j\rangle\}$ be two orthonormal and complete sets of states for α and β, respectively. If $|\eta\rangle$ is the state vector for $\varepsilon = (\alpha, \beta)$, one can carry out a (double) development of it over the sets $\{|\Psi_i\rangle\}$ and $\{|\Phi_j\rangle\}$ and write

$$|\eta\rangle = \sum_{ij} c_{ij}|\Psi_i\rangle|\Phi_j\rangle, \tag{5.8}$$

where the c_{ij}'s constitute a set of generally complex coefficients.

It is important to observe that in absence of superselection rules every state $|\eta\rangle$ of the previous form is a possible state vector for ε. This is a consequence of the superposition principle. In fact, every possible vector $|\Psi_i\rangle|\Phi_j\rangle$ is a conceivable state vector for ε, given the fact that it describes α in the state $|\Psi_i\rangle$ and β in the state $|\Phi_j\rangle$. Therefore, also every linear combination of such vectors, such as (5.8), is a possible state vector for ε.

Therefore the coefficients c_{ij} are totally unrestricted, except for the rather trivial normalization condition

$$\sum_{ij} |c_{ij}|^2 = 1.$$

In particular there is no way, if the superposition principle is to be of general validity, to restrict the c_{ij}'s to factorable numbers of the type

$$c_{ij} = x_i y_j. \tag{5.9}$$

Therefore there is no way to guarantee that not only ε but also α and β are in a well-defined state. *The factorization condition* (5.9) *is in fact the necessary and sufficient condition for having α and β in a quantum state.* In order to see this assume that α is in the state $|\Psi\rangle$ and β in the state $|\Phi\rangle$. Developing $|\Psi\rangle$ over the set $\{|\Psi_i\rangle\}$ and $|\Phi\rangle$ over the set $\{|\Phi_j\rangle\}$ one can then write

$$|\Psi\rangle = \sum_i x_i|\Psi_i\rangle,\ |\Phi\rangle = \sum_j y_j|\Phi_j\rangle. \tag{5.10}$$

Remembering from elementary quantum theory that if α is in the state $|\Psi\rangle$ and β in the state $|\Phi\rangle$, then $\varepsilon = (\alpha, \beta)$ is necessarily in the state $|\Psi\rangle\,|\Phi\rangle$, one obtains (5.8) and (5.9) from (5.10). Conversely, if the c_{ij}'s satisfy (5.9) one can write $|\eta\rangle$ as $|\Psi\rangle|\Phi\rangle$, where $|\Psi\rangle$ and $|\Phi\rangle$ are defined by (5.10), and are therefore vectors in the Hilbert spaces of α and β, respectively. But this means that they are state vectors for α and β.

Concluding, the factorization condition (5.9) is the only possibility for having α and β in some quantum state, but the superposition principle forces us to consider vectors $|\eta\rangle$, of the type (5.8), for which (5.9) does not hold. This implies that some states for ε are such that neither α nor β are in a quantum state.

States for $\varepsilon = (\alpha, \beta)$ of the latter kind are called state vectors of the second type, while state vectors of the first type are those for which (5.9) *holds.* One can also say that state vectors of the first (second) type are the factorizable (nonfactorizable) state vectors.

Given the well-known fact that the state vector is the only link between the quantum mechanical formalism and microphysical reality, one can also say that *quantum theory does not ascribe any separate reality to the systems* α *and* β *whose combination* (α, β) *is described by a state vector of the second type*: An embryo of the EPR paradox is already visible in this strange fact.

A general discussion of the EPR paradox first given by Furry shortly after the publication of the EPR paper (Furry, 1936) will be presented next. The starting point is a theorem proved by von Neumann according to which the state vector (5.8) can always be written in the form

$$|\eta\rangle = \sum_i c_i |\Psi_i\rangle |\Phi_i\rangle, \tag{5.11}$$

if the complete orthonormal sets $\{|\Psi_i\rangle\}$ and $\{|\Phi_j\rangle\}$ are suitably chosen. If $|\eta\rangle$ is of the first (second) type only one (more than one) of the coefficients c_i will be different from zero.

Let these two new sets of state vectors form sets of eigenstates of two linear Hermitian operators, A and B, respectively, so that

$$\begin{aligned} A|\Psi_i\rangle &= a_i|\Psi_i\rangle, \\ B|\Phi_i\rangle &= b_i|\Phi_i\rangle, \end{aligned} \tag{5.12}$$

these relations being valid for all values of the index i. One can thus say that A (B) represents an observable of the system α (β) and that the possible values of such an observable are given by the eigenvalues a_i (b_i).

The state vector (5.11) predicts a strict correlation of the measured values of the two observables, even if measurements are performed at different times. Thus, if a first observer measures A at time t_1 on the α component of a certain pair and finds $A = a_k$, then a second observer will necessarily find $B = b_k$ (the index k is the same!) for a measurement made at time $t \geqslant t_1$ on the β component of the same pair.

Suppose that this prediction has been checked many times and always found to be correct. This means that from the time t_1 onwards the outcome of the B measurement is certain (and equal to b_k) for the system β.

Can one apply the quantum mechanical axioms to this kind of knowledge? Many people would answer with a straightforward 'yes' to this question, believing that the application of quantum theory is unlimited. Actually a positive answer implies that quantum mechanics is complete, and completeness, as we saw, is a far from trivial feature of quantum theory. This problem will however be left for Sections 5.6 and 5.7, where it will be discussed at length. Let us continue with Furry's reasoning.

Above it was concluded that from the time t_1 onwards the outcome of the B measurement is certain and equal to b_k. But the two systems can be assumed to be very far from one another, so that no perturbation of the state of β can have taken place as a consequence of the measurement performed on α. Therefore the result of an eventual measurement of B on β must have been fixed and equal to b_k already before the measurement on α was performed. Quantum theory has a very precise way to describe a physical situation (state) of an atomic system for which the result of a measurement of a physical quantity is known *a priori*. It assigns as state vector the eigenvector of the operator corresponding to the known value of the physical quantity under consideration. Recalling Equation (5.12) it is obvious that the state of β must have been $|\Phi_k\rangle$ already before the time t_1. Given the predicted correlation of values for the observables A and B (the value $A = a_k$ is found necessarily for α whenever $B = b_k$ is found for β) one must conclude that the state vector for (α, β) already before time t_1 was

$$|\Psi_k\rangle|\Phi_k\rangle.$$

Repeating the above argument for a statistical ensemble of pairs (α, β) one concludes that the state vectors are

$$\begin{aligned}
|\eta_1\rangle &= |\Psi_1\rangle|\Phi_1\rangle \text{ in } |c_1|^2 \text{ percent of the cases,}\\
|\eta_2\rangle &= |\Psi_2\rangle|\Phi_2\rangle \text{ in } |c_2|^2 \text{ percent of the cases,}\\
&\vdots \\
|\eta_k\rangle &= |\Psi_k\rangle|\Phi_k\rangle \text{ in } |c_k|^2 \text{ percent of the cases,}\\
&\vdots
\end{aligned} \tag{5.13}$$

We started from the assumption that all the pairs of the considered statistical ensemble had the state vector $|\eta\rangle$ and concluded, instead, that the different state vectors $|\eta_1\rangle, |\eta_2\rangle, \ldots, |\eta_k\rangle, \ldots$ applied, with the stated frequencies, to different pairs.

Now, (5.11) and (5.13) are obviously different *mathematical* descriptions of the ensemble. Could it be that they are nevertheless equivalent descriptions for all practical purposes?

Furry could show that the answer is negative. His argument, slightly simplified, is the following. Consider two new observables A' for α and B' for β and suppose, furthermore, that

$$[A', A] \neq 0 \neq [B', B], \tag{5.14}$$

so that the knowledge of the value of A for α (of B for β) precludes the knowledge of the value of A' (of B') for the same system.

Now, suppose that

$$\begin{aligned}
A'|\Psi'_r\rangle &= a'_r|\Psi'_r\rangle,\\
B'|\Phi'_s\rangle &= b'_s|\Phi'_s\rangle,
\end{aligned}$$

are the eigenvalue relations for A' and B', and that $\{|\Psi'_r\rangle\}$ and $\{|\Phi'_s\rangle\}$ are the complete orthonormal sets of their respective eigenfunctions. It is now possible to calculate the probability of measuring: A' on α with the result a'_r and B' on β with the result b'_s. According to the axioms of quantum mechanics such a probability is given by

$$\Omega_1(a'_r, b'_s) = |\langle \eta | \Psi'_r \Phi'_s \rangle|^2$$

in the case the ensemble is described by the state vector (5.11), and by

$$\Omega_2(a'_r, b'_s) = \sum_k |c_k|^2 |\langle \eta_k | \Psi'_r \Phi'_s \rangle|^2$$

in the case it is described by the mixture (5.13).

But the two latter expressions are numerically different since it follows from (5.11) and from the definition of the vectors $|\eta_k\rangle$ given in (5.13) that

$$\Omega_1(a'_r, b'_s) = \left| \sum_k c_k^* \langle \eta_k | \Psi'_r \Phi'_s \rangle \right|^2,$$

which differs from the given expression for Ω_2 because of the presence of the interference term

$$\sum_{k \neq k'} c_k^* c_{k'} \langle \eta_k | \Psi'_r \Phi'_s \rangle \langle \eta_{k'} | \Psi'_r \Phi'_s \rangle^*,$$

which can be shown in most cases to be different from zero as a consequence of (5.14). A paradoxical conclusion has thus been obtained, because it is not possible for the same probability $\Omega(a'_r, b'_s)$ to have simultaneously the two different values Ω_1 and Ω_2!

More recently, other elegant ways expressing the incompatibility at the empirical level between state vectors of the second type and mixtures have been found. Fortunato called attention to the importance of the operator

$$P_\eta = |\eta\rangle\langle\eta| \tag{5.15}$$

for a discussion of the EPR paradox (Fortunato, 1976). P_η is a projection operator, which, being Hermitian, can be assumed to correspond to an observable. Its expectation value over the state (5.11) is obviously unity:

$$\langle \eta | P_\eta | \eta \rangle = 1. \tag{5.16}$$

The expectation value of the same observable over the mixture (5.13) is, by contrast, given by

$$\langle P_\eta \rangle = |c_1|^2 \langle \Psi_1 \Phi_1 | P_\eta | \Psi_1 \Phi_1 \rangle + |c_2|^2 \langle \Psi_2 \Phi_2 | P_\eta | \Psi_2 \Phi_2 \rangle + \\ + \cdots + |c_k|^2 \langle \Psi_k \Phi_k | P_\eta | \Psi_k \Phi_k \rangle + \cdots.$$

Since it is a simple matter to show that

$$\langle \Psi_k \Phi_k | P_\eta | \Psi_k \Phi_k \rangle = |c_k|^2,$$

it follows that

$$\langle P_\eta \rangle = |c_1|^4 + |c_2|^4 + \cdots + |c_k|^4 + \cdots.$$

But the sum of the fourth powers of the moduli of numbers whose *squared* moduli add up to 1, will certainly be less than 1, if there are at least two numbers different from zero. But the latter condition precisely characterizes a state vector of the second type, as we pointed out after Equation (5.11). Therefore the observable P_η has an expectation value equal to 1 [less than 1] over the state vector of the second type $|\eta\rangle$ [over the mixture of state vectors of the first type (5.13)].

A theorem allowing one to construct an observable whose expectation value can distinguish between a given state vector of the second type and all the possible mixtures of state vectors of the first type has been found by Cufaro-Petroni (1977). Suppose $|\eta\rangle$ is a given state vector of the second type for two correlated quantum systems. Let $\{|\eta_i\rangle\}$ be a complete orthonormal set of states for the correlated pairs (α, β) with $|\eta_1\rangle$ chosen in such a way that

$$|\eta_1\rangle \equiv |\eta\rangle,$$

the nature (first or second type) of the remaining vectors being irrelevant, and $\{\lambda_k\}$ be a set of nonzero real numbers satisfying only the condition

$$\lambda_1 > \lambda_k,$$

for all $k \neq 1$. It is then possible to construct the operator

$$\Gamma \equiv \sum_k \lambda_k |\eta_k\rangle\langle\eta_k|, \tag{5.17}$$

which is obviously a linear Hermitian operator and represents an observable. Its expectation value over the original state vector of the second type $|\eta\rangle$ is given by

$$\langle\eta|\Gamma|\eta\rangle = \lambda_1. \tag{5.18}$$

Consider next an arbitrary state vector of the first type $|y\rangle$. The expectation value of Γ over this state $|y\rangle$ is

$$\langle y|\Gamma|y\rangle = \sum_k \lambda_k |\langle y|\eta_k\rangle|^2,$$

as is easy to check from the definition (5.17) of Γ. Since, on the other hand, $|y\rangle$ is a normalized vector and $\{|\eta_k\rangle\}$ a *complete* set of vectors, one has

$$1 = \langle y|y\rangle = \left\langle y \left| \left\{ \sum_k |\eta_k\rangle\langle\eta_k| \right\} \right| y \right\rangle = \sum_k |\langle y|\eta_k\rangle|^2,$$

whence it follows that

$$\langle y | \Gamma | y \rangle \leqslant \lambda_1 \sum_k \langle y | \eta_k \rangle|^2 = \lambda_1. \tag{5.19}$$

Comparing (5.19) with (5.18) one thus sees that the average of the observable Γ allows one to discriminate experimentally between the state vector of the second type $|\eta\rangle$ and an arbitrary state vector of the first type. The sign $\leqslant$ in (5.19) can in fact become an equality only if $|y\rangle$ is orthogonal to all $|\eta_k\rangle$ with $k \neq 1$. This means however that it must be parallel to $|\eta_1\rangle$, which is not possible for a state vector of the first type.

5.4. Spin States for Two Particles

In the present section the general quantum mechanical description of two correlated quantum systems will be specialized to the case of spin-$\frac{1}{2}$ particles, in order to introduce the well-known 'singlet' state to which it is useful, for simplicity reasons, to refer the presentation of the EPR paradox.

In the third section of Chapter 2 we introduced spin states for spin-$\frac{1}{2}$ particles. That formalism will now be extended to *pairs* of correlated particles.

As was done in the previous section, given two independent quantum systems α and β, suppose that α is described by the wave function Ψ_1 and β by the wave function Ψ_2; then the compound system $\varepsilon = (\alpha, \beta)$ is described quantum mechanically by

$$\Psi(x_1, x_2) = \Psi_1(x_1)\Psi_2(x_2),$$

that is, by the product of the wave functions for the separate systems. One reason that the product is adopted stems from the factorizability of the position probabilities when α and β can be treated as independent systems. Let P_1 (P_2) be the probability for finding α (β) inside the volume V_1 (V_2). Then P_1 (P_2) is given as the volume integral over V_1 (V_2) of the squared modulus of ψ_1 (ψ_2). If the two systems are independent, as we have assumed, the probability P for finding α in V_1 *and* β in V_2 is given by $P = P_1 P_2$, as is well known from elementary probability calculus. If P is assumed to be given by the squared modulus of $\psi(x_1, x_2)$, then the factorizability of P into the product of P_1 and P_2 can be deduced directly from the factorizability of $\psi(x_1, x_2)$.

Factorizable states can be introduced also when one is dealing with spin. Let $|u_+\rangle$ and $|u_-\rangle$ be the eigenvectors, corresponding to the eigenvalues $+1$ and -1, respectively, of the matrix representing the third component of the spin angular momentum for the particle α; and let $|v_+\rangle$ and $|v_-\rangle$ be the corresponding eigenstates for the particle β. In other words, if σ_3, as given in Equation (2.4) is the matrix representing the third component of the spin for particle α, one has

$$|u_+\rangle = \begin{pmatrix} 1 \\ 0 \end{pmatrix}, \qquad |u_-\rangle = \begin{pmatrix} 0 \\ 1 \end{pmatrix}. \tag{5.20}$$

Now notice that the only factorizable spin states one can construct from $|u_+\rangle, |u_-\rangle, |v_+\rangle, |v_-\rangle$ are

$$|u_+\rangle|v_+\rangle, |u_+\rangle|v_-\rangle, |u_-\rangle|v_+\rangle, |u_-\rangle|v_-\rangle,$$

where the first function obtains when the spin vectors of both particles α and β point along the positive z direction, and so on.

A spin state of great interest for our purposes is the so-called *singlet state*, which is given by

$$|\eta_S\rangle = 2^{-1/2}(|u_+v_-\rangle - |u_-v_+\rangle) \tag{5.21}$$

and has four very important properties:

P1. *It is not a factorizable state.*
P2. *It predicts opposite results for measurements of the third component of the spins of particles* α *and* β.
P3. *It predicts the result zero for a measurement of the total squared spin of particles* α *and* β.
P4. *It is rotationally invariant.*

Some comments about these properties follow:

Property P1 is not difficult to prove, since, to begin with, the most general spin state for α is

$$|u\rangle = \begin{pmatrix} a \\ b \end{pmatrix} = a\begin{pmatrix} 1 \\ 0 \end{pmatrix} + b\begin{pmatrix} 0 \\ 1 \end{pmatrix} = a|u_+\rangle + b|u_-\rangle, (|a|^2 + |b|^2 = 1),$$

where a and b are some constants; similarly, the most general spin state for β is

$$|v\rangle = c|v_+\rangle + d|v_-\rangle, (|c|^2 + |d|^2 = 1),$$

where c and d are some other constants. Obviously, then, the most general factorizable spin state for the combined system is

$$|uv\rangle = ac|u_+v_+\rangle + ad|u_+v_-\rangle + bc|u_-v_+\rangle + bd|u_-v_-\rangle.$$

Now, since $|u_+v_+\rangle$ does not enter into $|\eta_S\rangle$, $|uv\rangle$ can equal $|\eta_S\rangle$ only if $ac = 0$. Thus $a = 0$, which implies that also $|u_+v_-\rangle$ is to be deleted from the expression for $|uv\rangle$, and/or $c = 0$, which implies a similar vanishing of $|u_-v_+\rangle$. It is therefore impossible, by any choice of $|u\rangle$ and $|v\rangle$, to satisfy $|\eta_S\rangle = |uv\rangle$. Since $|uv\rangle$ is the most general factorizable spin state, it follows that $|\eta_S\rangle$ is not factorizable; we have thus proven P1.

As for property P2, its validity is clearly suggested by the very structure of $|\eta_S\rangle$, which contains only $|u_+v_-\rangle$ and $|u_-v_+\rangle$. But the reader who wishes to rigorously confirm the property in question may introduce the Pauli matrices σ_1, σ_2, σ_3 for particle α and τ_1, τ_2, τ_3 for particle β and proceed to check that $|\eta_S\rangle$ is an eigenstate of $\sigma_3 + \tau_3$ with eigenvalue zero, that is,

$$(\sigma_3 + \tau_3)|\eta_S\rangle = 0.$$

From the physical interpretation of quantum states, it then follows that measurements of the third component of the spins of α and β must always give opposite results.

Also P3 can be verified by introducing the total squared spin operator, defined by

$$\Sigma^2 = (\sigma_1 + \tau_1)^2 + (\sigma_2 + \tau_2)^2 + (\sigma_3 + \tau_3)^2 = 6 + 2\boldsymbol{\sigma} \cdot \boldsymbol{\tau}$$

(all squared Pauli matrices equal unity), and checking that

$$\Sigma^2 |\eta_S\rangle = 0,$$

whence it follows, invoking quantum mechanics, that a measurement of the observable corresponding to Σ^2, on a system described by the state $|\eta_S\rangle$, will certainly give the result zero.

The fourth fundamental property of $|\eta_S\rangle$ (rotational invariance) entails essentially the following: If $|u_\pm\rangle$ and $|v_\pm\rangle$, which are eigenstates of σ_3 and τ_3, respectively, are replaced by $|u^n_\pm\rangle$ and $|v^n_\pm\rangle$, which denote eigenstates of $\boldsymbol{\sigma} \cdot \hat{\mathbf{n}}$ and $\boldsymbol{\tau} \cdot \hat{\mathbf{n}}$, respectively ($\hat{\mathbf{n}}$ being an arbitrary unit vector), then $|\eta_S\rangle$ can be shown to transform into

$$|\eta_S\rangle = 2^{-1/2}(|u^n_+ v^n_-\rangle - |u^n_- v^n_+\rangle), \tag{5.22}$$

which has the same structure as (5.21), albeit involving different states.

Another state that is central to our discussion is the *triplet* state, given by

$$|\eta_t\rangle = 2^{-1/2}(|u_+v_-\rangle + |u_-v_+\rangle), \tag{5.23}$$

which differs from $|\eta_S\rangle$ only by the plus sign replacing the minus sign.

One can prove that $|\eta_t\rangle$ shares with $|\eta_S\rangle$ the properties P1 and P2, but not P4; it is *not* rotationally invariant. Moreover, in place of P3, $|\eta_t\rangle$ has the following property: *Any measurement of the total squared spin of two particles described by* $|\eta_t\rangle$ *will give the result* $2\hbar^2$.

It is possible to write the states $|u_+v_-\rangle$ and $|u_-v_+\rangle$ as linear combinations of $|\eta_S\rangle$ and $|\eta_t\rangle$:

$$\begin{aligned} |u_+v_-\rangle &= 2^{-1/2}(|\eta_t\rangle + |\eta_S\rangle), \\ |u_-v_+\rangle &= 2^{-1/2}(|\eta_t\rangle - |\eta_S\rangle), \end{aligned} \tag{5.24}$$

as can be proven simply by adding and subtracting (5.21) and (5.23). Furthermore, on invoking the quantum mechanical interpretation of superpositions, one sees that:

Measurements of the total squared spin of two particles described by $|u_+v_-\rangle$ *will produce with equal probability the result zero and the result* $2\hbar^2$*; the same is true for* $|u_-v_+\rangle$.

It therefore follows that a large ensemble of pairs of particles described by $|\eta_S\rangle$ is observably different from a similar ensemble in which some pairs are described by $|u_+v_-\rangle$ and the remaining pairs by $|u_-v_+\rangle$. In fact, when measuring Σ^2 on pairs of the first ensemble, one will invariably find the result zero; while measurements of the same quantity on pairs of the second ensemble will furnish the results zero and unity with equal frequency.

This large observable difference, between an ensemble that is an arbitrary mixture of the states (5.24) and an ensemble whose elements are all described by $|\eta_S\rangle$, is the basis of a modern formulation of the EPR paradox, originally expounded by D. Bohm in his *Quantum Theory* (Bohm, 1951), which we shall discuss in the next sections.

5.5. Reality and Separability

The EPR formulation of the reality criterion has been quoted verbatim at the beginning of the present chapter. It centers on the idea that, if it is possible to predict with certainty the outcome of an act of measurement performed on an atomic system, then there must be a physical property ('an element of reality') belonging to that atomic system. Some points of elaboration are:

1. The element of physical reality is thought of as existing already before the act of measurement is concretely carried out. From a philosophical point of view, this means that a step into ontology has been taken. Strict positivists would, of course, have much to say against this hypothesis, even though it seems a very natural one to most ordinary people — who rightly tend to believe, for example, that our planet existed long before man appeared on it (in fact, for 99.9% of its history!) and that the galaxies exist also when no telescope is observing them.

2. The element of physical reality is viewed as belonging to the measured object and *not* to the measuring apparatus. The latter has been checked to work properly, so that it can register different values of the measured physical quantity. If one can predict that this apparatus will give an *a priori* known result by interacting with the object under investigation, this implies that the decisive factor producing the result of the measurement resides in the object. This does not mean that an active role of the apparatus, so characteristic of quantum measurements, is denied in general. The EPR reality criterion applies in fact only to those very special situations in which the outcome of an act of measurement is predictable with certainty.

3. The postulated element of reality is considered to be the *cause* of the precisely predictable outcome of the act of measurement. Realism and causality are therefore strictly tied together in the EPR reality criterion.

Notice that the element of reality is not necessarily coincident with the result of a measurement. Given, for instance, an electron with wave function

$$\psi(x) = \psi_0 \exp\left(\frac{i\mathbf{p}\cdot\mathbf{x}}{\hbar}\right), \tag{5.25}$$

one can predict with certainty that an apparatus measuring the particle's momentum will furnish the result $\mathbf{p}$. There then exists an element of reality associated with the electron's momentum, and this conclusion is emphatically not the same as saying that the electron *has* a well-defined momentum before one measures it. The EPR reality criterion postulates only the existence of a definite (but, in general, unknown) cause underlying the exactly predictable result that will emerge from a measurement of the particular (and perhaps only conventional) physical quantity that we call momentum and which we are at least able to define operationally.

4. Suppose that past experience has confirmed our ability to predict with certainty the outcome of a measurement under certain experimental conditions. For example, it could happen that electrons prepared with the wave function (5.25) have been checked many times to give invariably the result $\mathbf{p}$ in a measurement of momentum. In such a case we would be justified in concluding that the EPR element of reality leading to the predicted outcome of the measurement exists also when the measurement is not carried out, if the preparation of the state has been the same. Thus in the case of electrons in the state (5.25) we would be justified in saying that they possess an element of reality, which is essentially the physical property of leading necessarily to the result: momentum = $\mathbf{p}$, if and when a measurement of momentum is carried out. This is of course the regular scientific procedure. Physics is practiced for *learning* the properties of the material world, and it would be foolish, for instance, to say that the magnetic moment of the electron is known only for those electrons for which this quantity is actually measured! In fact, once the value of the magnetic moment has been measured accurately, it has been used in many physical situations in which no further observation of this quantity was made, such as bound states, superconductivity, and so on. With this natural assumption one precludes a future act of measurement from retroactively influencing a present element of reality. In other words, we ensure that causal connections operate from the past toward the future and not the other way around, i.e., that there exists an 'arrow of time'. In Section 5.10 a line of thought will be discussed that tries instead to solve the EPR paradox by assuming perfect time symmetry and therefore also the existence of retroactive causality.

We conclude with a simple example illustrating the practical meaning of all the previous four points.

Given an electron with wave function (5.25), suppose that we can predict *with certainty* the value of its momentum, either because our reasoning is based on a theory we trust, or because the correctness of the prediction has been checked many times, or both. Applying the EPR reality criterion, we thus attribute *to the electron* an objectively real property (that is, an element of reality), of unknown nature, which is *the cause* of the result 'momentum = **p**' for any momentum measurement carried out in the future, and which exists also if such a measurement is not made at all.

The reader may feel unsatisfied with these vague, qualitative, and philosophical conclusions. We beg for patience: 50% of a formidable theoretical tool has already been shaped; the remaining 50% will be discussed next.

In a first approximation, we may speak of *separability* as the reasonable assumption that two arbitrary objects interact very little when their mutual separation is very large. Already in this rough definition of separability we intentionally avoid assuming that the interaction between the two objects can be made to vanish completely; because not only does a nonzero, if very small (e.g., gravitational), interaction persist also at large distances, but the *total* absence of interaction is not a necessary condition for establishing the EPR paradox.

One may wonder what is to be understood under a 'very large distance'. Experience teaches us that such a distance allows no universal definition, but that it can none the less be identified quantitatively from case to case in all known instances. Thus, while 10^{-10} cm is already a 'very large' distance for nuclear interactions, which decrease exponentially with a characteristic length of 10^{-13} cm, much larger distances need to be assumed before electromagnetic and gravitational effects (which decrease inversely with the distance) can be neglected.

All that is important from the point of view of the EPR paradox is that a 'very large' distance, definable in the foregoing sense, exists in every known case. A more precise definition of the assumption of separability therefore is the following: *Given two physical systems, a minimal distance between them exists beyond which their mutual interactions become negligibly small.*

Again, one could wonder about the meaning of 'negligibly small interactions', but also here it is not difficult to agree that with this expression we more specifically imply: *unable to create the elements of reality of either system.* The restriction contained in the latter definition, being absolute, could perhaps be considered too strong, but a weaker version of it suffices for our purposes and leads to the modified definition: *capable of creating the elements of reality of either system at most in* 1% *of the cases.* Thus, in summary, a version of the idea of separability which seems impeccable is the following:

> *Separability is the assumption that, given any two physical systems, a minimal distance between them exists beyond which their mutual interaction becomes so small that it is capable of creating the elements of reality belonging to either system of the pair at most for* 1% *of such pairs.*

The hypothesis of separability can be applied also to quantum systems.

True, these systems are characterized only by their wave functions, which can fill broad regions of space. There are, however, also situations in which the spatial localization is very sharp. Take, for instance, two particles α and β with wave functions $\psi_\alpha(x_1)$ and $\psi_\beta(x_2)$, respectively. The wave function of the two particles, considered as a single system, is then

$$\psi(x_1, x_2) = \psi_\alpha(x_1)\psi_\beta(x_2).$$

Suppose, furthermore, that $\psi_\alpha(x_1)$ is a function of Gaussian form with modulus appreciably different from zero only in a region of width Δ_1, centered around the point x_{10}. Similarly, let $\psi_\beta(x_2)$ be a Gaussian function of width Δ_2 centered around x_{20} (see Figure 5.1). We shall regard the systems α and β as separable in the previously defined sense if the distance $|x_{20} - x_{10}|$ is much larger than either Δ_1 or Δ_2.

If now particles α and β are supposed to move to the left and the right, respectively, so that the distance between the centers of the two wave packets increases linearly with time, it is not difficult to show that Schrödinger's equation allows for situations in which the spreading of the wave packets is less than linear with time and consequently such that the conditions for separability improve with time. It might be objected that a Gaussian wave function actually fills the whole of space and not just a finite volume, so that there is after all a nonvanishing probability of finding α and β near to each other. This is of course true, but the point is that this probability can be made arbitrarily small (say, less than 10^{-100}!), given the increasingly rapid decrease of any Gaussian with the increase of distance from its midpoint.

We thus see that reasonable quantum situations exist in which any two quantum systems are effectively separated in space, and thus the conditions for the fulfillment of the above separability hypothesis are satisfied. In developing the EPR paradox in the next two sections, we shall confine ourselves exclusively to such situations as far as the spatial part of the wave function is concerned.

The importance of the separability idea was stressed by Albert Einstein, who, for instance, wrote:

> But on one supposition we should, in my opinion, absolutely hold fast: the real factual situation of the system S_2 is independent of what is done with the system S_1, which is spatially separated from the former. (Schilpp, 1970, p. 86)

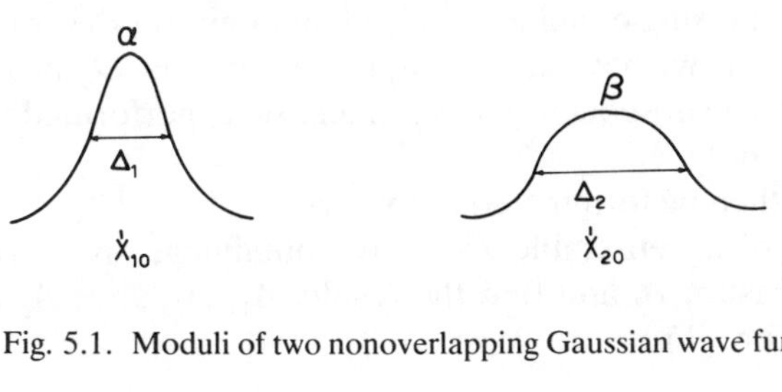

Fig. 5.1. Moduli of two nonoverlapping Gaussian wave functions.

Also Erwin Schrödinger recognized the strange nature of the quantum mechanical description of two correlated particles. In 1935 he asserted:

> When two systems of which we know the states by their respective representations enter into a temporary physical interaction due to known forces between them and then, after a time of mutual influence, the systems separate again, then they can no longer be described in the same way as before viz. by endowing each of them with a representative state vector. I would not call that *one* but rather *the* characteristics of quantum mechanics. (E. Schrödinger, 1935.)

In recent times many authors have discussed the idea of separability from different points of view. Hooker (1970) stressed that, in order to have a 'robust' physical basis for the quantum mechanical violations of separability, one should have some form of energy associated with the physical unity of the two quantum systems. Instead the total energy of the two systems is simply the numerical sum of the component energies. Wightman (1976) pointed out that the EPR paradox arises from the quantum mechanical prediction that the state of one of two correlated systems sometimes depends on what experiment is chosen to be done on the other, even though it may happen that there is no time for a light signal to travel from one system to the other to communicate the choice. Penetrating discussions of the idea of separability were also given by Eberhard (1978), who distinguished four different types of locality and showed that quantum theory contradicted three of them, and by Heywood and Redhead (1983), who stressed the difference between 'ontological locality' and 'environmental locality', the latter implying that no interaction between the system of interest and its surroundings exists which can take place before the act of measurement and alter the values of the physical properties of the system.

5.6. The EPR Paradox: Quantum Mechanics Complete

Let a system ε be given that is capable of decaying into two spin-$\frac{1}{2}$ systems α and β. There are several concrete examples in which it must be assumed that the spin state vector for the system (α, β) must be $|\eta_S\rangle$ as given by (5.8). We limit our attention to one of these examples and suppose furthermore that the spatial part of the total wave function is separable in the sense established in the previous section.

Consider a large ensemble E composed of N decays $\varepsilon \rightarrow \alpha + \beta$ of the system ε into α and β. The singlet state, which by assumption describes all these (α, β) pairs, implies, as we have seen, that, if the observer O_α measures σ_3 and finds $+1$ (-1), then a subsequent measurement of τ_3 performed by O_β will give -1 $(+1)$ with certainty.

The EPR paradox can then be formulated as follows:

1. On the systems α of an ensemble $E_1 \subset E$, containing, say, ℓ (α, β) pairs, assume that O_α measure σ_3 and find the results $A_1, A_2, \ldots, A_\ell$ (each of which equals either $+1$ or -1).

2. *Correlation.* The results of *subsequent* measurements of τ_3 performed by O_β on all the ℓ systems β of E_1 can then be predicted with certainty to be $-A_1, -A_2, \ldots, -A_\ell$. This prediction can be checked, e.g., on an ensemble $E_2 \subset E_1$. We assume that quantum mechanics is correct, so that every conceivable check of its predictions will give a positive answer.

3. *Reality.* The certainty arrived at in the previous step allows us to use the EPR reality criterion and to attribute to the system β an element of reality, call it t_3, which determines the result of a future measurement of τ_3 on β for all $\beta \in E_2$.

4. *Time arrow.* Assuming that the element t_3 is not generated retroactively in time by the future measurement on β, we conclude that t_3 belongs to all $\beta \in E_1$.

5. *Separability.* If systems α and β are separated by a very large distance (in the sense discussed in the previous section), we assume that the element of reality t_3 of β could not have been created by the measurement performed on α. Consequently, t_3 exists also if no measurement on α is carried out and, in fact, can be attributed to all $\beta \in E$.

6. *Completeness.* Assuming that quantum mechanics is complete, the element of reality t_3 must have a counterpart in the quantum theoretical description of the system β. But the only state vector describing as certain, and equal to -1 (or $+1$), the outcome of a measurement of τ_3 on β is v_- (or v_+). One of these two descriptions must therefore apply to all β of E. Actually v_+ and v_- should be attributed to E with equal weight ($= \frac{1}{2}$).

7. *Total spin.* Quantum mechanics predicts in the present case that the third component of the *total* spin of the system $\alpha + \beta$ is zero in all cases [recall property P2 of Section 5.4].

8. *Mixture.* The only state vectors of $\alpha + \beta$ describing β by either v_- or v_+ and giving zero for the third component of the total spin of the two atoms are u_+v_- or u_-v_+, respectively. These vectors must occur with equal statistical weights in the total ensemble E, which is thus described quantum mechanically by means of a mixture of factorizable state vectors.

The conclusion of point 8 above contradicts in an observable manner the description provided by $|\eta_S\rangle$, as was shown in Section 5.4. The contradiction is expressed by the different values of the average total squared spin of the two systems α and β obtained for $|\eta_S\rangle$ and for the mixture.

We thus arrived at an absurd conclusion. Now, assuredly one can rule out the occurrence of nonsensical conclusions strictly within the framework of quantum mechanics, since not only has this theory always been in harmony with experiments, but its logical structure has survived half a century of careful scrutiny, as well. This means that among the above numbered items there are some which contain elements foreign to and incompatible with quantum mechanics. A quick check will convince the reader that items 1, 2, 7, and 8 are strict consequences of quantum theory. The foreign elements must therefore have entered our reasoning via points 3 (*reality*), 4 (*time arrow*), 5 (*separability*), and 6 (*completeness*). Consequently, at least one of these faces repudiation if the paradox is to disappear — unless one is willing to admit that quantum mechanics is incorrect.

Einstein held the opinion that what should be questioned was the assumed completeness of quantum mechanics (point 6) and that the elimination of this tenet would leave us with a theory free of contradictions.

However, the majority of the scientists now working on the EPR paradox rather believe that the solution should be found in an intelligent negation of *Einstein locality*, which we define here as the set of the three assumptions 3 (*reality*), 4 (*time arrow*), and 5 (*separability*) above. The reason is simple: modern research has shown that the incompatibility between Einstein locality and the empirical predictions of quantum mechanics exists also for an *incomplete* theory, which means that the problem of completeness is totally irrelevant from the logical point of view.

Consequently, the EPR paradox has today become much more acute than its authors implied in their paper, since a very general and natural physical point of view, like that of Einstein locality, has been shown to be flatly incompatible with existent quantum theory.

The proof of the incompatibility between Einstein locality and the empirical predictions of quantum theory will be given in the following section for the case that quantum mechanics is instead assumed to be incomplete from the start. Here we shall rather show that the paradox can be expressed in a different way (by means of Bell's inequality) already in the case of a complete quantum theory. This will allow us to treat the two cases in a totally symmetrical way.

Consider again an ensemble formed by a very large number N of decays $\varepsilon \to \alpha + \beta$ and suppose that the observer O_α measures on α the dichotomic observable $A(a)$, while in a distant region of space a second observer O_β measures on β another dichotomic observable $B(b)$. The dichotomy of the observables means that they can assume only two values, which we take for simplicity to be ± 1. Notice that the observables $A(a)$ and $B(b)$ have been assumed to depend on the arguments a and b, respectively, which are assumed to be experimental parameters that are fixed in the structure of the apparatuses in any given experiment, but possibly variable over different experiments. Possible examples of such dichotomic observables are those corresponding to the spin matrices $\boldsymbol{\sigma} \cdot \hat{\mathbf{a}}$ and $\boldsymbol{\tau} \cdot \hat{\mathbf{b}}$ where $\sigma_1, \sigma_2, \sigma_3$ (τ_1, τ_2, τ_3) denote the Pauli matrices for α (β); here the experimental parameters are the unit vectors $\hat{\mathbf{a}}$ and $\hat{\mathbf{b}}$, which could be fixed experimentally, for example, by the directions of the inhomogeneous magnetic fields of two Stern–Gerlach apparatuses.

In practice any physical quantity can be used to define a dichotomic observable: for instance, one could say that $A(a) = -1$ $(+1)$ if the energy of an atom is below (above) a certain value a.

In any event, when measurements of such observables are made on all the N pairs of the given ensemble, O_α will obtain a set of experimental results

$\{A_1, A_2, \ldots, A_N\}$, while O_β will collect a similar set $\{B_1, B_2, \ldots, B_N\}$, all relative to fixed values of the parameters a and b. The results of the two sets are correlated in the sense that A_1 and B_1 pertain to the particles α and β, respectively, arising from the first decay; A_2 and B_2 are similarly associated with the second decay; and so on. By hypothesis, these results in every case equal ± 1.

The quantum mechanical *correlation function* $P(a, b)$ of the measurements A_i and B_i is defined as the *average product* of the results obtained by O_α and O_β for the same decays:

$$P(a, b) = N^{-1} \sum_{i=1}^{N} A_i B_i. \tag{5.26}$$

Since every product $A_i B_i$ is either $+1$ or -1, it follows that

$$-1 \leqslant P(a, b) \leqslant +1. \tag{5.27}$$

As a first application of definition (5.26), we proceed to calculate the quantum mechanical correlation function for two spin-$\frac{1}{2}$ particles in the singlet state when the observables are $\boldsymbol{\sigma} \cdot \hat{\mathbf{a}}$ and $\boldsymbol{\tau} \cdot \hat{\mathbf{b}}$. In quantum mechanics, the observable corresponding to the product of two observables like the foregoing is represented by an operator that can be written $\boldsymbol{\sigma} \cdot \hat{\mathbf{a}} \otimes \boldsymbol{\tau} \cdot \hat{\mathbf{b}}$, where the symbol $\otimes$ indicates a kind of *formal product*, different from the ordinary matrix product and requiring every matrix to act only on the spinors of the corresponding particle.

The quantum mechanical correlation function in the case of the singlet state is then

$$P(a, b) = \langle \eta_S | \boldsymbol{\sigma} \cdot \hat{\mathbf{a}} \otimes \boldsymbol{\tau} \cdot \hat{\mathbf{b}} | \eta_S \rangle.$$

Substituting Equation (5.21) into this expression one obtains, after a simple calculation,

$$P(a, b) = -\hat{\mathbf{a}} \cdot \hat{\mathbf{b}}. \tag{5.28}$$

This result is simple and elegant but it is incompatible with the great idea of Einstein locality, as we shall see in this and in the next section.

A mathematical property of the quantum mechanical correlation function (5.28) which will be very useful later on is the following: Consider two orthogonal unit vectors $\hat{\mathbf{a}}$ and $\hat{\mathbf{a}}'$ associated with particle α and, likewise, two orthogonal unit vectors $\hat{\mathbf{b}}$ and $\hat{\mathbf{b}}'$ pertaining to particle β, and suppose that the relative orientations of these vectors is as shown in Figure 5.2.

Now define the quantity

$$\Delta = |P(a, b) - P(a, b')| + |P(a', b) + P(a', b')|. \tag{5.29}$$

Recalling (5.28) we see that quantum mechanics predicts, for the choice of

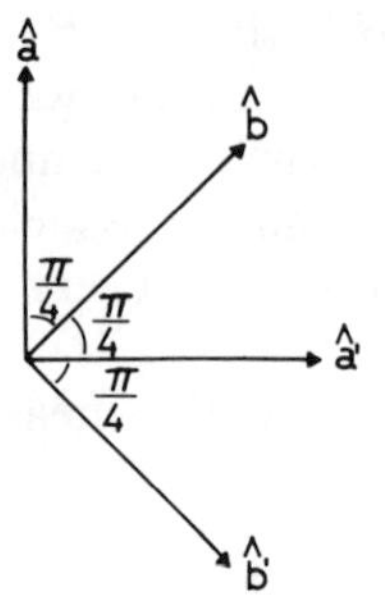

Fig. 5.2. Relative orientations of unit vectors used in the proof of $\Delta = 2\sqrt{2}$.

Figure 5.2

$$\Delta = |\hat{\mathbf{a}} \cdot \hat{\mathbf{b}} - \hat{\mathbf{a}} \cdot \hat{\mathbf{b}}'| + |\hat{\mathbf{a}}' \cdot \hat{\mathbf{b}} + \hat{\mathbf{a}}' \cdot \hat{\mathbf{b}}'| = 2\sqrt{2}.$$

It can moreover be shown that $2\sqrt{2}$ is the maximum value of Δ for all conceivable orientations of the vectors $\hat{\mathbf{a}}, \hat{\mathbf{a}}', \hat{\mathbf{b}}, \hat{\mathbf{b}}'$.

This last result is of great interest, because, as we shall see next, Δ is a physical quantity to which a reasoning based on Einstein locality attributes a maximum value of 2. The inequality $\Delta \leqslant 2$ usually goes under the name of *Bell's inequality*. It has been called 'the most profound discovery of science' (Stapp, 1977).

That Bell's inequality is connected with the EPR paradox we shall prove in the following by showing that *every mixture of factorizable state vectors* leads to its validity (Capasso *et al.*, 1973). The reasoning based on Einstein locality carried out at the beginning of the present section showed that pairs of correlated spin-$\frac{1}{2}$ systems should necessarily be described by a 50—50 mixture of the factorizable state vectors $|u_+v_-\rangle$ and $|u_-v_+\rangle$. We shall thus show that Einstein locality leads to the validity of Bell's inequality, at least in the presently examined case of a complete quantum theory (the opposite case will form the subject of the next section).

Consider an ensemble E of pairs of quantum systems (α, β) of any type and assume that they are described by the factorizable states $|\Psi_k\rangle|\Phi_k\rangle$ with the probability (frequency in the ensemble) p_k, where $k = 1, 2, \ldots$. Therefore, if E contains N pairs, there will be

$$\begin{aligned} &Np_1 \text{ pairs with state vector } |\Psi_1\rangle|\Phi_1\rangle,\\ &Np_2 \text{ pairs with state vector } |\Psi_2\rangle|\Phi_2\rangle,\\ &\vdots\\ &Np_k \text{ pairs with state vector } |\Psi_k\rangle|\Phi_k\rangle,\\ &\vdots \end{aligned} \tag{5.30}$$

Suppose that the dichotomic observables to be measured on α and β are described quantum mechanically by the operators $A(a)$ and $B(b)$, respectively, so that the operator corresponding to the joint measurement on the two systems is $A(a) \otimes B(b)$. The correlation function predicted by quantum theory is precisely the average of the latter operator over the mixture (5.30), so that

$$P(a, b) = \sum_k p_k \langle \Psi_k | \langle \Phi_k | \{A(a) \otimes B(b)\} | \Phi_k \rangle | \Psi_k \rangle. \tag{5.31}$$

The four correlation functions entering into Bell's inequality can then be written

$$\begin{aligned} P(a, b) &= \sum_k p_k \bar{A}_k \bar{B}_k, \quad P(a, b') = \sum_k p_k \bar{A}_k \bar{B}'_k, \\ P(a', b) &= \sum_k p_k \bar{A}'_k \bar{B}_k, \quad P(a', b') = \sum_k p_k \bar{A}'_k \bar{B}'_k, \end{aligned} \tag{5.32}$$

where

$$\begin{aligned} \bar{A}_k &= \langle \Psi_k | A(a) | \Psi_k \rangle, \qquad \bar{A}'_k = \langle \Psi_k | A(a') | \Psi_k \rangle, \\ \bar{B}_k &= \langle \Phi_k | B(b) | \Phi_k \rangle, \qquad \bar{B}'_k = \langle \Phi_k | B(b') | \Phi_k \rangle. \end{aligned} \tag{5.33}$$

An important property of these four averages is that they must have a modulus not exceeding 1, since they are expectation values of operators having ± 1 as their only possible eigenvalues. Therefore

$$|\bar{A}_k| \leqslant 1, |\bar{A}'_k| \leqslant 1, |\bar{B}_k| \leqslant 1, |\bar{B}'_k| \leqslant 1, \tag{5.34}$$

the previous inequalities being true for all values of k.

Inserting (5.32) in (5.29), one easily obtains

$$\Delta = \sum_k p_k \Delta_k, \tag{5.35}$$

where

$$\Delta_k \equiv |\bar{A}_k \bar{B}_k - \bar{A}_k \bar{B}'_k| + |\bar{A}'_k \bar{B}_k + \bar{A}'_k \bar{B}'_k|. \tag{5.36}$$

Remembering (5.34), one can then immediately deduce

$$\Delta_k \leqslant |\bar{B}_k - \bar{B}'_k| + |\bar{B}_k + \bar{B}'_k|, \tag{5.37}$$

whence it follows that $\Delta_k \leqslant 2$, since any two real numbers x and y such that $|x| \leqslant 1$ and $|y| \leqslant 1$ always satisfy $|x - y| + |x + y| \leqslant 2$. If $\Delta_k \leqslant 2$ is inserted in (5.35), one gets finally

$$|P(a, b) - P(a, b')| + |P(a', b) + P(a', b')| \leqslant 2, \tag{5.38}$$

since $\Sigma_k\, p_k = 1$. The inequality (5.38) is precisely Bell's inequality.

The previous results allow us to state the EPR paradox in a new way:

EPR PARADOX (1). *Given a statistical ensemble of pairs of spin-$\frac{1}{2}$ objects described quantum mechanically by the singlet state, Bell's inequality must be violated by more than 40% for suitable choices of the experimental parameters. Injecting into the theory the assumption of Einstein locality* (points 3, 4, 5 on p. 205) *and that of completeness* (point 6) *one concludes instead that Bell's inequality must always be satisfied.*

A study of generalized mixtures of factorizable state vectors for which the observables under study are not dichotomic, as assumed above, but multichotomic, meaning that they can assume any finite number of discrete values, has been carried out by Baracca *et al.* (1976). If the maximum value of these multichotomic observables is M_α (M_β) for those belonging to particle α (β), then the inequalities (5.34) must be replaced by

$$|\bar{A}_k| \leqslant M_\alpha, |\bar{A}'_k| \leqslant M_\alpha, |\bar{B}_k| \leqslant M_\beta, |\bar{B}'_k| \leqslant M_\beta. \tag{5.39}$$

As a consequence the (generalized) Bell's inequality becomes

$$|P(a, b) - P(a, b')| + |P(a', b) + P(a', b')| \leqslant 2M_\alpha M_\beta. \tag{5.40}$$

It has been possible to prove also in the present case that state vectors of the second type lead to violations of the latter inequality if the observables are suitably chosen (Baracca *et al.*, 1976; Bergia and Cannata, 1982).

The possibility of mixtures of *nonfactorizable* state vectors for dichotomic observables has been discussed by Horlick-Jones (1981), who found *rare* violations of Bell's inequality if the mixtures are generated at random with the help of a computer.

Finally it should be stressed that Capasso *et al.* (1973) could prove that all state vectors of the second type lead to violation of Bell's inequality, at least for two-dimensional Hilbert spaces.

5.7. The EPR Paradox: Quantum Mechanics not Complete

In the present section we shall carefully repeat the EPR argument by applying it to the singlet state of two spin-$\frac{1}{2}$ objects and showing that it leads to the validity of Bell's inequality, *even if one drops the assumption of completeness.*

Our starting point is the rotational invariance of the singlet state vector, that is, Equation (5.22) above, valid for an arbitrary orientation of the unit vector $\hat{\mathbf{n}}$. Consider a large ensemble E composed of N decays $\varepsilon \to \alpha + \beta$ of the system ε into the two spin-$\frac{1}{2}$ objects α and β and assume that the final pair (α, β) is described quantum mechanically by η_S. Use of this state vector implies, as we have seen, that, if O_α measures $\boldsymbol{\sigma} \cdot \hat{\mathbf{n}}$ and finds +1 (−1), then a subsequent measurement of $\boldsymbol{\tau} \cdot \hat{\mathbf{n}}$ performed by O_β will give −1 (+1) with certainty.

The EPR paradox can then be formulated as follows (the first five points duplicate corresponding points in the previous section, the only change being the observable considered):

1. On the systems α of an ensemble $E_1 \subset E$, containing, say, ℓ (α, β) pairs, assume that O_α measure $\boldsymbol{\sigma} \cdot \hat{\mathbf{n}}$ and find the results $A_1, A_2, \ldots, A_\ell$, all of which equals ± 1.

2. *Correlation.* The results of subsequent measurements of $\boldsymbol{\tau} \cdot \hat{\mathbf{n}}$ performed by O_β on the ℓ systems β of E_1 can then be predicted *with certainty* to be $-A_1, -A_2, \ldots, -A_\ell$. This prediction can obviously be checked, e.g., on an ensemble $E_2 \subset E_1$. If quantum mechanics is correct, as we assume, every conceivable check will always give a positive answer.

3. *Reality.* The certainty arrived at in the foregoing step allows us to apply the EPR reality criterion and to attribute to β an element of reality t_n which determines the result of a future measurement of $\boldsymbol{\tau} \cdot \hat{\mathbf{n}}$ on β for all $\beta \in E_2$.

4. *Time arrow.* Assuming that t_n is not generated retroactively in time by the future measurement on β, we conclude that t_n belongs to all $\beta \in E_1$.

5. *Separability.* The singlet state remains the same at arbitrary distances separating α and β and is time independent. Therefore, if the distance is very large, we assume that the element of reality t_n of β cannot have been created by the measurement performed on α. Consequently, t_n can be attributed to β for all $\beta \in E$.

6. *Lack of completeness.* Assuming that quantum mechanics is *not* complete, we need not describe the element of reality with a state vector. For our present purposes it is enough to have concluded that such an element of reality exists for all $\beta \in E$.

7. *Different elements of reality.* The reasoning of points 1 through 5 has led to the conclusion that t_n belongs to β even if no measurement on either α or β is carried out. In this way *the assignment of an element of reality has been completely disentangled from all acts of measurement.* Given the symmetry of the problem in α and β and given the invariant nature of η_S under rotations, this allows us to introduce an arbitrary number of elements of reality for α and β. In particular, one can attribute to α two elements of reality s_a and $s_{a'}$ corresponding to the observables $\boldsymbol{\sigma} \cdot \hat{\mathbf{a}}$ and $\boldsymbol{\sigma} \cdot \hat{\mathbf{a}}'$, respectively. Simultaneously, one can attribute to β two elements of reality t_b and $t_{b'}$ corresponding to the observables $\boldsymbol{\tau} \cdot \hat{\mathbf{b}}$ and $\boldsymbol{\tau} \cdot \hat{\mathbf{b}}'$, respectively. This can be done for all pairs (α, β) of E. See Figure 5.3.

8. *Values of the observables.* As seen above, every observable can be used for the discovery of a corresponding element of reality *belonging to the measured system*: Therefore, the symbols a, a', b, b' for the parameters of the experimental apparatus have been attached to the elements of reality only to recall the observables whose values they fix and which allowed their

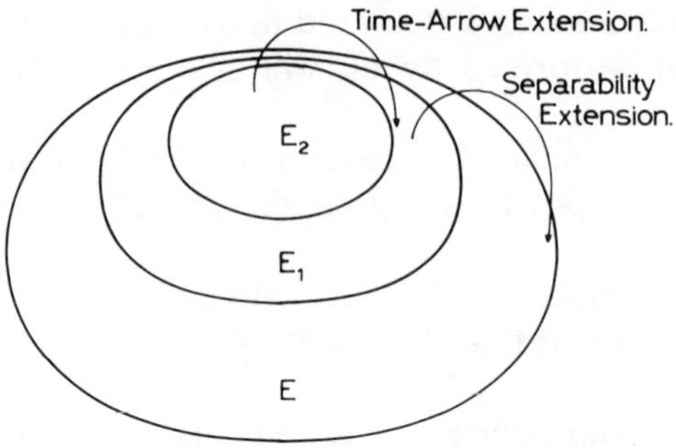

Fig. 5.3. *Structure of proof of the EPR paradox.* E_2: Set of (α, β) pairs for which both $\sigma_n(\alpha)$ and $\sigma_n(\beta)$ are measured; the EPR reality criterion attributes an element of reality to all α's of this set. E_1: Set of (α, β) pairs for which only $\sigma_n(\beta)$ is measured; the 'time-arrow' assumption attributes an element of reality to all α's of this set. E: Set of all (α, β) pairs for which, in general, neither $\sigma_n(\alpha)$ nor $\sigma_n(\beta)$ are measured; the separability assumption extends the element of reality to all α's of this set.

discovery. It could even happen that we are always discovering the same element of reality, which fixes all the observables. This would only mean that our notation is redundant, not that it is wrong! Let us introduce the following new symbols:

$$\begin{array}{lll} \Sigma(s_a), & \text{the measured value of the observable} & \boldsymbol{\sigma}\cdot\hat{\mathbf{a}}, \\ \Sigma(s_{a'}), & \text{the measured value of the observable} & \boldsymbol{\sigma}\cdot\hat{\mathbf{a}}', \\ T(t_b), & \text{the measured value of the observable} & \boldsymbol{\tau}\cdot\hat{\mathbf{b}}, \\ T(t_{b'}), & \text{the measured value of the observable} & \boldsymbol{\tau}\cdot\hat{\mathbf{b}}'. \end{array} \tag{5.41}$$

Obviously:

$$\Sigma(s_a) = \pm 1, \Sigma(s_{a'}) = \pm 1, T(t_b) = \pm 1, T(t_{b'}) = \pm 1, \tag{5.42}$$

since our observables are dichotomic with values ± 1. One can see from these equalities something that should have been clear already from the way the elements of reality were introduced. Each of them is a *variable*, assuming at least two values, possibly two groups of values, such that the first one (the second one) determines the value $+1$ (-1) of the corresponding observable.

9. *The correlation function.* The statistical ensemble E of N pairs (α, β) can be subdivided into subensembles having fixed values of all our four elements of reality. Calling

$$n(s_a, s_{a'}, t_b, t_{b'})$$

the number of pairs with fixed values of s_a, $s_{a'}$, t_b, $t_{b'}$ one must obviously have

$$\sum n(s_a, s_{a'}, t_b, t_{b'}) = N, \tag{5.43}$$

where the sum is taken over all the possible values of the four variables.

From the definition of the correlation function (average product of results of correlated measurements on α and β), it follows that

$$P(a, b) = N^{-1} \sum n(s_a, s_{a'}, t_b, t_{b'}) \Sigma(s_a) T(t_b), \tag{5.44}$$

where the sum is taken again over all the possible values of the four variables. Expressions similar to (5.44) can easily be written down for the other three correlation functions that enter into Bell's inequality.

10. *Bell's inequality.* From (5.44) and (5.42), it follows easily:

$$|P(a, b) - P(a, b')| \leqslant N^{-1} \sum n(s_a, s_{a'}, t_b, t_{b'}) |T(t_b) - T(t_{b'})|,$$

$$|P(a', b) + P(a', b')| \leqslant N^{-1} \sum n(s_a, s_{a'}, t_b, t_{b'}) |T(t_b) + T(t_{b'})|.$$

On summing these inequalities, while using the obvious relation

$$|T(t_b) - T(t_{b'})| + |T(t_b) + T(t_{b'})| = 2,$$

which is a consequence of (5.42), one obtains Bell's inequality (5.38).

The results of the present section allow us to state the EPR paradox in the following way:

EPR PARADOX (2). *Given a statistical ensemble of pairs of spin-$\frac{1}{2}$ objects described quantum mechanically by the singlet state, Bell's inequality must be violated by more than* 40% *for suitable choices of the experimental parameters. Injecting into the theory the assumption of Einstein locality* (points 3, 4, 5 on p. 211) *and that of lack of completeness of the theory* (point 6), *one concludes instead that Bell's inequality must always be satisfied.*

It is easy to check that the EPR paradox (2), given above, and the EPR paradox (1), given in the previous section, exhibit only one difference, the assumed completeness or lack of completeness of quantum theory. Since the same conclusion has been obtained under opposing assumptions about the completeness of the theory, this can only mean that the latter problem represents a redundant variable, from the strictly logical point of view, which can thus be eliminated. This observation allows us to forget, once and for all, about the question of completeness and to cast the EPR paradox in its final form:

EPR PARADOX. *Given a statistical ensemble of pairs of spin-$\frac{1}{2}$ objects described quantum mechanically by the singlet state, Bell's inequality must be violated by more than* 40% *for suitable choices of the experimental parameters. Injecting into the theory the assumption of Einstein locality* (defined as the set of three ideas: reality, time arrow, and separability), *one concludes instead that Bell's inequality must always be satisfied.*

Formulated in the foregoing manner, the EPR paradox can be viewed as a theorem of incompatibility between Einstein locality and the empirical predictions of quantum theory. This means either that quantum mechanics is incorrect or that the world cannot be analyzed in terms of separable elements of reality. To get rid of the EPR paradox — which is an absolute logical necessity — one is forced to choose only one of these alternatives. Denying the existence of separable elements of reality would amount to abandoning with immense consequences for our world view, one of the most important strongholds of natural philosophy. On the other hand, inferring that quantum mechanics is wrong would mean that a scientific revolution in fundamental physics were required. Thus either choice between the two alternatives posed by modern physics would amount to a revolutionary development. Moreover, such a choice can be made on the basis of experimental physics by measuring the correlation functions appearing on the left-hand side of Bell's inequality.

The proof of Bell's inequality given above is similar, albeit not identical, to the proof found by Wigner (1970). Another important point is that no particular interpretation of quantum theory has been adopted in this or in the previous section. All the interpretations that agree on the empirical predictions used in the previous discussions (essentially: correlation of spin values for the two particles and violations of Bell's inequality) are therefore incompatible with Einstein locality. Thus the opinion expressed by Margenau (1936) that the EPR paradox is only a question of interpreting the theory correctly appears untenable.

At this point, no better comment can be found than the one made by Shimony:

> I hope that the rigor and beauty of the argument of EPR is apparent. If one does not recognize how good an argument it is — proceeding rigorously from premises which are thoroughly reasonable — then one does not experience an adequate intellectual shock when one finds out that the experimental evidence contradicts their conclusion. The shock should be as great as the one experienced by Frege when he read Russell's theoretical paradox and said, "Alas, arithmetic totters!" (Shimony, 1978.)

Comments on the experimental situation will be made in future sections of this and of the following chapter, and it will be seen that the situation for Einstein locality is not at all as bad as is usually thought.

An interesting consequence of Einstein locality has recently been found by Garg and Mermin (1982) who were able to deduce Bell-type inequalities for two particles of arbitrary spin-j. They could also prove that the singlet state for two spin-j particles leads to violations of Einstein locality for arbitrarily large values of j, right up to and beyond the threshold of the classical world, in which we know that it is possible to assign *a priori* well-defined values of all the observable quantities. This result of Garg and Mermin is bad news for the coherence and rationality of existing quantum theory, which thus seems to extend its 'magic' predictions also into the macroscopic domain, where classical physics has successfully ruled without magic approaches!

We conclude the present section with a discussion of a macroscopic situation that violates the Bell inequality (Aerts, 1982). In place of the two spin-$\frac{1}{2}$ objects studied before, let us now consider two vessels V_1 and V_2 having the form of cubes of sidelength 20 cm. Each vessel is assumed to contain 8 liters of water, and the two vessels are connected by a tube T containing 16 liters of water, so that the system $V_1 + V_2 + T$ holds 32 liters of water (see Figure 5.4). Two dichotomic observables are defined for each of the two vessels; specifically, we agree to measure the observables

$$A(a) \text{ and } A(a') \text{ on } V_1 \text{ and } B(b) \text{ and } B(b') \text{ on } V_2,$$

with the following definitions:

DEFINITION OF $A(a)$. This measurement checks whether the volume of water that can be extracted from V_1 is more than 10 liters. We perform this experiment by extracting water from V_1 with a siphon and collecting it in a reference vessel V_0. We give the answer 'yes' ($A(a) = 1$) if the water flowing from V_1 to V_0 exceeds 10 liters, otherwise we give the answer 'no' ($A(a) = -1$).

DEFINITION OF $A(\mathrm{a}')$. This measurement checks whether the water is transparent. We take a glass of water from V_1 and hold it up against a light source. If the light gets through, we give the answer 'yes' ($A(a') = 1$), otherwise we give the answer 'no' ($A(a') = -1$).

DEFINITION OF $B(b)$. Here we check whether or not the depth of the water in V_2 is more than 15 cm. If it is, our answer is 'yes' ($B(b) = 1$), if not, our answer is 'no' ($B(b) = -1$).

DEFINITION OF $B(b')$. Here we test whether or not the water is fit to drink. We perform the experiment by taking a spoon of water from V_2 and

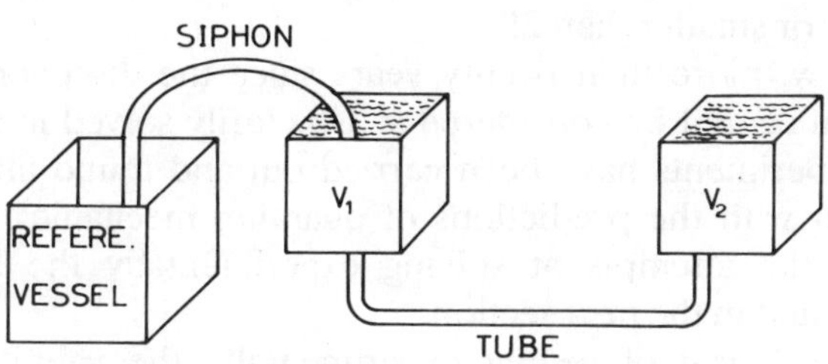

Fig. 5.4. Aerts' example of a macroscopic situation in which Bell's inequality is violated.

drink it. We wait five minutes, and, if not ill by then, we answer 'yes' ($B(b') = 1$), otherwise we give the answer 'no' ($B(b') = -1$).

The order of the experiments is important, and we shall assume that the measurement on V_1 is made first. The key of the idea is that the result of $B(b)$ depends on the measurement made in V_1. If water is extracted, one will of course find $A(a) = +1$, because it is possible to extract 10 liters from V_1, but the level in V_2 will decrease from 20 cm to 10 cm, and one will then find $B(b) = -1$. If instead the transparency of the water in V_1 is checked, the answer will be $A(a') = +1$, but the level in V_2 will have no practical change and one will find $B(b) = +1$.

If we assume that all our water is drinkable, we shall of course find $B(b') = +1$, so that the correlation functions have the following values:

$$P(a, b) = A(a)\,B(b) = (+1)(-1) = -1,$$

$$P(a, b') = A(a)\,B(b') = (+1)(+1) = +1,$$

$$P(a', b) = A(\text{a}')\,B(b) = (+1)(+1) = +1,$$

$$P(a', b') = A(a')\,B(b') = (+1)(+1) = +1.$$

Substitution of these results into (5.29) shows that the left-hand side assumes the value 4, in violation of Bell's inequality. This result is of course a consequence of the nonlocal nature of the apparatus considered. Without the tube T connecting the two vessels, there would not be any violation of the inequality!

5.8. From Theory to Practice

In the two previous sections the EPR paradox has been reduced to a very simple conceptual alternative: either the Bell inequality is always true in nature and the presently existing quantum theory must be modified, or the Bell inequality is violated and Einstein locality has to be rejected. Clearly then, one must only measure the value of the left-hand side of the Bell inequality in a situation for which quantum theory predicts its violation and find out whether it is larger or smaller than 2!

Unfortunately, even now, more than twenty years since the discovery of the inequality, the problem cannot be considered satisfactorily solved in spite of the fact that several experiments have been carried out and found almost always to be in agreement with the predictions of quantum mechanics. The very instructive story of the attempts at solving experimentally the EPR paradox will be told in this and in the next section.

At first sight, a practical way of testing experimentally the validity of (5.38) might appear to be the following. First of all *we build* the source in such a way that the decays $\varepsilon \rightarrow \alpha + \beta$ lead to emissions from a screen surrounding the physical source only when the system α (β) flies to the right (to the left), where a first (a second) *two-channel analyzing apparatus* can direct it either on a first or on a second channel, depending on its physical properties. The dichotomic choice forced in this way upon the atomic systems can then be used to define Bell's dichotomic observables by saying that $A(a) = \pm 1$ ($B(b) = \pm 1$), depending on the channel, first or second, chosen by the system α (β).

When the problem of *detecting* which channel was chosen by either system is considered, one discovers however that detectors are often not very efficient. Therefore, the logical element of certainty, so essential for the development of the EPR paradox, is lost as soon as one goes from the world of theory to that of practical experimental research. Of course this could have been anticipated: no absolute certainty is ever likely to be achieved in any real experiment!

On thinking more carefully, however, all we need is not perfection but reasonably good accuracy, since the discrepancy expressed in the EPR paradox is between the empirical predictions $\Delta \leqslant 2$ and $\Delta = 2\sqrt{2}$, which quantities are different by about 41%! Even 'good accuracy' is difficult to obtain, unfortunately.

In 1969 Clauser, Horne, Shimony and Holt (CHSH) suggested the use of pairs of optical photons coming from atomic cascades. For such photons they *assumed that the binary choice was the one between transmission and absorption by a polarizer.* For every choice of the polarizers' orientations a and b, they introduced four probabilities $T(a_\pm, b_\pm)$, where $T(a_+, b_-)$, for instance, is the probability that observer O_α will find $A(a) = +1$ (photon α transmitted through polarizer with axis a) and that O_β find $B(b) = -1$ (photon β absorbed by polarizer with axis b). The correlation function can then be written

$$P(a, b) = T(a_+, b_+) - T(a_+, b_-) - T(a_-, b_+) + T(a_-, b_-), \tag{5.45}$$

since the product of the results obtained by O_α and O_β is $+1$ [-1] in the cases of $T(a_+, b_+)$ and $T(a_-, b_-)$ [in the cases of $T(a_+, b_-)$ and $T(a_-, b_+)$]. See Figure 5.5 for the experimental set up.

The sum of the probabilities of the four possible destinies of the pairs of photons must of course be 1:

$$T(a_+, b_+) + T(a_+, b_-) + T(a_-, b_+) + T(a_-, b_-) = 1. \tag{5.46}$$

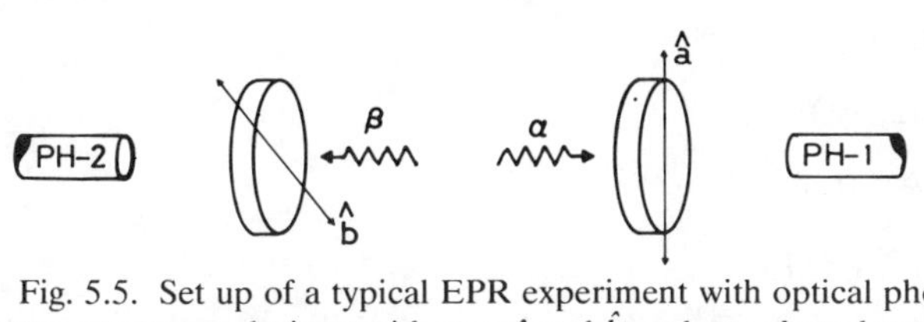

Fig. 5.5. Set up of a typical EPR experiment with optical photons (α and β). There are two one-way polarizers with axes $\hat{a}$ and $\hat{b}$, and two photodetectors PH-1 and PH-2.

Considering furthermore the case in which the second polarizer has been removed (the symbol ∞ is used to denote this) it is clear that the second photon will be transmitted with certainty through empty space, so that

$$T(a_+, b_+) + T(a_+, b_-) = T(a_+, \infty). \tag{5.47}$$

If instead the first polarizer has been removed one will get, with a similar notation,

$$T(a_+, b_+) + T(a_-, b_+) = T(\infty, b_+). \tag{5.48}$$

Finally, if both polarizers have been removed both photons will certainly be transmitted, so that

$$T(\infty, \infty) = 1. \tag{5.49}$$

Using now (5.46), (5.47) and (5.48) in order to eliminate $T(a_+, b_-)$, $T(a_-, b_+)$ and $T(a_-, b_-)$ from $P(a, b)$, it is a simple matter to show that

$$P(a, b) = 4T(a_+, b_+) - 2T(a_+, \infty) - 2T(\infty, b_+) + 1. \tag{5.50}$$

In the latter expression only cases of double transmission appear, which should be nearer to experimental observations (see Figure 5.5). At this point one must face a very important problem: can one *measure* experimentally the right-hand side of (5.50) with an error of, say, a few percent? The only way to know that a photon has been transmitted through a polarizer is to detect it, but photon detectors have efficiencies of 10—20%, i.e., far below the required accuracy!

This problem has traditionally been 'solved' by means of *ad hoc* assumptions concerning the nature of the detection process. These will be considered critically in the next section and in the following chapter, so that we can momentarily forgo any critical remarks. The additional assumption made by CHSH (Clauser *et al.*, 1969) is, for example:

> *Given that a pair of photons emerge from two regions of space where two polarizers can be located, the probability of their joint detection by two photomultipliers is independent of the presence and of the orientation of the polarizers.*

Denoting with the letter D the probabilities for double detection, one can translate the previous sentence into the following relations

$$\begin{aligned} D(a, b) &= D_0 T(a_+, b_+), \\ D(a, \infty) &= D_0 T(a_+, \infty), \\ D(\infty, b) &= D_0 T(\infty, b_+), \\ D(\infty, \infty) &= D_0 T(\infty, \infty), \end{aligned} \tag{5.51}$$

where D_0 is a positive constant, $D(a, b)$ is the probability for detecting both photons with both polarizers in place and having orientations a and b,

$D(a, \infty)$ is the probability for detecting both photons with the second polarizer removed and the first one oriented along a, and so on.

The rates of double detections depend of course on the number N_0 of photon pairs entering per second into the solid angles defined by the optical apparatuses. Using the letter R to denote rates, one has

$$\begin{aligned} R(a, b) &= N_0 D_0 T(a_+, b_+), \\ R(a, \infty) &= N_0 D_0 T(a_+, \infty), \\ R(\infty, b) &= N_0 D_0 T(\infty, b_+), \\ R_0 &= N_0 D_0 T(\infty, \infty) = N_0 D_0, \end{aligned} \tag{5.52}$$

where $R(\infty, \infty)$ has been called R_0 following the usual notation, and the meaning of the new symbols is obvious.

If the T functions are obtained from (5.52) and substituted into (5.50), one gets

$$P(a, b) = 4\,\frac{R(a, b)}{R_0} - 2\,\frac{R(a, \infty)}{R_0} - 2\,\frac{R(\infty, b)}{R_0} + 1. \tag{5.53}$$

Only coincidence rates enter into (5.53). By virtue of the CHSH additional assumption the correlation function has therefore become measurable!

Equation (5.53) allows us to transform the Bell inequality into a more directly measurable expression. In fact, from (5.38) it is easy to deduce

$$-2 \leqslant P(a, b) - P(a, b') + P(a', b) + P(a', b') \leqslant +2.$$

Substituting in these inequalities expressions of the type (5.53) for the four correlation functions, one obtains

$$-1 \leqslant \frac{R(a, b)}{R_0} - \frac{R(a, b')}{R_0} + \frac{R(a', b)}{R_0} + \frac{R(a', b')}{R_0} - \\ - \frac{R(a', \infty)}{R_0} - \frac{R(\infty, b)}{R_0} \leqslant 0. \tag{5.54}$$

Only coincidence rates enter into the previous inequalities, which can therefore be checked experimentally. Historically, the second of the inequalities (5.54) was obtained in the CHSH paper while the first one first appeared in print in the 1972 letter reporting on the Freedman—Clauser experiment (Freedman and Clauser, 1972).

A useful simplification of (5.54) can be obtained if one accepts two qualitative predictions of quantum theory which contain nothing paradoxical and which can anyway be directly checked in experiments; they are:

1. The prediction that $R_1 \equiv R(a', \infty)$ does not depend on a' and that $R_2 \equiv R(\infty, b)$ does not depend on b.
2. The prediction that every T, D, R function should depend only on the relative angle between the orientations of the polarizers (of course,

when both polarizers are inserted); for example,

$$R(a, b) = R(a - b)$$

and so on.

Adopting these two simplifications, one gets, from (5.54),

$$-1 \leqslant \frac{R(a-b)}{R_0} - \frac{R(a-b')}{R_0} + \frac{R(a'-b)}{R_0} + \frac{R(a'-b')}{R_0} - \frac{R_1}{R_0} - \frac{R_2}{R_0} \leqslant 0. \tag{5.55}$$

The axes of the polarizers can be taken to be coplanar. It is obviously possible to choose them in such a way that

$$a - b = a' - b = a' - b' = \phi, a - b' = 3\phi. \tag{5.56}$$

In fact, it is enough to take clockwise steps of size ϕ from a to b, from b to a', from a' to b', in order to have all the relations (5.56) satisfied. From (5.55) and (5.56), it follows that

$$-1 \leqslant \frac{3R(\phi)}{R_0} - \frac{R(3\phi)}{R_0} - \frac{R_1 + R_2}{R_0} \leqslant 0. \tag{5.57}$$

There is a further interesting step that can be taken. Consider the previous inequalities for $\phi = 22\frac{1}{2}°$ and for $\phi = 67\frac{1}{2}°$, where the maximal quantum mechanical violations take place. One can write, respectively,

$$-1 \leqslant \frac{3R(22\frac{1}{2}°)}{R_0} - \frac{R(67\frac{1}{2}°)}{R_0} - \frac{R_1 + R_2}{R_0} \leqslant 0, \tag{5.58}$$

and

$$-1 \leqslant \frac{3R(67\frac{1}{2}°)}{R_0} - \frac{R(22\frac{1}{2}°)}{R_0} - \frac{R_1 + R_2}{R_0} \leqslant 0. \tag{5.59}$$

The inequalities (5.58) are obvious. Instead, (5.59) are consequences of the fact that three times $67\frac{1}{2}°$ is $202\frac{1}{2}°$, which is physically equivalent to $22\frac{1}{2}°$. From (5.58) and (5.59) one can easily obtain the so-called Freedman inequality

$$\delta \equiv \left| \frac{R(22\frac{1}{2}°)}{R_0} - \frac{R(67\frac{1}{2}°)}{R_0} \right| - \frac{1}{4} \leqslant 0, \tag{5.60}$$

which does not involve R_1 or R_2.

It is perhaps not useless to repeat that all the new results deduced from (5.51) to (5.60) *have become possible only because the CHSH assumption has been made*; without its intervention much weaker inequalities would have been obtained. One should therefore guard against confusing the original Bell inequality with the much stronger inequalities which have now been deduced, and which should more properly be called strong inequalities, as will be done systematically next.

We adopt thus the following definitions:

1. *Weak inequality*: The inequality deduced only under the assumption of Einstein locality and violated by quantum mechanics in the case of nearly perfect instruments.

2. *Strong inequality*: The inequality deduced from Einstein locality *and* from *ad hoc* additional assumptions such as the CHSH hypothesis stated above and the CH hypothesis which will be written down later in this section.

The strong and the weak inequalities can assume different forms. We have shown, for instance, that the weak inequality can be expressed in the more or less equivalent ways (5.54), (5.55), (5.57), and (5.60).

In the previous considerations, *probabilities* have entered in an essential way: for instance, double detection probabilities were introduced in (5.51). This is of course a necessity, since the deterministic scheme considered originally by EPR and by Bell cannot apply to a concrete physical situation. However, the previous deductions were based also on the Bell inequality, of which only a deterministic proof was given.

A more systematic probabilistic approach was proposed by Clauser and Horne (1974) (CH). They started by characterizing pairs of photons with a common variable λ representing their physical state and introduced probabilities for transmsision through the polarizer *and* for detection. Thus: $p(a, \lambda)$ denotes the probability that the system α crosses the polarizer with axis a and that it is subsequently detected; $q(b, \lambda)$ is a similar probability for the system β; $D(a, b, \lambda)$ signifies the probability that both systems α and β will cross their respective polarizers with axes a and b and be detected. Clauser and Horne furthermore proposed that the very definition of the locality condition should be written

$$D(a, b, \lambda) = p(a, \lambda)\, q(b, \lambda). \tag{5.61}$$

It is of course not obvious that this definition should exhaust all possible forms of locality, but this important problem will not be discussed here, since it will receive considerable attention in the next chapter.

Supposing that the variable λ can vary in the set Λ with density $\rho(\lambda)$, it is possible to write

$$p(a) = \int d\lambda\, \rho(\lambda) p(a, \lambda),$$

$$q(b) = \int d\lambda\, \rho(\lambda) q(b, \lambda), \tag{5.62}$$

$$D(a, b) = \int d\lambda\, \rho(\lambda) p(a, \lambda)\, q(b, \lambda)$$

for the ensemble probabilities expressed as weighted averages of the individual probabilities.

In order to deduce inequalities from their definition of locality Clauser and Horne considered the following simple algebraic theorem. Given the six real numbers x, x', y, y', X, Y, such that

$$0 \leqslant x, x' \leqslant X, 0 \leqslant y, y' \leqslant Y,$$

one must always have

$$-XY \leqslant xy - xy' + x'y + x'y' - x'Y - Xy \leqslant 0. \tag{5.63}$$

The proof of (5.63) is straightforward, since the intermediate quantity in (5.63) is linear in each of the four variables x, x', y, y' so that its extremes must be looked for on the boundary of the variables. The boundary is obviously defined by 16 points which are analyzed separately in Table 5.1, wherein we have set

$$U = xy - xy' + x'y + x'y' - x'Y - Xy. \tag{5.64}$$

We have thus proved that (5.63) holds without exception.

That inequality one can now apply to the EPR paradox by making the identifications

$$\begin{aligned} x &= p(a, \lambda), \\ x' &= p(a', \lambda), \\ y &= q(b, \lambda), \\ y' &= q(b', \lambda). \end{aligned} \tag{5.65}$$

Table 5.1

x	x'	y	y'	U
0	0	0	0	0
X	0	0	0	0
0	X	0	0	$-XY$
0	0	Y	0	$-XY$
0	0	0	Y	0
X	X	0	0	$-XY$
X	0	Y	0	0
X	0	0	Y	$-XY$
0	X	Y	0	$-XY$
0	X	0	Y	0
0	0	Y	Y	$-XY$
X	X	Y	0	0
X	X	0	Y	$-XY$
X	0	Y	Y	$-XY$
0	X	Y	Y	0
X	X	Y	Y	0

If one substitutes these quantities into Equation (5.63), multiplies the result by $\rho(\lambda)$, and integrates over λ, the outcome is

$$-XY \leqslant D(a,b) - D(a,b') + D(a',b) + \\ + D(a',b') - p(a')Y - Xq(b) \leqslant 0. \qquad (5.66)$$

What is the correct value of X and Y in the inequalities (5.66)? The straightforward answer is of course $X = Y = 1$, since the probabilities in the right-hand sides of (5.65) could reach the value 1 for some value of λ. This leads to the inequality of the strong type (no additional assumption)

$$-1 \leqslant D(a,b) - D(a,b') + D(a',b) + \\ + D(a',b') - p(a') - q(b) \leqslant 0, \qquad (5.67)$$

which could also be deduced directly from Bell inequality (5.38).

The problem with the latter inequality is the usual one: for real detectors it is not violated by the quantum mechanical predictions. To overcome this unpleasant situation, Clauser and Horne (1974) proposed to use the following additional hypothesis:

> *For every photon in the state λ, the probability of a detection with a polarizer in place on its trajectory is less than or equal to the detection probability with the polarizer removed.*

In practice, this new assumption is equivalent to the following four inequalities

$$\begin{aligned} p(a,\lambda) &\leqslant p(\infty,\lambda), \\ p(a',\lambda) &\leqslant p(\infty,\lambda), \\ q(b,\lambda) &\leqslant q(\infty,\lambda), \\ q(b',\lambda) &\leqslant q(\infty,\lambda), \end{aligned} \qquad (5.68)$$

where the symbol ∞ indicates, as before, that the polarizer has been removed. The new relations (5.68) allow us to use (5.63) and (5.65) with

$$X = p(\infty,\lambda),\ Y = q(\infty,\lambda). \qquad (5.69)$$

Substituting Equations (5.65) and (5.69) into (5.63), multiplying the result by $\rho(\lambda)$, and integrating over λ, one obtains

$$-D(\infty,\infty) \leqslant D(a,b) - D(a,b') + D(a',b) + \\ + D(a',b') - D(a,\infty) - D(\infty,b) \leqslant 0. \qquad (5.70)$$

This is an inequality of the strong type, deduced with the help of additional assumptions and much stronger than the weak-type inequality (5.67). In fact, (5.70) contains only double-detection probabilities, and the quantum mechanical predictions can be shown to violate this equation for suitable choices of the polarizer axes. Since the ratio of double-detection probabilities coincides with the corresponding ratio of detection rates, one can easily demonstrate that (5.70) agrees with (5.54), from which observation it follows that all the results deduced in the CHSH approach from (5.54) are valid also in the present CH approach.

Up to the present point we have considered applications of the EPR paradox to experiments in which a *polarizer* is placed on the trajectory of each photon. This is however not a very convenient configuration, since the dichotomic choice is between the transmission of the photon and its absorption. Now, of course, an absorption cannot be detected, and a considerable amount of information is therefore lost inside the polarizer, where the photon absorption takes place. A better experiment would be one in which a truly binary choice is made and where the two alternatives are both detectable.

In 1981 Garuccio and Rapisarda (GR) studied an experiment in which a piece of calcite, monitored by two detectors following the ordinary and extraordinary rays, respectively, was used as analyzing system for each one of the two photons. The apparatus which they considered is shown in Figure 5.6.

While the theoretical approach of Garuccio and Rapisarda was similar to that of Clauser and Horne, with the variable λ and with factorizable probabilities, an important difference is that one must now deal with four simultaneously measurable coincidence rates. Denoting a photon detection on the ordinary or the extraordinary ray with a_+ and a_-, respectively, one has, instead of (5.62),

$$D(a_\pm, b_\pm) = \int d\lambda\, \rho(\lambda) p(a_\pm, \lambda) q(b_\pm, \lambda), \tag{5.71}$$

where $p(a_+, \lambda)$ is the probability that the photon α emerges and is detected in the ordinary beam when the axis of the calcite has the orientation a, and so on.

Garuccio and Rapisarda proposed an entirely new definition of a correla-

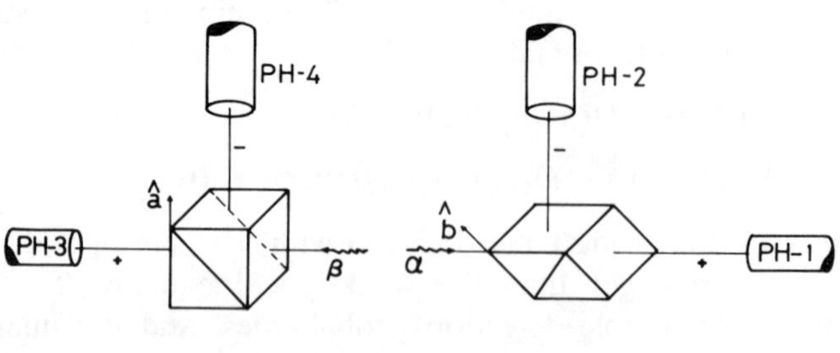

Fig. 5.6. Set up of the Rapisarda experiment. There are two two-way polarizers with axes $\hat{a}$ and $\hat{b}$ and four photodetectors PH-1 . . . PH-4.

tion function, based on all the available experimental information, by writing

$$E(a, b) \equiv \frac{D(a_+, b_+) - D(a_+, b_-) - D(a_-, b_+) + D(a_-, b_-)}{D(a_+, b_+) + D(a_+, b_-) + D(a_-, b_+) + D(a_-, b_-)}. \tag{5.72}$$

Substituting (5.71) into the latter expression, one easily finds

$$E(a, b) = \frac{\int d\lambda\ \rho(\lambda) f(a, \lambda) g(b, \lambda)}{\int d\lambda\ \rho(\lambda) F(a, \lambda) G(b, \lambda)}, \tag{5.73}$$

where

$$\begin{aligned} f(a, \lambda) &\equiv p(a_+, \lambda) - p(a_-, \lambda), \\ g(b, \lambda) &\equiv q(b_+, \lambda) - q(b_-, \lambda), \\ F(a, \lambda) &\equiv p(a_+, \lambda) + p(a_-, \lambda), \\ G(b, \lambda) &\equiv q(b_+, \lambda) + q(b_-, \lambda). \end{aligned} \tag{5.74}$$

The problem with the definition (5.72) is again that no inequality violated by quantum theory can be obtained for actually feasible experiments. As a consequence, GR were also forced to introduce an additional assumption, which can be formulated as follows:

> *For every photon in the state λ, the sum of the detection probabilities in the 'ordinary' and in the 'extraordinary' beams emerging from a two-way polarizer does not depend on the polarizer's orientation.*

The practical implications of this assumption is that the function F does not depend on a, the function G does not depend on b, and the denominator of (5.73) does not depend on either a or b. A better notation is therefore the following:

$$\begin{aligned} F(\lambda) &= p(a_+, \lambda) + p(a_-, \lambda), \\ G(\lambda) &= q(b_+, \lambda) + q(b_-, \lambda), \\ H_0 &= \int d\lambda\ \rho(\lambda) F(\lambda) G(\lambda). \end{aligned} \tag{5.75}$$

The foregoing simplifications allow one to obtain new inequalities, of the CHSH type, which are violated by the quantum mechanical predictions for the case of real experiments. It is a simple matter to show that

$$\begin{aligned} &|E(a, b) - E(a, b') + E(a', b) + E(a', b')| \\ &\quad \leqslant H_0^{-1} \int d\rho\, \{|f(a, \lambda)|\, |g(b, \lambda) - g(b', \lambda)| + \\ &\qquad + |f(a', \lambda)|\, |g(b, \lambda) + g(b', \lambda)|\}, \end{aligned}$$

whence, using the obvious inequalities

$$|f(a,\lambda)| \leqslant F(\lambda),$$
$$|f(a',\lambda)| \leqslant F(\lambda),$$
$$|g(b,\lambda)| \leqslant G(\lambda),$$
$$|g(b',\lambda)| \leqslant G(\lambda),$$

one obtains

$$|E(a,b)-E(a,b')+E(a',b)+E(a',b')| \leqslant 2, \tag{5.76}$$

as a consequence of the fact that any two numbers g and g' that satisfy $|g| \leqslant G$ and $|g'| \leqslant G$ must also satisfy $|g-g'|+|g+g'| \leqslant 2G$. Garuccio and Rapisarda could show that the quantum mechanical predictions for the left-hand side of (5.76) did in fact imply large violations of this inequality, since values as large as 3 can be deduced for suitable choices of the parameters a, a', b, b'.

5.9. The Experimental Information

Actual experiments on the validity of the strong inequality have almost always been carried out with photons. The quantum mechanical treatment of photon polarization is similar to that of spin-$\frac{1}{2}$ in one important respect: both observables are dichotomic. In the case of photon polarization this property is actually due to the absence of a photon mass, a fact which has the practical effect of eliminating from the theoretical scheme the photons with longitudinal polarization. Only photons whose linear polarization is perpendicular to the direction of propagation are therefore left, a situation similar to that of classical electromagnetic waves whose transverse nature is well known. Since also states of circular polarization can be considered, we define

$|R\rangle$ state for right-handed circular polarization,

$|L\rangle$ state for left-handed circular polarization,

$|x\rangle$ state for linear polarization along the x axis,

$|y\rangle$ state for linear polarization along the y axis.

These two sets of states are not unrelated. Elementary textbooks on quantum mechanics show that

$$|R\rangle = \frac{\{|x\rangle + i|y\rangle\}}{\sqrt{2}},$$
$$|L\rangle = \frac{\{|x\rangle - i|y\rangle\}}{\sqrt{2}}. \tag{5.77}$$

The existence of dichotomic observables for photons has the practical effect that Bell-type inequalities can be formulated also for correlated photons. Correspondingly, there are also situations where quantum theory gives a description of the polarization of two correlated photons in terms of nonfactorizable state vectors, analogous to the singlet state of two spin-$\frac{1}{2}$ objects, which imply violations of Einstein locality.

In the case of photons the parity quantum number plays an important role and it is necessary to distinguish, for instance, the $J^P = 0^+$ from the $J^P = 0^-$ state, represented respectively by the state vectors

$$
\begin{aligned}
|0_+\rangle &= \frac{\{|R_\alpha\rangle|R_\beta\rangle + |L_\alpha\rangle|L_\beta\rangle\}}{\sqrt{2}}, \\
|0_-\rangle &= \frac{\{|R_\alpha\rangle|R_\beta\rangle - |L_\alpha\rangle|L_\beta\rangle\}}{\sqrt{2}}.
\end{aligned}
\tag{5.78}
$$

These states can also be expressed in terms of linear polarization states by using relations (5.77) both for photon α and photon β. One obtains

$$
\begin{aligned}
|0_+\rangle &= \frac{\{|x_\alpha\rangle|x_\beta\rangle + |y_\alpha\rangle|y_\beta\rangle\}}{\sqrt{2}}, \\
|0_-\rangle &= \frac{\{|x_\alpha\rangle|y_\beta\rangle - |y_\alpha\rangle|x_\beta\rangle\}}{\sqrt{2}},
\end{aligned}
\tag{5.79}
$$

The basis states with respect to which the polarization is expressed are arbitrary. Using the rotated x' and y' axes one obtains results identical to (5.79) for both states, with x', y' in place of x, y. This property expresses the invariance of zero angular momentum states for rotations about the z axis.

The reasoning leading to the EPR paradox could easily be repeated for the above two photon states, starting for instance from the observation that elements of reality corresponding to polarization measurements along the x and x' axes can be assigned to the photon β from measurements performed (or thought to be performed) on the first photon. All the inequalities (of the weak and of the strong type) found in the previous sections clearly apply also to correlated photon pairs, since they were deduced merely from the assumed dichotomic nature of the measured quantities.

Several experiments have been performed in the past fifteen years in order to check the validity of these inequalities. They are collected in Table 5.2, where also the atom used, the type of cascade, and the wavelengths of the emitted photons are shown.

The most widely used cascade is the $(J = 0) \rightarrow (J = 1) \rightarrow (J = 0)$ cascade of calcium. The quantum mechanical predictions following from the

Table 5.2 Experiments to test the strong inequality

Reference	Atom	Cascade	λ_1	λ_2
Freedman—Clauser, 1972	^{40}Ca	$4p^2\,{}^1S_0 \to 4p4s\,{}^1P_1 \to 4s^2\,{}^1S_0$	5513	4227
Holt—Pipkin, 1973	^{198}Hg	$9\,{}^1P_1 \to 7\,{}^3S_1 \to 6\,{}^3P_0$	5676	4047
Clauser, 1976a	^{202}Hg	$9\,{}^1P_1 \to 7\,{}^3S_1 \to 6\,{}^3P_0$	5676	4046
Clauser, 1976b	^{202}Hg	$9\,{}^1P_1 \to 7\,{}^3S_1 \to 7\,{}^3P_0$	5676	4046
Fry—Thompson, 1976	^{200}Hg	$7\,{}^3S_1 \to 6\,{}^3P_1 \to 6\,{}^1S_0$	4358	2537
Aspect *et al.*, 1981	^{40}Ca	$4p^2\,{}^1S_0 \to 4p4s\,{}^1P_1 \to 4s^2\,{}^1S_0$	5513	4227
Aspect *et al.*, 1982a	^{40}Ca	$4p^2\,{}^1S_0 \to 4p4s\,{}^1P_1 \to 4s^2\,{}^1S_0$	5513	4227
Aspect *et al.*, 1982b	^{40}Ca	$4p^2\,{}^1S_0 \to 4p4s\,{}^1P_1 \to 4s^2\,{}^1S_0$	5513	4227
Falciglia *et al.*, 1983	^{40}Ca	$4p^2\,{}^1S_0 \to 4p4s\,{}^1P_1 \to 4s^2\,{}^1S_0$	5513	4227

state $|0_+\rangle$ applying to this case are, for the quantities appearing in (5.57),

$$\frac{R(\phi)}{R_0} = \frac{1}{4}\,\varepsilon^1_+\varepsilon^2_+ + \frac{1}{4}\,\varepsilon^1_-\varepsilon^2_- F_1(\theta)\cos 2\phi, \tag{5.80}$$

$$\frac{R_1}{R_0} = \frac{1}{2}\,\varepsilon^1_+, \quad \frac{R_2}{R_0} = \frac{1}{2}\,\varepsilon^2_+, \tag{5.81}$$

where $F_1(\theta)$ is a function of the half-angle θ subtended by the primary lenses representing a depolarization due to noncollinearity of the two photons and where

$$\begin{aligned} \varepsilon^1_\pm &= \varepsilon^1_M \pm \varepsilon^1_m, \\ \varepsilon^2_\pm &= \varepsilon^2_M \pm \varepsilon^2_m. \end{aligned} \tag{5.82}$$

Here ε^1_M (ε^1_m) is the transmittance of the first polarizer for light polarized parallel (perpendicular) to the polarizer axis; similar notation has been used for the second polarizer. All these transmittances are usually very near to the ideal case, with ε^i_M close to unity and ε^i_m close to zero ($i = 1, 2$). The values of these parameters are collected in Table 5.3 for the indicated experiments. The notation used in Table 5.3 is the usual one, in the sense that the index 1 (2) on an optical parameter refers to the first (second) photon of the cascade, with wavelength λ_1 (λ_2) in Table 5.2. We should add two important observations. First, the experiment by Aspect *et al.* (1982a) was of the GR type and used as polarizers two polarizing cubes made of two prisms with suitable dielectric thin films on the sides that are stuck together. They are two-way prisms and work similarly to the calcite prisms considered by GR. For such polarizers the notation of Table 5.3 is not the usual one, and one

Table 5.3 Optical transmittances of the two polarizers

Reference	ε_M^1	ε_m^1	ε_M^2	ε_m^2
Freedman—Clauser, 1972	0.97 ± 0.01	0.038 ± 0.004	0.96 ± 0.01	0.037 ± 0.004
Holt—Pipkin, 1973	0.910 ± 0.001	$< 10^{-4}$	0.880 ± 0.001	$< 10^{-4}$
Clauser, 1976	$\simeq 0.965$	$\simeq 0.011$	$\simeq 0.972$	$\simeq 0.008$
Fry—Thompson, 1976	0.98 ± 0.01	0.02 ± 0.005	0.97 ± 0.01	0.02 ± 0.005
Aspect *et al.*, 1981	0.971 ± 0.005	0.029 ± 0.005	0.968 ± 0.005	0.028 ± 0.005
Aspect *et al.*, 1982a	0.950 ± 0.005	0.007 ± 0.005	0.930 ± 0.005	0.007 ± 0.005
Aspect *et al.*, 1982b	0.96 ± 0.01	0.005 ± 0.005	0.93 ± 0.01	0.007 ± 0.005
Falciglia *et al.*, 1983	0.9095 ± 0.0023	0.0044 ± 0.0002	0.7625 ± 0.0024	0.0041 ± 0.0003

should therefore make the identifications

$$T_1^{\|} = \varepsilon_M^1,\ T_1^{\perp} = \varepsilon_m^1,\ T_2^{\|} = \varepsilon_M^2,\ T_2^{\perp} = \varepsilon_m^2.$$

Similar considerations and identifications hold for the Catania experiment (Falciglia *et al.*, 1983), which uses calcite polarizers.

The second observation is that the experiment by Aspect *et al.* (1982b) used time-varying analyzers and that each one of these switching devices was followed by *two* polarizers in two different orientations. Therefore the optical parameters for this particular experiment should actually be doubled for each photon. In practice, however, it happens that the two 'large' transmittances are identical for each photon and that the two 'small' ones are different by insignificant amounts within the quoted errors. It follows that the parameters shown in Table 5.3 are good for the two polarizers following every switching device.

As one can see from Table 5.2, some experiments used mercury isotopes and cascades of the type $(J = 1) \rightarrow (J = 1) \rightarrow (J = 0)$. Since in the initial atomic state there are three levels with different J_3 values, the state of the emitted photon pair is not unique and the calculation is slightly more complicated than in the case of the calcium cascade. The result is however simple if it is assumed that the three different initial atomic levels are equally populated and incoherent. One then obtains for the $(J = 1) \rightarrow (J = 1) \rightarrow (J = 0)$ cascade

$$\frac{R(\phi)}{R_0} = \frac{1}{4}\,\varepsilon_+^1\varepsilon_+^2 - \frac{1}{4}\,\varepsilon_-^1\varepsilon_-^2 F_2(\theta)\cos 2\phi, \tag{5.83}$$

$$\frac{R_1}{R_0} = \frac{1}{2}\,\varepsilon_+^1,\ \frac{R_2}{R_0} = \frac{1}{2}\,\varepsilon_+^2. \tag{5.84}$$

In (5.83), $F_2(\theta)$ is a new function of the half-angle θ subtended by the primary lenses and represents again the depolarization due to noncollinearity of the two photons.

We next review the published experiments on the EPR paradox which were performed by using atomic photon pairs.

1. *Freedman and Clauser* (1972). In this experiment the $3d4p\ ^1P_1$ state of calcium in a beam was excited by radiation from a deuterium arc lamp. About 10% of the atoms go to the $4p^2\ ^1S_0$ state which is the initial state of the (0, 1, 0) cascade emitting photons of wavelengths 551.3 nm and 422.7 nm. Since the natural calcium used in this experiment was an almost pure sample of the isotope with zero nuclear spin, there was no significant reduction expected in the polarization correlation due to hyperfine structure. On each side of the source the photons were collected and collimated by a lens, then passed through a filter and a linear polarizer to a photomultiplier. Freedman and Clauser used pile-of-plates polarizers each of which was

about 1 m in length and consisted of ten glass sheets inclined nearly at the Brewster angle.

The photomultiplier pulses were fed to a coincidence circuit and coincidence measurements were made for 100s time intervals, the intervals during which all the plates were removed alternating with intervals in which the plates were inserted. The results obtained, as the relative orientation of the transmission axes of the polarizers was varied, were found to be in good agreement with the quantum mechanical predictions. The strong inequality (5.47) ('Freedman's inequality') can be written

$$\eta \leqslant 0.250 \tag{5.85}$$

if

$$\eta = \left| \frac{R(22\tfrac{1}{2}°)}{R_0} - \frac{R(67\tfrac{1}{2}°)}{R_0} \right|. \tag{5.86}$$

The results for $R(22\tfrac{1}{2}°)$ and for $R(67\tfrac{1}{2}°)$, combined with those with both polarizers removed for R_0, gave $\eta = 0.300 \pm 0.008$, in clear violation of Freedman's strong inequality and in agreement with the quantum-mechanical prediction $\eta_{qm} = 0.301 \pm 0.007$.

2. *Holt and Pipkin* (1974). In the second experiment the 567.6 nm and 404.7 nm photons emitted in the (1, 1, 0) cascade of the zero nuclear spin isotope ^{198}Hg of mercury were observed. Since the final cascade level is not the ground state of the atom, no precautions had to be taken to avoid the effects of resonance trapping observed in the Freedman—Clauser experiment (Freedman, 1972). To produce the required radiation, mercury vapor was excited to the $9\,^1P_1$ state by a 100 eV electron beam, both the beam and the vapor being contained in an encapsulated source made from Pirex glass. Calcite polarizers were used. These polarizers have a much better extinction ratio than pile-of-plates polarizers, but the values of ε_M are somewhat low (see Table 5.3).

Experimentally it was found that $\eta = 0.216 \pm 0.013$, a result which disagrees with the quantum mechanical prediction $\eta_{qm} = 0.266$ and clearly does *not* violate the strong inequality. This discrepancy has never been completely explained. Proponents of local realistic theories have suggested (Corchero, 1986) there may be some significance to be attached to the use of calcite polarizers.

3. *Clauser* (1976a). Clauser repeated the Holt—Pipkin experiment using the same cascade but in the ^{202}Hg isotope of mercury. Also, instead of calcite polarizers, he used pile-of-plates polarizers. This experiment gave $\eta = 0.2885 \pm 0.0093$, violating Freedman's strong inequality and in close agreement with the quantum mechanical prediction $\eta_{qm} = 0.2841$.

In an extension of the previous experiment, Clauser (1976b) measured the *circular* polarization correlation by inserting quarter-wave plates between each linear polarizer and the source. The quarter-wave plates were obtained by applying pressure to bars of commercial grade quartz. Assuming ideal quarter-wave plates, quantum mechanics predicts that the zero angular momentum state vectors (5.78) should remain unmodified after the two photons have crossed the plates. Therefore also (5.85) remains a valid form of Freedman's strong inequality. From the experimental results Clauser found $\eta = 0.235 \pm 0.025$ while, taking into account the transmission efficiencies of the polarizers and the assumed lack of stability of the quarter-wave plates, he obtained from the theory $\eta_{qm} = 0.252$ (which almost does not violate (5.78)). Within the limit of experimental errors, these circular-polarization results were in agreement with quantum mechanics, but the agreement was not very satisfactory.

4. *Fry and Thompson* (1976). These authors used the 435.8 nm and 253.7 nm photons emitted in the (1, 1, 0) cascade in the zero nuclear spin isotope ^{200}Hg of mercury (see Table 5.2). The 7 $^3S_1$ state of a mercury beam was populated in a two-step process with electron bombardment excitation of the 6 $^3P_2$ metastable state followed downstream, where all short-lived states had decayed, by absorption of resonant radiation from a tunable dye laser. The laser band width was narrow enough that the ^{200}Hg isotope could be selectively excited. The polarizers used in this experiment were of the pile-of-plates variety, and the magnetic field in the interaction volume was reduced to less than 5 mG.

Since the initial state of the cascade had $J = 1$, it was necessary to take into account the possibility of unequal population of, and coherence between, the initial Zeeman sublevels, which Fry and Thompson did by measuring the polarization of the 435.8 nm radiation. Allowing for such effects and considering the transmission efficiencies of Table 5.3, it was predicted $\eta_{qm} = 0.294 \pm 0.007$, while experiment gave $\eta = 0.296 \pm 0.014$, in agreement with quantum theory but in violation of Freedman's strong inequality.

5. *Aspect et al.* (1981). Aspect, Grangier, and Roger (AGR) used the 551.3 nm and 422.7 nm photons from the (0, 1, 0) cascade of calcium. In their case the calcium atoms were excited to the $4p^2\ ^1S_0$ state by a non-resonant two-photon absorption process using a krypton-ion laser beam of wavelength 406 nm and a dye laser beam tuned to 581 nm, both laser beams being at right angles to the calcium atomic beam emitted from a tantalum oven. The laser beams were focused at the interaction region to provide a source about 60 m in diameter by 1 mm long. The density varied between 3×10^{10} cm^{-3} and 10^{11} cm^{-3}, resulting in cascade rates equal or higher than 4×10^7 s^{-1}. Selective excitation of the ^{40}Ca isotope of calcium prevented the polarization correlation from being reduced by hyperfine-structure effects. The photons from the cascade were analyzed by polarizers of the pile-of-plates type and filters in much the same way as in previous experiments.

The high atomic density of the source generated coincidence rates of up to 100 s^{-1}, allowing a 1% statistical accuracy in only 100 s counting time.

Measuring $R(22\frac{1}{2}°)$, $R(67\frac{1}{2}°)$, and R_0, AGR obtained $\eta = 0.3072 \pm 0.0043$, in agreement with the quantum mechanical prediction of $\eta_{qm} = 0.308 \pm 0.002$ but in violation of Freedman's (strong) inequality by more than 13 standard deviations.

6. *Aspect et al.* (1982a). In 1982 AGR performed an experiment, originally suggested and analyzed by Garuccio and Rapisarda (1981) using two-channel polarizers instead of the previous one-channel pile-of-plates type. Each polarizer was a polarizing cube, built using the properties of dielectric thin films and antireflection coated and was rotatable about the observation axis. The arrangement allowed the quantity $E(a, b)$ defined in Equation (5.59) to be measured directly in a single run, using a fourfold coincidence technique for each of the four relative orientations of the polarizers: (a, b), (a, b'), (a', b) and (a', b').

If the left-hand side of the (strong) inequality (5.63) is called S, AGR found experimentally $S = 2.697 \pm 0.015$, in violation of (5.63) itself but in full agreement with the quantum mechanical prediction $S_{qm} = 2.70 \pm 0.05$.

7. *Aspect et al.* (1982b). In all experiments described so far the transmission axes of the polarizers were held fixed during every set of measurements. Thus, there was the possibility of an exchange of information between the two polarizers with a velocity not exceeding that of light. Such a possibility, although very unlikely, given the known nature of interactions, could be ruled out if the settings of the polarizers were changed in a time shorter than the time of flight of the photons from the source to the polarizers. In the experiment performed by Aspect, Dalibard, and Roger (ADR) an optical switch rapidly redirected the light incident from the source to one of two polarizing cubes on each side of the source. In contrast to the previous experiment only the transmitting channels of the polarizing cubes were used. The switching of the light was obtained by a Bragg reflection from an ultrasonic standing wave in water. The light was completely transmitted when the amplitude of the standing wave was zero and was almost fully deflected through 10 mrad when the amplitude was a maximum. Switching between the two channels occurred about once every 10 ns, and since this time, as well as the 5 ns lifetime of the intermediate level of the cascade, was smaller than L/c (40 ns), where $L = 12$ m was the separation between the two switches and c the speed of light, a detection event on one side and the corresponding change of orientation on the other side were separated by a space-like interval.

In the ADR experiment the coincidence rates were only a few per second, with an accidental background of about 1 per second. If U is the intermediate quantity in the (strong) inequality (5.57), ADR found experimentally $U = 0.101 \pm 0.020$, in clear violation of (5.57) itself but in agreement with the quantum mechanical prediction $U_{qm} = 0.112$.

Although the switching was in practice periodic rather than random, the switches on the two sides were driven by two different generators at different frequencies, and it was *assumed* that they functioned in an uncorrelated way. That this situation can instead hide a conceptual difficulty has been shown by Zeilinger (1986).

Finally it should be noted that some criticism (Selleri, 1984; Pascazio, 1985) of the AGR and ADR experiments has been raised on the grounds that there may have been significant resonance trapping due to the high density of the calcium source. A reply to these criticisms has been given by Aspect and Grangier (1985).

8. *Perrie et al.* (1985). Perrie, Duncan, Beyer, and Kleinpoppen (PDBK) measured for the first time the polarization correlation of the two photons emitted simultaneously by metastable atomic deuterium in a true second-order decay process. Single-photon decay from the $2S_{1/2}$ state of deuterium is forbidden, and the main channel for the de-excitation is by the *simultaneous* emission of two photons which can have any wavelength consistent with energy conservation for the pair. However, because of the absorption in oxygen, the observation window was in practice limited between 185 nm and 355 nm.

In the PDBK experiment a 1 keV metastable atomic deuterium beam of density about 10^4 cm^{-3} was produced by charge exchange, in cesium vapor, of deuterons extracted from a radiofrequency ion source. Electric field pre-quench plates upstream from the observation region allowed the $2S_{1/2}$ component of the beam to be switched on and off by Stark mixing the $2S_{1/2}$ and $2P_{1/2}$ states and, at the end of the apparatus, the beam was fully quenched so that the resulting Lyman-signal could be used to normalize the two-photon coincidence signal. The two-photon radiation was collected and collimated by a pair of lenses and the polarizers were of the pile-of-plates type.

Measuring $R(22\frac{1}{2}°)$, $R(67\frac{1}{2}°)$ and R_0, PDBK obtained $\eta = 0.268 \pm 0.010$, in agreement with the quantum theoretical prediction $\eta_{qm} = 0.272 \pm 0.008$, but in violation of Freedman's strong inequality by slightly more than two standard deviations.

In an extension of the previous experiment (Duncan, 1986) the circular polarization correlation was measured by placing achromatic quarter-wave plates in each detection arm between linear polarizer and source (see Figure 5.7). The obtained results did *not* violate Freedman's inequality and would have disagreed with the quantum theoretical predictions *if* the quarter-wave plates had been assumed to be perfectly achromatic. With the introduction of a considerable degree of achromaticity, PDBK could reconcile their observations with theory, considering also imperfect parallelism of the incoming photons. It is surprising however that the only two measurements of circular polarizations in EPR experiments (the present one and the already discussed Clauser (1976b) experiment) led to results of difficult interpretation.

9. *Hassan et al.* (Duncan, 1988). These authors inserted in the apparatus of the previous experiment a $\lambda/2$ plate in one detection arm between

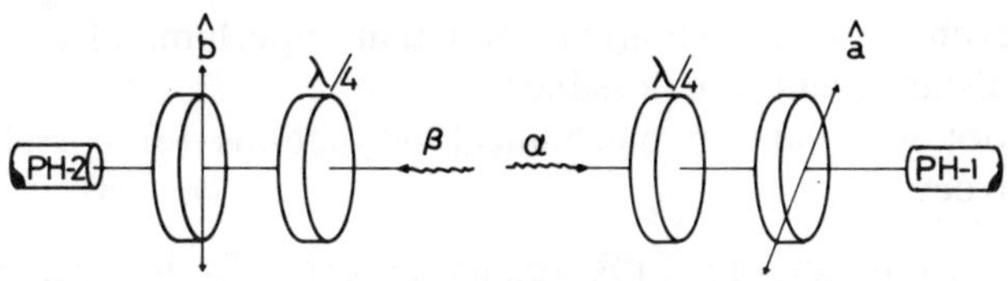

Fig. 5.7. EPR experiment for the detection of circular-polarization correlations. Each photon crosses a $\lambda/4$ plate and a polarizer. Axes of $\lambda/4$ plates must be parallel.

polarizer and photomultiplier. By rotating the fast axis of the half-wave plate through half the angle of rotation of the transmission axis of the linear polarizer, it was possible to ensure that the planes of polarization of the two photons were always parallel just prior to detection in the photomultipliers. The results of this experiment led to $\eta = 0.271 \pm 0.021$, in violation of Freedman's strong inequality and in agreement with the quantum mechanical prediction $\eta_{qm} = 0.272 \pm 0.008$.

Garuccio and Selleri (1984) proposed a mechanism for the violation in nature of the CHSH additional assumption based on the idea of polarization-dependent 'enhanced' detection of the photon pairs. The results of this experiment are inconsistent with the Garuccio–Selleri mechanism, which is therefore ruled out. There remain other possible mechanisms, as shown in the final section of the next chapter, where deterministic local models completely in agreement with quantum mechanical predictions for not too high detector efficiency are reviewed.

In a further experiment an additional linear polarizer was inserted in one arm of the detection system: In this way the first photon crossed *two* polarizers (a, a'), while the second one crossed just one polarizer (b). The orientation of polarizer a was held fixed, while polarizer b was rotated through an angle $a - b$ in a clockwise sense and polarizer a' through an angle $a' - a$ in the opposite sense. The ratio $R(a - b;\, a' - a)/R(a - b;\, \infty)$ was measured, where $R(a - b;\, a' - a)$ is the coincidence rate with all polarizers in place and $R(a - b;\, \infty)$ the coincidence rate with polarizer a' removed. The results are in good agreement with the quantum mechanical predictions, thus indicating once more, that a polarization-dependence of 'enhanced' photon detection was not responsible for violations of the strong inequalities.

5.10. Solution 1: Modifying the Past

We saw that the solution of the EPR paradox must necessarily be looked for in one of the following alternatives:

1. Deny the validity of the EPR criterion of physical reality.
2. Accept signals propagating from the future toward the past.

3. Assume that events are not separable and that superluminal connections between distant events are possible.
4. Assume that not all quantum mechanical predictions for correlated systems are correct.

The simplest way to reject the EPR reality criterion is to accept an idealistic philosophy that denies the existence of an external reality independent of man and of its observations. This is consistent with Bohr's assumption that the purpose of a scientific description of nature is not to disclose the real essence of physical phenomena but only to track down, as far as possible, the relations between different aspects of our experience. The debate on what the task of physics should be goes back to the birth of modern science and can be schematized in two extreme and opposite ways. First there is the optimistic view that science actually allows us to discover the real behavior of phenomena. Galilei, Maxwell, Boltzmann, and Einstein can be considered champions of this view. The second position is positivistically oriented and stresses that the *only* role of science is to create some rational ordering among our otherwise chaotic sense impressions. This current of thought is represented, for instance, by Mach, Ostwald, Jordan and, to some extent, Bohr. That this debate can now be terminated and the second school of thought declared the winner seems to me highly unlikely, considering not only the long history of this discussion, but also the fact that if physics is presently under the influence of Bohr's views rather that Einstein's, there are modern sciences (astrophysics, geophysics, molecular biology, etc.) which have greatly progressed by embracing the optimistic conception of science. This could perhaps be the reason for the little favor that Bohr's proposed solution of the EPR paradox is encountering among contemporary scientists — with rare exceptions (e.g., D. Canals-Frau, 1982).

A second way to reject the EPR criterion of physical reality is to stress that it is unduly strong, as it requires for its very use that predictions *with certainty* be possible. Thus it commits the defender of reality to a choice far more restrictive than simple objective causality. It has however been possible to show (Selleri, 1988) that the reality criterion can be generalized in such a way as to avoid the previous objection and that, nevertheless, the EPR paradox still obtains. Therefore this is not a possible way out of the paradox. These ideas will be discussed in more detail in the next chapter.

In the present section the idea of nonlocality will be discussed by reviewing those proposed solutions of the paradox that deny the validity of point 2 above.

A champion of the idea that retroactions from the future can modify past events is Olivier Costa de Beauregard. He noted that at least twice in prequantum physics contradictions were introduced between *factlike* irreversibility of physical processes and *lawlike* reversibility of the equations of a physical theory. The first time was when Boltzmann deduced the second law of thermodynamics from his statistical mechanics and tried therefore to deduce irreversibility from a time-symmetric theory, such as Newtonian mechanics. This led to the discovery of well-known paradoxes by Loschmidt and by Zermelo. The second time was in classical electrodynamics when the principle of retarded waves was used to exclude one half of the mathematically permissible solutions of the wave equations.

But mathematics should be granted a more important role as a way to discover fundamental properties of the physical world. If all our physical theories are time-symmetric, this must mean that time symmetry must play an objective role in our physical world. Therefore retroactions should be possible. Another fact pointing in the same direction is Aristotle's definition of *information* as a word connoting not only gain of knowledge, as it is understood today, but also, symmetrically, an *organizing power*. The examples given by Aristotle were the craftsman or the artist's work, and also biological ontogenesis. A different way of seeing a *final* cause at work is to consider modern cybernetics which, according to Costa de Beauregard, without having searched for it, has hit upon the intrinsic symmetry of the two facets of Aristotle's conception of information. In computers and other information processing machines the chain

$$\text{information} \overset{\text{(i)}}{\rightarrow} \text{negentropy} \overset{\text{(ii)}}{\rightarrow} \text{information}$$

means that a concept is coded and sent as a message, before being decoded and received. Negentropy is of course entropy with a minus sign. Step (ii) above is the *learning transition*, where information shows up as gain in knowledge, while step (i) is the *willing transition*, where information acts as an organizing power.

In the theoretical framework (*de jure*) there is a complete symmetry between the two transitions. In spite of this there is a dissymmetry in practice (*de facto*) because irreversibility is generated by several factors (misprints in the coding, noise along the line, mistakes in decoding, and so on).

The practical relationship between the variation of negentropy ΔN and the variation of information ΔI is given by

$$\Delta N = (k \ln 2)\Delta I. \tag{5.87}$$

If N and I are both expressed in 'practical' units, it turns out that the factor multiplying ΔI is of the order of 10^{-16}. Therefore, concludes Costa de Beauregard, it is very difficult to produce important increases of negentropy (decreases of entropy!). If one lets $k \rightarrow 0$, one loses completely the second aspect of Aristotle's information and obtains a situation where gaining knowledge is absolutely without cost but producing order is utterly impossible. Consciousness is thus made totally passive; it registers what goes on outside itself, and that is all.

If the roots of Costa de Beauregard's conceptions sink deep into classical physics, it is in quantum theory that he thinks the most important effects of retroaction can be seen. Again, he stresses, the theory is completely time symmetrical, but only until the idea of the *collapse of the wave function* is introduced. At this point quantum theory commits itself to the philosophy of the retarded waves. In Costa de Beauregard's opinion, this happened because 'the Copenhagen school has forgotten the hidden face of Aristotle's information'.

It is precisely in the discussion of the EPR paradox that this 'hidden face' shows up again. In order to understand the essence of the paradox, consider the mathematical apparatus of quantum theory. The problem today is, according to Costa de Beauregard, to tailor the wording of the EPR situation after the mathematics, because the birth of quantum physics has given rise to an irreversible victory of formalism over modelism. From this *starting point* he obtains conclusions upsetting our familiar notions concerning space and time. In fact, he deduces that when an EPR pair, for instance two photons described by one of the state vectors (5.78), is measured by two observers in two widely separated spacetime regions, then it is the very observation that produces, *in the past* of the acts of measurement, the right physical properties of the photon pair. Each observer is thus capable of *telediction* plus *teleaction*, by taking, so to speak, a trip into the past, specifically into the source that emitted the two photons.

The outcome of Costa de Beauregard's theory is that the element of reality introduced in the formulation of the EPR paradox of Section 5.5 is accepted, but that it is viewed as created by *actually performed* future acts of observation. From the discussion carried out in that section, it is clear that Costa de Beauregard's proposal represents undoubtedly a solution of the paradox.

As we saw at the end of Chapter 3, a similar proposal of retroaction, though not devoted to the EPR paradox, was made by Wheeler.

A third author who introduces retroactions in time is Rayski (1985) who discusses well-known quantum mechanical paradoxes, such as de Broglie's box paradox and Schrödinger's cat paradox, to conclude that the quantum mechanical wave function 'does not refer to the physical reality directly, but represents rather our knowledge about it'. In the case of the EPR paradox, Rayski concludes that quantum mechanics is incomplete, i.e., that it does not describe fully the physical reality, because one is pushed to admit that any particle of an EPR pair possesses well defined values of *all* of its spin components. The obvious conclusion at this point would be that Bell's inequality can be expected to be valid in nature, but Rayski avoids discussion of this point. Rather he comments that another and inequivalent way out of the EPR paradox exists if one admits propagation backward in time as if anticausality were on equal footing with the ordinary causality. In this case the action at a distance exhibited in the EPR paradox can be regarded as only apparent, whereas the information obtained from a direct measurement made on the first one of the two widely-separated particles in reality propagates backward in time to the instant of their emission from a common source and — passing from the first to the second particle — propagates together with it forward in time. This point of view is natural, according to Rayski, if one considers that Einstein's special theory of relativity is invariant not only under proper, but also under improper Lorentz transformations, i.e., that it is invariant also under time reversal.

A similar point of view has been expressed by Rietdijk (1978, 1980, 1981), who stressed that the natural idea of reality of the theory of relativity implies that events that are future for one observer may be present — and therefore *already real* — for a second observer moving relative to the first. Since the second observer's description of reality is as good as that of the first, we are pushed to accept a four-dimensional reality of the universe, in which past, present, and future are on an equal footing — given, real, and in all cases 'already existing'. Since some thought experiments proposed by Rietdijk imply that there must be retroactive influences from the 'already existing' future to the present, it is possible to imagine a solution of the EPR paradox along these lines.

The four-dimensional reality of the universe being established, *events* rather than objects become the actual stuff of the universe. This conception must modify also the most relevant metric for describing our world. For, just as it is natural in the case of *objects* to measure distances by means of other objects (measuring rods), so, in the case of *events* or occurrences, is it natural to measure their 'distances' apart by means of units of 'occurrence', that is, by establishing how much 'occurring' or action is needed to get from event B to event A. In this way the quantum of action h naturally appears as the (indivisible) atom of occurrence. One discovers in this way, on analyzing some concrete physical situations, that the 'distance' separating some events, which is considered very large if one uses ordinary three-dimensional spatial distances, can actually be *exactly zero* when measured in terms of action. This holds in particular for some EPR situations. What happens, on considering the action $Et - \mathbf{p} \cdot \mathbf{x}$ as a primary quantity and E, $\mathbf{p}$, t, $\mathbf{x}$ as derived ones, is that one step beyond relativity has been taken: The unification of spacetime and energy-momentum into action (reminiscent of the less specific joining of space and time into spacetime by relativity), leaves the Minkowski scheme merely as a theoretical construct (as earlier was the fate of the Euclidean metric). This outcome can give rise to discrepancies and contradictions if one imposes a Minkowski space description on micro-situations in which the quantum of action becomes important. Precisely such discrepancies give rise to the EPR paradox according to Rietdijk.

As a final comment on proposals regarding retroaction in time, which we have attempted to present above as objectively as possible, is that if, on the one hand, they are quite natural from the perspective of a theoretical physicist used to dealing with formal structures invariant under time reversal, they cannot avoid, on the other hand being somewhat reminiscent of science fiction. But if one insists that the best possible approximation to the truth about nature is *in all cases* given by the presently available theories, then acceptance of retroactions seems almost inescapable. The macroscopic world that we inhabit exhibits irreversibility. This fact is hard to reconcile with purely time-symmetric theories, and I believe that Prigogine is right in stressing that irreversibility should be given a more fundamental role in our theories. However, if this is done, the possibility of introducing retroactions is immediately lost.

5.11. Solution 2: Superluminal Connections

Several authors have considered the possibility of a concrete existence of a certain type of action at a distance as a way out of the EPR paradox (this is the third alternative of Section 5.10).

According to Bohm, the essential new feature implied by quantum theory is nonlocality. A system cannot be analyzed into parts whose basic properties do not depend on the state of the whole system, no matter how 'well separated' in space these parts might appear to be. He believes, with Hiley, that the well-known experiments on Bell's inequality reveal in an especially clear way the nonlocal nature of quantum phenomena (Bohm and Hiley, 1978). However, nonlocality

> is involved in an essential way in every manifestation of a many-body system, as treated by Schrödinger's equation in a $3N$-dimensional configuration space. (Bohm and Hiley, 1978, p. 94.)

In the case of two particles with mass m, Schrödinger's equation is

$$i\hbar \frac{\partial \Psi}{\partial t} = -\frac{\hbar^2}{2m}[\nabla_1^2 + \nabla_2^2]\Psi + V\Psi, \tag{5.88}$$

where

$$\Psi = \Psi(\mathbf{x}_1, \mathbf{x}_2, t) \tag{5.89}$$

is the wave function of the two particles, $V(\mathbf{x}_1, \mathbf{x}_2)$ is the potential acting on them, and ∇_1^2 and ∇_2^2 refer to particles 1 and 2, respectively. Writing

$$\Psi = R \exp\left\{\frac{iS}{\hbar}\right\} \tag{5.90}$$

and

$$R^2 = \Psi^*\Psi, \tag{5.91}$$

one can obtain, from (5.88),

$$\frac{\partial R^2}{\partial t} + \boldsymbol{\nabla}_1 \cdot \left(R^2 \frac{\boldsymbol{\nabla}_1 S}{m}\right) + \boldsymbol{\nabla}_2 \cdot \left(R^2 \frac{\boldsymbol{\nabla}_2 S}{m}\right) = 0 \tag{5.92}$$

and

$$\frac{\partial S}{\partial t} + \frac{(\nabla_1 S)^2}{2m} + \frac{(\nabla_2 S)^2}{2m} + V + Q = 0, \tag{5.93}$$

where

$$Q = Q(\mathbf{x}_1, \mathbf{x}_2, t) = -\frac{\hbar^2}{2m}\left(\frac{\nabla_1^2 R}{R} + \frac{\nabla_2^2 R}{R}\right). \tag{5.94}$$

Evidently, Equation (5.92) describes the conservation of probability with density $R^2 = \Psi^*\Psi$ in the configuration space of the two particles. Equation (5.93) is instead a Hamilton—Jacobi equation for the system of the two particles, acted on not only by the classical potential V but also by the quantum potential $Q(\mathbf{x}_1, \mathbf{x}_2, t)$. There are two strikingly new features of this quantum potential:

(i) In general it does not produce a vanishing interaction between the two particles as $|x_1 - x_2| \to \infty$.
(ii) It cannot be expressed as a universally determined function of the coordinates $\mathbf{x}_1$ and $\mathbf{x}_2$. Rather, it depends on $\Psi(\mathbf{x}_1, \mathbf{x}_2)$ and therefore on the 'quantum state' *of the system as a whole*.

It is the latter feature that brings about the nonlocal nature of quantum phenomena. Bohm has suggested also a sort of ontological model of quantum nonlocality which tries to provide a general framework in which the new phenomena might look less unnatural than it is usually thought. This he has done by introducing the new notion of 'unbroken wholeness' that characterizes two correlated quantum systems (e.g., see Bohm, 1988). He considers the interesting example of a hologram and stresses that the different parts of the object are not in correspondence with different parts of its hologram, but rather each of the latter parts individually is somehow expressing the whole object. Accordingly, if one illuminates any part of the hologram, one gets information about the entire object, even if less detailed and from fewer angles. Similarly, Bohm thinks that what appears to us as two separated quantum objects might in actuality only be a manifestation of a truly interconnected wholeness. The hologram of two spheres, for instance, stores the information of each ball over the entire hologram. It can therefore be said that *in the hologram* the two spheres are really in a way amalgamated and impossible to separate. Bohm views this as an example of the true physical situation giving rise to the EPR paradox: in space there is only an 'unbroken wholeness', which sometimes can give rise to manifestations that *appear* as two separate objects.

As a further example, Bohm considers a device consisting of two concentric glass cylinders, separated by a highly viscous fluid such as glycerine, which is so designed that the outer cylinder can be turned very slowly, thus causing the diffusion of the viscous fluid to be negligible. If one puts a droplet of insoluble ink into this viscous fluid and slowly turns the outer cylinder, the ink will become strung out into a thread that finally becomes invisible. When the rotation of the cylinder is reversed, the ink will suddenly become visible again. Now, it could be said that the ink particles have been *enfolded* into the glycerine, like an egg is folded into a cake. But, while one can in no way unfold the egg, the material of which undergoes irreversible diffusive mixing, it is possible to *unfold* the ink droplet from the glycerine, because in this case the mixing is viscous, diffusion being prevented by the slow turning of the outer cylinder.

Now suppose that, instead of a single ink droplet, one starts out with two spatially separated droplets. When the outer cylinder is turned, each droplet will, as before, become enfolded, until finally particles from both droplets will appear to intermingle at random. But, of course, despite appearances, the ink particles will retain their identification by origin, because they are correspondingly drawn out into two threads that remain separate and distinct, and indeed in such a way that, on reversal of the fluid motions, the threads will unfold to ultimately reconstitute the respective droplets from which they originated. With the help of this analogy, Bohm draws a distinction between what he calls the *enfolded order* and the undisguised, or unfolded order, which constitutes our ordinary description of reality.

In a situation where the threading out of 'droplets', as above, exists, one must make a distinction between wholes according to where the process of unfolding would produce a droplet or whether it would produce two droplets, and so on (different states of the vacuum). It is possible to have situations where what is going to be visible is only a very small part of the enfolded order, and one therefore introduces a distinction between what is manifest and what is not. Reality might enfold and become nonmanifest or it might unfold, to reveal a manifest order, and then enfold again. Bohm claims that, whereas the fundamental movement of Descartes is one that crosses space in time, a localized entity moving from one place to another, his own fundamental movement is marked by folding and unfolding. In this theory it is obviously possible to have interactions between two entities which manifest themselves as two separated objects, and in principle this view could solve the EPR paradox.

A second author willing to abandon separability is Stapp, who believes that the quantum mechanical predictions have been accurately confirmed under experimental conditions essentially equivalent to those needed for the EPR argument. Hence he says that it may now be safely concluded that the world we live in is nonlocal (Stapp, 1985). However, he does not believe that the results obtained by Bell (1965) and/or by Clauser and Horne (1974) are sufficient for establishing the need for nonlocality since, in his opinion, these authors made very strong assumptions about microscopic reality that are not compatible with the orthodox quantum thinking. The denial of these 'strong' assumptions of realism does not imply, however, any retreat to idealism or subjectivism. It is in fact possible to replace them with weaker realistic assumptions that are compatible with an 'informal' Copenhagen interpretation of quantum phenomena. Stapp distinguishes in fact a *strict* Copenhagen interpretation, in which nothing at all is said about any reality other than our observations, from an *informal* interpretation, in which one accepts the common sense idea of a *macroscopic* reality that exists independently of our observations and can be described, at least approximately, with the concepts of classical physics. The 'informal' interpretation is by Stapp at least partly related to Heisenberg's well-known idea of a transition from the 'possible' to the 'actual' taking place during the act of measurement:

> . . . orthodox thinking, at least at the informal level, suggests that objective reality is built upon myriads of macroevents which, rising from a sea of micro-level potentialities, create or actualize attributes that weave together to form the fabric of a macroscopic spacetime reality that is describable in terms of the concepts of classical physics. (Stapp, 1988.)

In this way the microcosm is made only of a chaotic '*sea of micro-level potentialities*'. This failure of the ideal picture of a 'well-defined' microworld, however, does not jeopardize the general idea of scientific realism; for, according to Stapp, one can always fall back on the idea that such a type of reality applies only at the macroscopic level.

A model theory proposed by Stapp contains certain 'hidden variables' d which represent all the deterministic and stochastic quantities that characterize the unified organic world and which are *not* used to provide the basis for a Clauser—Horne factorization structure of probabilities. They do not reflect ideas of separation, localization, or microscopic structure. Stapp writes $d = (d', d'')$, where d' is strictly predetermined and d'' is any stochastic variable.

Furthermore Stapp states that every act of measurement involves a *choice*. This choice 'picks the actual from among what had previously been mere possibilities: the choice renders fixed and settled something that had prior to the choice been undetermined'. A 'choice' variable c is also introduced and written $c = (c_1, c_2)$, where c_1 and c_2 represent the choices of experiment in the regions R_1 and R_2, respectively, where two correlated observations of the EPR type are made. The 'choices' c_1 and c_2 are considered independent free variables. Each of them can assume an infinite number of different values.

Suppose that there are two observables A_1 and B_1 that can be measured in R_1 and other two, A_2 and B_2, that can be measured in R_2. *The choice variable picks one observable before an act of measurement is made*. More precisely, the chosen observable is, in R_1,

$$A_1 \quad \text{if } c_1 \in \Gamma_1(A),$$

$$B_1 \quad \text{if } c_1 \in \Gamma_1(B),$$

where $\Gamma_1 = \Gamma_1(A) \cup \Gamma_1(B)$ is the set of possible values of c_1. Furthermore, in R_2 the chosen observable is

$$A_2 \quad \text{if } c_2 \in \Gamma_2(A),$$

$$B_2 \quad \text{if } c_2 \in \Gamma_2(B),$$

where $\Gamma_2 = \Gamma_2(A) \cup \Gamma_2(B)$ is the set of possible values of c_2.

Depending on the values of c_1 and c_2, there are thus four possible experiments that one can choose to perform in R_1 and R_2, corresponding to the four pairs of observables

$$(A_1, A_2), (A_1, B_2), (B_1, A_2), (B_1, B_2).$$

Now, the results of the measurements of whatever observables have been chosen are assumed to be

$$r_1(c_1, c_2, d) \quad \text{in } R_1,$$

$$r_2(c_1, c_2, d) \quad \text{in } R_2,$$

in a general nonlocal theory, and

$$r_1(c_1, d) \quad \text{in } R_1,$$

$$r_2(c_2, d) \quad \text{in } R_2,$$

if instead locality is assumed. Stapp could easily prove that the local choice contradicts the empirical predictions of quantum theory and concludes:

> The proof given above shows that neither determinism, nor counterfactual definiteness, nor any idea of reality incompatible with orthodox quantum thinking need be assumed in order to prove the incompatibility of rudimentary predictions of quantum theory with the EPR idea that no influence can propagate faster than light. (Stapp, 1988.)

The remark about the absence of a 'counterfactual definiteness' is of course justified by the important fact that, in Stapp's theory, the fixing of c_1 and c_2 and d fixes the value *only of the observable that is actually measured*: the values of the other three observables remain instead completely indefinite.

A detailed discussion of the EPR paradox has been given by Paty (1982). He makes some interesting comments about 'the degree of certainty that one should grant to the conclusion of observations' and points out that researches are conceivable both in the direction of eventual defects of the experimental methods used for testing the inequalities of Einstein locality and in the direction of particular, very specific models that might simulate a quantum-like behavior while saving locality. Paty also states that no experiment can be 'crucial', since it always rests on a complex organization of facts and theories. In spite of this opinion Paty concludes that local models for EPR-type experiments look like alternatives 'without a serious future', since the experiments done after 1970 'have clearly shown' that the choice must be in favor of quantum mechanics and that Bell-type inequalities cannot be true in nature.

A further interesting way of accepting nonlocality has been proposed by Ne'eman (1986), who is convinced that recent experiments on the EPR paradox have provided an 'overwhelming refutation' of the local deterministic alternatives. He also considers recent trends in particle physics as showing, with high probability, that gauge theories could be 'the only physical theories relevant to the physical world'. Furthermore, Ne'eman assumes that gauge theories should be regarded as specific geometric constructs, namely fiber bundle manifolds. A principal fiber bundle

$$\mathscr{P}(\mathscr{M}, G, \pi, \cdot)$$

is a manifold defined by four elements:

(i) a base space $\mathscr{M}$ (spacetime, for instance);
(ii) a structure group G (like $U(1)$ in quantum electrodynamics);
(iii) a 'vertical' projection π,

$$\pi : p(x, \Gamma) \rightarrow x, \quad p \in \mathscr{P}, \quad x \in \mathscr{M}, \quad \Gamma \in G;$$

(iv) the 'right action' $(\cdot)$ of G on $\mathscr{P}$ [defined by three algebraic properties; see Equation (1a) of Ne'eman (1986)].

One can show that the principal fiber bundle $\mathscr{P}$ is a strongly correlated manifold, because the group G and its Lie algebra have *global* characteristics. All *local* algebraic generators are assured to be correlated at all different points $x, x' \in \mathscr{M}$. Therefore, in an important sense, operators are nonlocalized. This can provide, according to Ne'eman, the nonlocal solution of the EPR paradox. What makes the quantum mechanical predictions look 'unreal' would in this view be only our habit to visualize experimental situations in ordinary Galilean space (or, at best, in a Riemannian 4-space), while this should only be considered a *base space* and 'the true arena' of events would consist of fiber bundle manifolds.

A nonlocal solution of the EPR paradox has been proposed also by Rastall (1982), who believes that there is now good experimental evidence that nonlocal systems exist and that they can be described by quantum mechanics. He distinguishes two types of nonlocal influences, those that can be used for transmitting *signals* and those which cannot. The latter type of nonlocal influences are compatible with causality, according to Rastall, who adds the following example. Given two spatially separated regions R_1 and R_2, suppose that a sequence of random numbers is generated in R_1 and another such sequence is generated in R_2. If it turns out that the two sequences are exactly the same, then one concludes that the random-number generators are malfunctioning, or that an improbable coincidence has occurred, or that there is a nonlocal effect. If one checks the random-number generators and repeats the experiment several times, with the same result of two identical sequences, one must conclude that the mysterious correlation between the numbers is indeed a nonlocal effect. However the correlation cannot be used to send signals from one region to the other and the nonlocal influence is not noncausal.

In concluding this section, we shall discuss the nonlocal model of Vigier and collaborators (Vigier, 1979). They adopt the idea, first presented by Dirac (1952), that, in the light of quantum mechanics, the *aether* is no longer ruled out by relativity, provided that the velocity distribution of the particles constituting the aether has a constant value over the hyperboloid

$$v_0^2 - v_1^2 - v_2^2 - v_3^2 = 1.$$

In such a case, in fact, the velocity distribution looks the same to all observers, and the aether does not produce any physical effect on moving bodies. In Vigier's model, this aether-like physical vacuum is made of extended *rigid* particles that can support within their interiors signals with superluminal velocity. The statistical properties of quantum objects (particles) reflect then nothing else than the real random fluctuations of the aether.

This theory is different from the usual approaches based on some analogy between quantum phenomena and Brownian motion, also because the necessary existence of undulatory phenomena is explicitly recognized. There are in fact waves which propagate as real physical collective excitations (i.e., as density waves) on the top of the foregoing thermostat (Dirac's aether).

In this way, information originating on the boundary of the ψ wave (such as the opening or closing of a slit in the double-slit interference experiment or the observation of one of the two particles forming an EPR pair) reacts with superluminal velocity (via the quantum potential) on the particle motions that propagate with subluminal group velocities along the flow lines of the quantum mechanical ψ waves.

The existence of superluminal propagations does not necessarily imply a breakdown of causality if 'causality' is defined as follows:

(i) possibility to solve the two-particle problem in the forward (or backward) time direction as a Cauchy problem;
(ii) timelike nature of all particle trajectories;
(iii) invariance of the formalism under the Poincaré group of transformation.

The problem of causality has however been rediscussed by Cufaro-Petroni (1985) in an interesting and penetrating way from a physical point of view. Consider the following objection that can be raised against a theory, like that of Vigier and collaborators, in which superluminal connections are introduced. Two particles propagate in two widely separated regions of space R_1 (on the earth) and R_2 (in the Andromeda galaxy) and form an EPR pair. Their propagation takes place according to precise deterministic equations containing nonlocal potentials such as, Bohm's potential Q of Equation (5.94). Each particle 'knows' instantaneously what the other particle is doing, and reacts accordingly. It seems natural to conclude that the switching on and off of a magnetic field in R_1 must have instantaneous consequences on the particle located in R_2, because of the superluminal physical connection. The experiment can then be set up in such a way that the second particle enters a detector D_1 (D_2) if the magnetic field in R_1 is off (if it is on). Therefore, the observer on Andromeda can instantaneously learn what his fellow observer is doing on earth. Using ensembles of correlated EPR pairs, it then becomes possible to transmit instantaneous information from R_1 and R_2.

Cufaro-Petroni's answer to this objection is that we live in a completely deterministic world and that it does not make any sense to consider 'modifications' of its properties, such as the one introduced before through the switching on and off of a magnetic field:

> In a completely deterministic world there is no possible 'modification': The world IS and we cannot intervene from exterior in its tissue in order to modify it because we are IN the world. (Cufaro-Petroni, 1985.)

A signal always needs a free will that is external to the physical process considered and which, at a given time, decides to modify the regular evolution of the process in order to send a message. But if instead we assume, with Cufaro-Petroni, that the particles of the human brain are connected to all physical processes, obey the same equations, and therefore behave in a unique and strictly correlated way, then there are no 'signals' at all. At this point a statement made by Spinoza is recalled: "Men are wrong in considering themselves free, and this opinion consists only in the fact that they are conscious of their actions and unaware of the causes which determine them."

While the present writer has no doubt that there is some important truth in Spinoza's words, he nevertheless believes that it is very difficult to deny human beings even the freedom to push a button in order to switch on a field!

5.12. Solution 3: New Definitions of Probability

Given the very fundamental nature of the EPR paradox, it is not surprising that all the conceivable solutions are investigated, even those that might appear 'far fetched'. In the present section those attempted solutions of the paradox will be reviewed that try to introduce some fundamental modification of probability. We shall see that some authors have introduced negative and/or complex probabilities, while others have questioned the ergodic nature of quantum probability.

A 'negative-probability solution' of the EPR paradox has been proposed by Mückenheim (1982). The importance of this proposal is increased by the fact that the idea of negative probability has been entertained at different times by physicists such as Dirac and Feynman. In all cases we should be open to all types of solution of the EPR paradox: After all we live in times where it is normal to study the singularities in the complex angular momentum plane; so, why not negative probabilities?

During the Second World War Dirac expressed the opinion that

> Negative energies and probabilities should not be considered as nonsense. They are well-defined concepts mathematically, like a negative sum of money, since the equations which express the important properties of energies and probabilities can still be used when they are negative. Thus negative energies and probabilities should be considered simply as things which do not appear in experimental results. (Dirac, 1941.)

More recently, Feynman (1982) has stressed that the only difference between a probabilistic classical world and the equations of the quantum world 'is that somehow or other it appears as if the probabilities would have to go negative . . .'.

In order to understand the logical possibility to solve the EPR paradox by extending the range of variation of probabilities, we should remember that in all the proofs of the Bell inequality discussed in this chapter an implicit assumption that probabilities (and frequencies, in ensembles) are positive and not larger than 1 has been made. For example, in the proof given on p. 213 we used the assumption that the populations

$$n(s_a, s_{a'}, t_b, t_{b'})$$

are all positive. It is enough to relax this hypothesis to obtain a theoretical scheme in which the Bell inequality is not generally valid anymore. Similarly, the proof given on pp. 222 and 223 used in an essential way the inequalities

$$0 \leqslant x, x', y, y' \leqslant 1,$$

where x, x', y, y' were later to be identified with probabilities. If the above inequalities are not assumed, the validity of the Bell inequality no longer follows.

In view of these considerations, it is perhaps not surprising that Mückenheim could build a negative-probability local hidden-variable model that could reproduce all the predictions of quantum theory for the 'singlet' state of two spin-$\frac{1}{2}$ particles. Each of the two particles has a spin vector, $\mathbf{S}$ for the first one and $-\mathbf{S}$ for the second one, where $\mathbf{S}$ is assumed to have a random distribution over the sphere of radius $(\sqrt{3}/2)\hbar$ in a statistical ensemble of such pairs. The length $(\sqrt{3}/2)\hbar$ is chosen of course in such a way as to reproduce the quantum mechanical eigenvalue of $\mathbf{S}^2$, which is $3/4\hbar^2$. If $\hat{a}$ is a unit vector, the projection of $\mathbf{S}$ over $\hat{a}$ satisfies

$$-\frac{\sqrt{3}}{2}\hbar \leqslant \mathbf{S}\cdot\hat{a} \leqslant +\frac{\sqrt{3}}{2}\hbar. \qquad (5.95)$$

Next Mückenheim assumes that the probabilities $p_+(\hat{a}, \mathbf{S})$ and $p_-(\hat{a}, \mathbf{S})$ of measuring $\mathbf{S} \cdot \hat{a}$ and finding the positive and the negative eigenvalue, respectively, are linear functions of $\mathbf{S} \cdot \hat{a}$, so that the expressions satisfying

$$p_+(\hat{a}, \mathbf{S}) + p_-(\hat{a}, \mathbf{S}) = 1 \tag{5.96}$$

are given by

$$p_+(\hat{a}, \mathbf{S}) = 0.5 + \frac{\hat{a} \cdot \mathbf{S}}{\hbar} \quad \text{and} \quad p_-(\hat{a}, \mathbf{S}) = 0.5 - \frac{\hat{a} \cdot \mathbf{S}}{\hbar}. \tag{5.97}$$

Obviously these probabilities can assume negative values because of (5.95). In the case of an EPR pair, one can consider the case of correlated spin measurements along $\hat{a}$ and $\hat{b}$ for the first and second particle, respectively. The correlation function is given by

$$P(\hat{a}, \hat{b}) = \frac{1}{4\pi} \int \mathrm{d}\Omega \left[\frac{\hbar}{2} p_+(\hat{a}, \mathbf{S}) - \frac{\hbar}{2} p_-(\hat{a}, \mathbf{S}) \right] \times$$

$$\times \left[\frac{\hbar}{2} p_+(\hat{b}, -\mathbf{S}) - \frac{\hbar}{2} p_-(\hat{b}, -\mathbf{S}) \right].$$

Substituting (5.97) into the previous relation and carrying out the integration one obtains

$$P(\hat{a}, \hat{b}) = -\frac{\hbar^2}{4} \hat{a} \cdot \hat{b},$$

which coincides with the quantum mechanical correlation function for the singlet state. A local model is thus able to reproduce the quantum mechanical violations of the Bell inequality, if negative probabilities are introduced.

One can also show that the introduction of *complex* numbers in the EPR paradox can reconcile Einstein locality with the quantum mechanical predictions (Ivanovic, 1978). In such a case one can in fact deduce an inequality of the type

$$|P(\hat{a}, \hat{b}) - P(\hat{a}, \hat{b}')| + |P(\hat{a}', \hat{b}) + P(\hat{a}', \hat{b}')| \leqslant 2\sqrt{2},$$

which is less strict than the Bell inequality and is furthermore precisely the one satisfied by the quantum mechanical correlation functions.

Another way to 'solve' the EPR paradox is to question some fundamental assumptions that are usually made every time a probability is measured. A nonergodic interpretation of quantum phenomena has been proposed which could in particular lead to a solution of the EPR paradox. The basic idea is that a sequence of particles, even if separated by large time intervals from one another, do not behave independently in their interactions with the measuring apparatus (Buonomano, 1980, 1988).

Consider, for instance, the double-slit experiment. In the nonergodic approach one assumes that a particle passing through a slit knows if the other slit is open, because this information is recorded in the medium filling the space between the two screens: More exactly, it is assumed that the action of other particles coming from the second slit has modified the properties of space and given rise to the storage of the relative information. In this way particles interfere with other particles, but only indirectly through the medium. Furthermore, interference can only happen after a sufficiently large number of particles has crossed the apparatus and 'polarized' the medium. In this way one considers quantum systems simply as classical particles, and interference phenomena are attributed to the action of the medium.

Such an idea can be put to empirical test as follows. Consider a quantum experiment, repeated a large number of times, every repetition being called a 'run'. Let R represent the number of runs and N the number of particles in every run, assumed constant for simplicity. Let furthermore λ_{rn} represent the state of the nth particle in the rth run, and let Λ_{rn} represent the state of the experimental apparatus just before interacting with the nth particle of the rth run. The result Q_{rn} of the measurement is assumed to be completely fixed, once λ_{rn} and Λ_{rn} are given:

$$Q_{rn} = Q_{rn}(\lambda_{rn}, \Lambda_{rn}).$$

Starting from these quantities, three types of average are possible. In fact, consider the following definitions

$$\overline{Q}_r = \frac{1}{N} \sum_{n=1}^{N} Q_{rn},$$

$$\overline{Q}_n = \frac{1}{R} \sum_{r=1}^{R} Q_{rn},$$

$$\overline{Q} = \frac{1}{N} \frac{1}{R} \sum_{n=1}^{N} \sum_{r=1}^{R} Q_{rn}.$$

$\overline{Q}_r$ is called the *run average*, $\overline{Q}_n$ is the *ensemble average* at 'time' n, and $\overline{Q}$ is sometimes called the *grand-time average*.

From the point of view of the nonergodic approach, it is clear that the only possibility for avoiding the medium polarization effects is to keep the runs distant in time from one another and eventually also in different regions of space where no experiments have been previously carried out. Thus the ensemble average for $\overline{Q}_{n=1}$ should represent events collected under conditions where the medium does *not* act on the particles: There cannot be memory effects for $N = 1$, since no previous particles entered the apparatus in each one of the considered runs! Therefore $\overline{Q}_{n=1}$ should describe a situation in which no quantum phenomena show up and classical physics holds unreservedly. By contrast, $\overline{Q}_n$ for large n, $\overline{Q}_r$ for all r, and $\overline{Q}$ all describe the quantum mechanical situation. The case of $\overline{Q}_n$ for not too large values of n, but with $n \neq 1$, represents mixed situations where a transition between classical and quantum physics is taking place.

Buonomano observed that in real experiments it is always implicitly or explicitly assumed that

$$\overline{Q}_r = \overline{Q}_n = \overline{Q}$$

(*ergodic assumption*) but that such an assumption should really be checked with suitably designed experiments. An experiment of this type could, for instance, start as a regular double-slit experiment, but an ensemble average should be taken using many identical independent apparatuses, with exactly one photon measured by each apparatus, all of which have been prepared in the same initial state. Real laboratory averages are not of this type but are time averages and, of course, necessarily involve an ergodic type assumption for interpreting what has been measured at the same time as ensemble and as time average.

This nonergodic interpretation of quantum mechanics can in principle solve the EPR paradox, because it can explain the apparent violations of Einstein locality as due to nonergodic effects within a strictly local theory. Consider in fact the left-hand side of a polarization correlation experiment and divide the space between polarizer and source into M cells, numbering them from left to right. Thus the polarizer is in cell 1 and the source in cell M. Let $C_{rn}^{(m)}$ be the state of cell m after the nth photon has passed through it in run r, and let λ_{rn} and Λ_{rn} be defined as previously stated. Assume also that

$$C_{rn}^{(m)} = C_{rn}^{(m)}[\Lambda_{r,n-1}^{(m-1)}, \Lambda_{r,n-1}^{(m+1)}, \lambda_{r,n-1}].$$

This means that the state of cell m depends on the (previous) state of the two neighboring cells as well as on the state of the previous photon that passed through. It follows that, after the passage of one photon, the state of cell 2 depends on the state of the polarizer; after two photons have passed, cell 3 depends on the state of the polarizer; etc. Then after $N > M$ photons have passed through, cell M, that is, the source, depends on the state of the polarizer.

If the right-hand side of our polarization correlation experiment is assumed to behave in the same manner, one obtains a situation in which the source may produce pairs of photons in a state dependent on the configuration of the analyzing–detecting apparatus. As is well known, no Bell-type inequality can be obtained in such a case.

If the polarizer angle is changed, then the state of the emitted photons is changed accordingly, but only after a sufficient number of photons have passed through the apparatus and established a new equilibrium between the polarizer and the state of the emitted photons. The strong quantum mechanical correlations do not exist initially but develop after many pairs have traveled through the apparatus. These ideas can thus provide a solution of the EPR paradox, because all the previous considerations, including the 'birth' of the quantum mechanical correlations, had a strictly local nature, thanks to the introduction of the nonergodic assumption.

In the nonergodic approach the experimental apparatus has somehow a more important role than it should, even according to quantum theory, in defining the physical properties of the 'quantum systems' that one is studying. A very specific model of this type has been proposed by Scalera (1983), who considered an EPR experiment in which a source S emits two correlated beams of 'light' in opposite directions with the following property. A beam consists of a continuous ribbon which is helix-shaped and propagates in the direction of its axis. Passing through a polarizer with axis $\hat{a}$, there is only a modification of the ribbon, in that the directions $\pm\hat{a}$ are marked on it. It is next assumed that a photomultiplier placed behind the polarizer reacts only to the integral of energy contained in the ribbon, always starting from a marked direction and ending with the following marked direction. Thus a counter 'ticks' only immediately after the entrance in it of the second one of two successive marked directions. It is also assumed that this detector is blind to the initial piece of ribbon arriving before the first marked direction. Thus a 'detector' placed behind a 'polarizer' ticks regularly every $L/2c$ seconds, where L is the step of the helix and c the velocity of propagation of the helix (velocity of 'light').

Coming next to the EPR experiment, it is assumed that the correlation between the two ribbons emitted by the source in opposite directions is such that they satisfy a perfect specular symmetry: the two vectors perpendicular to the axis and lying on the ribbon are parallel if they are emitted at equal times. It follows that if the two polarizers have axes separated by an angle θ, then the second ribbon will be marked a distance $(\theta/2\pi)L$ behind its position simultaneously emitted to the one marked on the first ribbon. As a consequence, the second counter will tick a time $(\theta L/2\pi c)$ after the first counter. Are there coincidences? Clearly, the answer depends on the definition of the *coincidence time* τ. Coincidences are certainly present if

$$\tau \leqslant \left(\frac{\theta L}{2\pi c}\right). \tag{5.98}$$

Otherwise they are certainly absent. This means that the coincidence probability is a step function of θ, being equal to 1 (to 0) if (5.98) is (is not) satisfied. From this it follows that the left-hand side of the Bell inequality, that is the quantity Δ defined in (5.29), can assume values as high as 4, in clear violation of the inequality. A contradiction with well-established consequences of Einstein locality has therefore been generated from a model in which the "quantum" systems (here the ribbons) behave in a classical and strictly local way. How is this possible? The answer is that a source of nonlocality has been introduced by the electronic circuit that defines a 'coincidence'. Every pair of 'ticks' from the two counters is time-analyzed in the circuit, and the delay is compared with the left-hand side of (5.98) in order to decide whether they form a coincidence or not. In this way, information coming from one detector is used in order to decide the meaning of the tick of the other detector, and nonlocality is introduced into the experiment.

A disturbing feature of Scalera's model is that it uses detectors and electronics working in ways strictly similar to those of the apparatuses actually used in concrete experiments. Furthermore, it has been possible to show that the model has less to do with the nonergodic assumption than it might be thought (Scalera, 1984).

A second way of seeing that the coincidence counting electronics can be a 'source of nonlocality' has been discussed by Pascazio (1985), who considered the following model. Photons used in EPR experiments are assumed to be made of a point-like particle surrounded by a wave packet. The latter has a 'leading edge', and the distance x of the particle from it is supposed to vary in the statistical ensemble of similar photons with a density $\rho(x)$. Considering two photons of an EPR pair, they will be characterized by the two distances x_1 and x_2 for photon 1 and photon 2, respectively. The basic idea in Pascazio's model is that *there exists a dependence of the probability that the energy-carrying particle crosses the polarizer and is detected by the photomultiplier on the position* (x_1 or x_2) *of the particle in the wave*. If the previous probabilities are written $p_1(\hat{a}, x_1)$ and $p_2(\hat{b}, x_2)$ for the first and second photon crossing a polarizer with axis $\hat{a}$ ($\hat{b}$), then the double detection probability can be written

$$D(\hat{a}, \hat{b}) = \int_{X_1} \mathrm{d}x_1 \int_{X_2} \mathrm{d}x_2\, \rho(x_1, x_2) p_1(\hat{a}, x_1) p_2(\hat{b}, x_2),$$

where X_i are the x_i domains ($i = 1, 2$), $\rho(x_1, x_2)$ is the probability density for the pair of values x_1 and x_2, and where eventual additional variables have been ignored for simplicity. It is useful to change the space variables into time delays (everything is of course traveling at the speed of light). Let t_1 (t_2) be the time elapsed between the arrival of the leading edge of the first (second) wave packet and the arrival of the first (second) particle at the detector. With an obvious notation, one can then write

$$D(\hat{a}, \hat{b}) = \int_{T_1} \mathrm{d}t_1 \int_{T_2} \mathrm{d}t_2\, \rho(t_1, t_2) p_1(\hat{a}, t_1) p_2(\hat{b}, t_2),$$

where T_i are the t_i domains ($i = 1, 2$). Assuming for simplicity that the (eventual) extraction of a photoelectron is instantaneous, one can consider t_i ($i = 1, 2$) as the time elapsed between the arrival time of the ith wave packet leading edge and the extraction of a photoelectron. If τ_i is the time needed by the ith photomultiplier to turn the photoelectron into a macroscopically observable electrical pulse, plus the time necessary to 'shape' this pulse in the discriminator, then the time interval t_i^* separating the arrival of the leading edge of the wave packet on the photocathode from the appearance of a shaped electrical pulse in the measuring device is

$$t_i^* = t_i + \tau_i \quad (i = 1, 2).$$

Now, the coincidence electronics imposes the requirement that

$$t_2^* + T \leqslant t_1^* + \omega,$$

where T is the lifetime of the excited level that is going to emit the second photon of the cascade (T can also be interpreted as the time interval between the emissions of the leading edges of the two wave packets) and ω the so-called coincidence window. Obviously ω is *a priori* fixed in every coincidence experiment. It is, by definition, the maximum separation between two signals in order for them to be considered a coincidence.

From the two previous relations one easily obtains

$$t_2 \leqslant t_1 + \tau_1 - \tau_2 + \omega - T.$$

Therefore, the domain T_2 of the hidden variable t_2 is formally a function of properties of the first measuring apparatus, such as τ_1 and ω! This is a manifestly nonlocal feature of Pascazio's model which is introduced by the counting electronics.

Such an idea, like the one previously discussed by Scalera, is clearly an invitation to the experimentalists to pay great care in avoiding spurious sources of nonlocality when they perform EPR-type experiments.

5.13. Solution 4: Modifications of Quantum Theory

Already in 1936 it was clear that the EPR paradox had evidenced the existence of a striking

> ... disagreement between the results of quantum-mechanical calculations and those to be expected on the assumption that a system once freed from dynamical interference can be regarded as possessing independently real properties. (Furry, 1936.)

This conclusion is for some people very difficult to accept: that the old-fashioned trick of subjectivist philosophy ('nothing exists outside our thoughts') may become an unavoidable conclusion of modern science has led, and still leads, many physicists to the idea that something must be wrong with modern theoretical physics. Even Einstein entertained such a point of view:

... Einstein has (in a private communication) actually proposed such an idea; namely, that the current formulation of the many-body problem in quantum mechanics may break down when particles are far enough apart. (Bohm and Aharonov, 1957.)

The first organic examination of an eventual breakdown of quantum theory (with immediately negative conclusions) was made by Bohm and Aharonov in 1957. They considered the annihilation of a positron—electron pair into two energetic photons ('gamma rays') and showed that the quantum state produced is the one corresponding to the ket $|0^-\rangle$ of (5.79), so that each photon is always found with a state of linear polarization orthogonal to that of the other, no matter what may be the choice of axes with respect to which the state of polarization is expressed. Bohm and Aharonov calculated the ratio $R = \Gamma_1/\Gamma_2$, where Γ_1 is the rate of double scattering of the two photons through a fixed angle θ when the planes π_1 and π_2 formed by the lines of motion of the first and the second photon and their common original direction of motion are *perpendicular*; Γ_2 is the same rate when the planes π_1 and π_2 are *parallel*. The value of $R = \Gamma_1/\Gamma_2$ predicted by the $|0^-\rangle$ state is

$$R = \frac{(\gamma - 2\sin^2\theta)^2 + \gamma^2}{2\gamma(\gamma - 2\sin^2\theta)}, \tag{5.99}$$

where

$$\gamma = \frac{k_0}{k} + \frac{k}{k_0}.$$

Here k_0 is the wave number of the incident photon, k that of the final photon. Bohm and Aharonov considered an angle of 82° for which the ratio k_0/k can easily be calculated from Compton scattering kinematics and obtained $R = 2.85$. This figure cannot be compared directly with the experiment of Wu and Shaknov (1950), because photons were detected with an angular spread around the 'ideal' value of 82°. To such a concrete situation applies rather the prediction

$$R = 2.00,$$

obtained with a suitable angular average of (5.99), which agrees very well with the experimental result

$$R = 2.04 \pm 0.08.$$

Bohm and Aharonov could show that *the hypothesis of a breakdown of the* $|0^-\rangle$ *state vector with increasing distance between the two photons and of its substitution with mixtures of factorizable vectors* leads *necessarily to considerably smaller values of R*, such that

$$R \lesssim 1.5.$$

These results show that Wu's experiment is explained adequately by the existing quantum theory, which implies distant correlations, of the type leading to the EPR paradox, but not by any simple hypothesis implying a simple-minded breakdown of the quantum theory that could avoid the paradox. It would however not be correct to conclude that this experimental evidence constitutes an argument against Einstein locality, since there are well-known models capable of reproducing the quantum mechanical predictions for the experiments on double rescattering following electron—positron annihilations into two gamma rays.

One such model was proposed by Kasday (1971). Suppose there are two 'hidden vectors' $\hat{\lambda}_1$ and $\hat{\lambda}_2$ associated with photon 1 and photon 2, respectively, and let the photons ultimately scatter in the directions of these vectors. Then simply give $\hat{\lambda}_1$ and $\hat{\lambda}_2$ the same probability distribution as that of the momenta $\hat{k}_1$ and $\hat{k}_2$ of the scattered photons:

$$\rho(\hat{\lambda}_1, \hat{\lambda}_2) = F(\hat{k}_1, \hat{k}_2),$$

where $F(\hat{k}_1, \hat{k}_2)$ is the probability distribution of the momenta $\hat{k}_1$, $\hat{k}_2$ of the scattered photons. The assumption is clearly that the photons have 'decided in advance', at the time of annihilation, in which direction they would ultimately scatter. The model is clearly local: changing the position of 'detector 1' does not affeect the parameter $\hat{\lambda}_2$, for example, and therefore it does not change the response of 'detector 2'. Furthermore, the model reproduces the results of *all* measurements that can be made on the scattered photons. It does not matter much that the model is not really 'believable' from a physical point of view: if one local model is consistent with the data, there is every reason for believing that many more local models are possible and that also physically reasonable ones can be found. Anyway, there is a proof that locality is not ruled out, even by this old experimental evidence.

The previous 1957 discussion by Bohm and Aharonov, even if far from conclusive as far as Einstein locality is concerned, has nevertheless reached an important conclusion. *If one were willing to consider the possibility that the theory to be applied to EPR pairs at macroscopic distances between the two components were not the usual quantum theory, but a 'reduced theory', in which every state vector of the second type is in one way or another to be replaced by a mixture of state vectors of the first type, then the available experimental evidence would be decisive against such a possibility.*

This being the case, it is surprising that several important authors rediscovered and tried to apply the Bohm—Aharonov idea long after it had been discarded by its proponents.

Thus Jauch developed an ambitious approach to quantum theory based on an 'algebra of propositions', where he defined as 'mixtures of the 2nd kind' the quantum mechanical descriptions of EPR pairs based on nonfactorizable state vectors, that is, the quantum mechanical descriptions which lead to the EPR paradox. His idea was the following:

We may thus say the essence of our new notion of state is contained in the statement: *Mixtures of the* 2nd *kind do not exist.* (Jauch, 1971.)

Similarly de Broglie objected to Bell's proof of the incompatibility between Einstein locality and quantum mechanical predictions that the latter had been deduced from a formalism implying strong nonlocal correlations, unless the two particles were transported by one and the same 'wave train'. A unique wave train is however possible, according to de Broglie, only if the two particles are *near* to one another, near meaning in this context: distant less than the coherence length of the common wave packet. Therefore, the usual quantum formulas should break down at distances larger than a meter for photon pairs generated by an incoherent source, at distances larger than a micron for electron pairs, and so on. At larger distances de Broglie states that the 'singlet' state of two electrons, for instance, cannot be applied anymore and that ". . . the motion of each particle in its wave train becomes entirely independent of the motion that the other particle can have in its far-away wave train" (de Broglie, 1974).

Some steps in the direction of the Bohm–Aharonov hypothesis have recently been made by Piccioni and Mehlhop (1988). They started from the conviction that the 'singlet' state and other states leading to the EPR paradox cannot exist because they lead to unacceptable actions at a distance. Therefore, when the two components of an EPR pair are physically divided by a large distance, it is reasonable to think, according to Piccioni and Mehlhop, that a different quantum mechanical description applies, but not the one in terms of nonfactorizable state vectors. They note that the wavelength of the de Broglie waves of, e.g., two fermions composing an EPR pair can be of the order of fermis. Considering also the collimation of the two beams one can conclude in the following way:

Suppose that a source, for instance a machine in which neutrons are made to collide with protons, produces, at a certain rate, pairs of neutrons and protons which will separate from each other and yet remain in the [singlet] state (SS) when far apart. A key point is that, given the smallness of their wavelengths for realizable experiments, it is easy to see that inescapably each particle is confined to a wave packet of minimal dimensions, say less than a few centimeters, when they are a hundred meters apart. Thus we must assume that we know, for each pair produced, where the neutron will be at a certain time, and similarly for the proton. . . . The EPR is a paradox, because it demands that two particles far apart should behave as if they still interacted with each other. (Piccioni and Mehlhop, 1988.)

After having examined several objections and alternative possibilities, these authors conclude that renouncing the assumption that the singlet state is valid at large distances is the only available option. To this they also add the conviction that the source produces the EPR pairs in correlated spin eigenstates along some direction. Thus Piccioni and Mehlhop try to resurrect what essentially amounts to the Bohm–Aharonov hypothesis.

The empirical consequences of the Bohm–Aharonov hypothesis in the case of photon pairs from atomic cascades were examined by Fortunato *et al.* (1977), who considered the observable corresponding to the following linear Hermitian (projection) operator

$$\Gamma_0 = |0^-\rangle\langle 0^-|, \tag{5.100}$$

where $|0^-\rangle$ is the second-type state vector defined in (5.79). From that definition and from (5.100), it follows that

$$\Gamma_0 = \tfrac{1}{2}\{|x_\alpha\rangle\langle x_\alpha| \otimes |y_\beta\rangle\langle y_\beta| + |y_\alpha\rangle\langle y_\alpha| \otimes |x_\beta\rangle\langle x_\beta| - \\ - |y_\alpha\rangle\langle x_\alpha| \otimes |x_\beta\rangle\langle y_\beta| - |x_\alpha\rangle\langle y_\alpha| \otimes |y_\beta\rangle\langle x_\beta|\}.$$

If one puts

$$\begin{aligned} |x_k\rangle\langle x_k| &= \tfrac{1}{2}(1 + R_k), \\ |y_k\rangle\langle y_k| &= \tfrac{1}{2}(1 - R_k), \\ |x_k\rangle\langle y_k| &= \tfrac{1}{2}(P_k + iQ_k), \\ |y_k\rangle\langle x_k| &= \tfrac{1}{2}(P_k - iQ_k), \end{aligned} \tag{5.101}$$

with R_k, P_k, Q_k Hermitian operators ($k = \alpha, \beta$), one obtains

$$\Gamma_0 = \tfrac{1}{4}(1 - P_\alpha \otimes P_\beta - Q_\alpha \otimes Q_\beta - R_\alpha \otimes R_\beta). \tag{5.102}$$

The physical meaning of the operators P_k, Q_k, and R_k is easily obtained by noticing that their eigenstates and eigenvalues are given by

$$\begin{aligned} &R_k|x_k\rangle = |x_k\rangle, \\ &R_k|y_k\rangle = -|y_k\rangle, \\ &\frac{P_k(|x_k\rangle + |y_k\rangle)}{\sqrt{2}} = \frac{|x_k\rangle + |y_k\rangle}{\sqrt{2}}, \\ &\frac{P_k(|x_k\rangle - |y_k\rangle)}{\sqrt{2}} = -\frac{|x_k\rangle - |y_k\rangle}{\sqrt{2}}, \\ &\frac{Q_k(|x_k\rangle + i|y_k\rangle)}{\sqrt{2}} = \frac{|x_k\rangle + i|y_k\rangle}{\sqrt{2}}, \\ &\frac{Q_k(|x_k\rangle - i|y_k\rangle)}{\sqrt{2}} = -\frac{|x_k\rangle - i|y_k\rangle}{\sqrt{2}}. \end{aligned} \tag{5.103}$$

Therefore, the eigenvalues of P_k, Q_k, R_k ($k = \alpha, \beta$) are ± 1, and their eigenstates are linear polarization states along x and y (for R_k), linear polarization states along 45° and −45° (for P_k), and circular polarization states, clockwise and anticlockwise (for Q_k). It follows that the expectation value of the product of two such operators (one for each system of the EPR pair) is exactly what has been defined as a *correlation function* in previous sections. Adopting again the symbol $P(a, b)$ for correlation functions, we get from (5.102)

$$\Gamma_0 = \tfrac{1}{4}\{1 - P(0°, 0°) - P(45°, 45°) - P(\mathrm{RHC}, \mathrm{RHC})\} \tag{5.104}$$

where angles are measured starting from the x axis and RHC indicates a 'right-handed-circular' polarization measurement.

The expectation value of the operator Γ_0 in the state $|0^-\rangle$ is unity, as obviously follows immediately from the definition (5.100). From (5.104) a precise quantum mechanical prediction follows for an optical photon pair in the $|0^-\rangle$ state:

$$P(0°, 0°) + P(45°, 45°) + P(\mathrm{RHC}, \mathrm{RHC}) = -3. \tag{5.105}$$

If instead one assumes that a mixture of factorizable photon states should be adopted at large distances between the two photons, it has been shown (Fortunato *et al.*, 1977) that, *for all possible mixtures*, one must have

$$P(0°, 0°) + P(45°, 45°) + P(\mathrm{RHC}, \mathrm{RHC}) \geqslant -1. \tag{5.106}$$

This strikingly large disagreement between the two predictions should lead to rather easy experimental discriminations between the two possibilities. As a matter of fact, an experiment has been performed by Clauser (1976b) in order to check (5.105) against (5.106). The result has come out in favor of the quantum mechanical prediction, but the experiment was marred by difficulties connected with the quarter-wave plates.

A careful experiment will anyway probably agree with (5.105) for at least three orders of reasons:

(i) Bohm and Aharonov already showed that mixtures of nonfactorizable vectors lead to disagreements with empirical evidence in the case of gamma rays produced in positron—electron annihilation.
(ii) Classical models of the photon exist that are perfectly local but which agree with (5.105), as shown by Garuccio *et al.* (1977).
(iii) Also in the case of spin-$\frac{1}{2}$ particles it is impossible to find mixtures leading to the perfect anticorrelation condition so typical of certain quantum mechanical predictions (Bergia *et al.*, 1985).

Thus the 'reduced' quantum theory, in which all state vectors leading to the EPR paradox are assumed to 'decay' spontaneously into mixtures of factorizable state vectors when the two systems composing an EPR pair increase their mutual distance, appears to be only a mutilated theoretical scheme with no hope of representing the physical reality. In spite of this, the Bohm—Aharonov hypothesis keeps being 'rediscovered' from different points of view (Hawking, 1976; Ellis *et al.*, 1983).

A reformulation of quantum mechanics has been proposed by Ghirardi *et al.* (1976) in which time evolution is governed by the usual Schrödinger equation when the two systems are close together, while a continuous transition to a mixture of factorizable states takes place with increasing distance. Obviously, the generalized equation violates quantum mechanics for large separation of the subsystems: however, the dynamics of this model has been chosen in such a way that *only the correlations* between the subsystems turn out to be different.

A theory based on the C^* algebras of quantum mechanical operators has recently been put forward by Wan and McLean. This theory possesses features that are absent in conventional quantum mechanics, since:

(i) a pure state can evolve in time into a mixture;
(ii) two coherent pure states can evolve in time into two spatially disjoint states.

Some interesting results obtained by Wan and McLean (1984) concern the resolution of the de Broglie paradox and the possibility to develop a quantum theory of measurements free of contradictions as far as the 'reduction of the wave packet' is concerned. An interesting feature of their theory, which leads these authors to the hope of solving the EPR paradox as well, is the existence of *asymptotically localizing states.* Consider the following definitions: ϕ is an element of the Hilbert space of the physical system under consideration; ϕ_t is the time-evolved vector given by $\phi_t = U_t^0 \phi$, where $U_t^0 = \exp(H_0 t/i\hbar)$, H_0 being the free Hamiltonian of the system; $E_x([x_1, x_2])$ is the spectral measure of the position observable x relative to the interval $[x_1, x_2]$; $E_p([p_1, p_2])$ is the spectral measure of the momentum observable p relative to the interval $[p_1, p_2]$. Wan and McLean (1983) could then show that the following theorem holds:

$$\operatorname*{Lim}_{t \to \infty} \| E_x([v_1 t, v_2 t])\phi_t \|^2 = \| E_p([mv_1, mv_2])\phi \|^2, \tag{5.107}$$

where m is the mass of the considered quantum system. Since it is rather easy to choose the right-hand side of (5.107) equal to unity (it suffices to choose a state ϕ composed of the superposition of momentum states all included in the interval $[mv_1, mv_2]$), it follows that the corresponding ϕ_t will localize asymptotically (that is, for large t), in the sense that it will contract itself into the spatial interval $[v_1 t, v_2 t]$.

By building a theory based exclusively on spatially localizing states, it becomes possible to show that all interferences disappear asymptotically, so that no EPR paradox exists anymore (Wan and McLean, 1984). However this theory, just like the 'reduced quantum theory' of Bohm and Aharonov, falls into considerable and unresolved troubles when conservation laws are considered (Wan and Timson, 1985).

A nonlinear field theory of matter, based on general relativity (Sachs, 1980), has an asymptotic, low-energy limit that predicts the outcome of experimental tests of Bell's inequalities. The inequalities should be valid if correlated spin-$\frac{1}{2}$ particles, observed in coincidence, were a spacelike distance apart; they should be violated if they were separated by timelike distances. This approach is therefore not equivalent to the usual quantum theory, which predicts violations of Bell's inequalities for all types of distances. It does not appear convincing, however, to attribute mixtures of factorizable states to Bohr's point of view about the EPR paradox.

References

Books

Bitsakis, E., and Tambakis, N. (eds): 1985, *Determinism in Physics*, Gutenberg, Athens.
Bohm, D.: 1951, *Quantum Theory*, Prentice-Hall, Englewood Cliffs.
Bohr, N.: 1961, *Atomic Theory and the Description of Nature*, Cambridge University Press, Cambridge.
d'Espagnat, B. (ed.): 1971, *Foundations of Quantum Mechanics* (Italian Physical Society, Proceedings of the International School of Physics 'Enrico Fermi', Course IL), Academic Press, New York.
d'Espagnat, B.: 1983, *In Search of Reality*, Springer, New York.
Jammer, M.: 1974, *The Philosophy of Quantum Mechanics*, Wiley, New York.
Lahti, P., and Mittelstaedt, P. (eds): 1985, *Symposium on the Foundations of Modern Physics: 50 years of the Einstein—Podolsky—Rosen Gedanken-experiment*, World Scientific, Singapore.
Screaton, G. R. (ed.): 1961, *Dispersion Relations*, Oliver and Boyd, Edinburgh.
Selleri, F. (ed.): 1988, *Quantum Mechanics versus Local Realism: The Einstein, Podolsky, and Rosen Paradox*, Plenum, New York.
Tarozzi, G., and van der Merwe, A.: 1985, *Open Questions in Quantum Physics*, Reidel, Dordrecht.
Tarozzi, G., and van der Merwe, A.: 1988, *The Nature of Quantum Paradoxes*, Kluwer Academic Publishers, Dordrecht.
van der Merve, A., Selleri, F., and Tarozzi, G.: 1988, *Microphysical Reality and Quantum Formalism*, Kluwer Academic Publishers, Dordrecht.
Schilpp, P. A. (ed.): 1970, *Albert Einstein: Philosopher-Scientist* (3rd edn), Open Court, La Salle.

Articles

Aerts, D.: 1982, *Lett. Nuovo Cimento* **34**, 107.
Aspect, A., *et al.*: 1981, *Phys. Rev. Lett.* **47**, 460; 1982a, *Phys. Rev. Lett.* **49**, 91; 1982b, *Phys. Rev. Lett.* **49**, 180.
Aspect, A., and Grangier, P.: 1985, *Lett. Nuovo Cimento* **43**, 345.
Baracca, A., *et al.*: 1975, *Nuovo Cimento* **28B**, 4.
Baracca, A., *et al.*: 1976, *Int. J. Theor. Phys.* **15**, 473.
Bartell, L. S.: 1980, *Phys. Rev. D* **22**, 1352.
Barut, A. O., and Meystre, P.: 1984, *Phys. Rev. Lett.* **53**, 1021.
Bell, J. S.: 1965, *Physics* **1**, 195; 1975, *Epist. Lett.* **7**, 14.1; 1976, *Epist. Lett.* **9**, 17.0.
Bergia, S., and Cannata, F.: 1982, *Found. Phys.* **12**, 843.
Bergia, S., Cannata, F., and Monzoni, V.: 1985, *Found. Phys.* **15**, 145.
Bohm, D., and Aharonov, Y.: 1957, *Phys. Rev.* **108**, 1070.
Bohm, D., and Hiley, B. J.: 1978, *Found. Phys.* **8**, 93.

Bohm, D.: 1988, in A. van der Merwe *et al.* (eds); see book list.
Bohr, N.: 1935, *Phys. Rev.* **48**, 696.
Buonomano, V.: 1980, *Nuovo Cimento* **57B**, 146; 1988, in F. Selleri (ed.); see book list.
Canals-Frau, D.: 1982, *Ann. Fond. L. de Broglie* **7**, 217.
Carazza, B.: 1988, in G. Tarozzi and A. van der Merwe (eds); see book list.
Capasso, V., Fortunato, D., and Selleri, F.: 1973, *Int. J. Theor. Phys.* **7**, 319.
Clauser, J. F., *et al.*: 1969, *Phys. Rev. Lett.* **23**, 880.
Clauser, J. F., and Horne, M. A.: 1974, *Phys. Rev. D* **10**, 526.
Clauser, J. F.: 1976a, *Phys. Rev. Lett.* **37**, 1223; 1976b, *Nuovo Cimento* **33B**, 740.
Clauser, J. F., and Shimony, A.: 1978, *Rep. Progr. Phys.* **41**, 1881.
Corchero, E. S.: 1986, *Proposed Atomic Cascade Experimental Test of Symmetric Local Hidden-Variables Theories*, University of Santander preprint.
Costa de Beauregard, O.: 1976, *Found. Phys.* **6**, 539; 1977, *Nuovo Cimento* **42B**, 41; 1979, *Nuovo Cimento* **51B**, 267; 1980, *Found. Phys.* **10**, 513.
Cufaro-Petroni, N.: 1977, *Nuovo Cimento* **40B**, 235 and **40B**, 381; 1985, in G. Tarozzi and A. van der Merwe (eds); see book list.
Dalla Chiara, M. L., and Toraldo di Francia, G.: 1985, in P. Lahti and P. Mittelstaedt (eds); see book list.
de Broglie, L.: 1974, *C.R. Acad. Sc.* (*Paris*) **278**, 721.
de Muynck, W. M., and van Stekelenborg, J. T.: 1986, *Phys. Lett.* **116A**, 420.
Dirac, P. A. M.: 1941, *Proc. Roy. Soc.* **180A**, 1; 1952, *Nature* **169**, 702; 1975, in *Directions in Physics*, H. Hora and J. R. Shepanski (eds), Sidney.
Duncan, A.: 1986, in *Book of Invited Papers, Tenth International Conference on Atomic Physics*, H. Narumi and T. Shimamura (eds), Tokyo.
Duncan, A.: 1988, in F. Selleri (ed.); see book list.
Ellis, J., *et al.*: 1983, SLACPUB 3131/CERNTH 3619.
Eberhard, P. H.: 1978, *Nuovo Cimento* **46B**, 392.
Einstein, A., Podolsky, B., and Rosen, N.: 1935, *Phys. Rev.* **47**, 777.
Falciglia, F., Garuccio, A., Iaci, G., and Pappalardo, L.: 1983, *Lett. Nuovo Cimento* **37**, 66.
Feynman, R. P.: 1982, *Int. J. Theor. Phys.* **21**, 467.
Fine, A.: 1982, *Phys. Rev. Lett.* **48**, 291.
Freedman, S. J.: 1972, Ph.D. Thesis, University of California, Lawrence Berkeley Laboratory Report No. LBL-391.
Freedman, S. J., and Clauser, J. F.: 1972, *Phys. Rev. Lett.* **28**, 938.
Fortunato, D.: 1976, *Lett. Nuovo Cimento* **15**, 289.
Fortunato, D., Garuccio, A., and Selleri, F.: 1977, *Int. J. Theor. Phys.* **16**, 1.
Fry, E. S., and Thompson, R. C.: 1976, *Phys. Rev. Lett.* **37**, 465.
Furry, W. H.: 1936, *Phys. Rev.* **49**, 393; **49**, 476.
Garuccio, A., and Rapisarda, V.: 1981, *Nuovo Cimento* **65A**, 269.
Garuccio, A., Scalera, G. C., and Selleri, F.: 1977, *Lett. Nuovo Cimento* **18**, 26.
Garuccio, A., and Selleri, F.: 1984, *Phys. Lett.* **103A**, 99.
Garg, A., and Mermin, N. D.: 1982, *Phys. Rev. Lett.* **49**, 901; 1984, *Found. Phys.* **14**, 1.
Ghirardi, G. C., Rimini, A., and Weber, T.: 1976, *Nuovo Cimento* **31B**, 177; **36B**, 97.
Grib, A. A.: 1984, *Sov. Phys. Usp.* **27**, 284.
Hawking, S. W.: 1976, *Phys. Rev. D* **14**, 2460.
Heywood, P., and Redhead, M. L.: 1983, *Found. Phys.* **13**, 481.
Hiley, B. J.: 1977, *Contemp. Phys.* **18**, 411.
Holt, R. A., and Pipkin, F. M.: 1974, University of Harvard, preprint (unpublished).
Hooker, C. A.: 1970, *Am. J. Phys.* **38**, 851.
Horlick-Jones, T.: 1981, *Phys. Lett.* **85A**, 206.
Kemble, E. C.: 1935, *Phys. Rev.* **47**, 973.
Kraus, K.: 1985, in G. Tarozzi and A. van der Merwe (eds); see book list.

Kunstatter, G., and Trainor, L. E.: 1984, *Phys. Lett.* **103A**, 32; *Am. J. Phys.* **52**, 598.
Kasday, L.: 1971, in B. d'Espagnat (ed.); see book list.
Ivanovic, I. D.: 1978, *Lett. Nuovo Cimento* **22**, 14.
Jammer, M.: 1985, P. Lahti and P. Mittelstaedt (eds); see book list.
Jauch, J. M.: 1971, in B. d'Espagnat (ed.); see book list.
Margenau, H.: 1936, *Phys. Rev.* **49**, 240.
Marshall, T. W., Santos, E., and Selleri, F.: 1983, *Phys. Lett.* **98A**, 5.
Mermin, N. D., and Schwarz, G. M.: 1982, *Found. Phys.* **12**, 101.
Mermin, N. D.: 1985, *Physics Today* **38**, 38.
Mittelstaedt, P.: 1985, in P. Lahti and P. Mittelstaedt (eds); see book list.
Mückenheim, W.: 1982, *Lett. Nuovo Cimento* **35**, 300; 1986, *Phys. Rep.* **133**, 337.
Ne'eman, Y.: 1983, *Proc. Natl. Acad. Sci. USA* **80**, 7051; 1986, *Found. Phys.* **16**, 361.
Paty, M.: 1982, *Fundamenta Scientiae* **3**, 79.
Pascazio, S.: 1985, *Nuovo Cimento* **5D**, 23; *Phys. Lett.* **111A**, 339; 1986, *Phys. Lett.* **118A**, 47.
Perrie, W., Duncan, A. J., Beyer, H. J., and Kleinpoppen, H.: 1985, *Phys. Rev. Lett.* **54**, 1790.
Piccioni, O., and Mehlhop, W.: 1988, in A. van der Merwe *et al.* (eds); see book list.
Pipkin, F. M.: 1978, *Progr. Atomic Mol. Phys.* **14**, 281.
Popper, K. R.: 1985, in G. Tarozzi and A. van der Merwe (eds); see book list.
Rastall, P.: 1982, *Phys. Lett.* **87A**, 279.
Rayski, J.: 1985, in P. Lahti and P. Mittelstaedt (eds); see book list.
Rietdijk, C. W.: 1978, *Found. Phys.* **8**, 615; 1980, *Found. Phys.* **10**, 403; 1981, *Found. Phys.* **11**, 783.
Rosen, N.: 1985, in P. Lahti and P. Mittelstaedt (eds); see book list.
Ruark, A. E.: 1935, *Phys. Rev.* **48**, 466.
Sachs, M.: 1980, *Nuovo Cimento* **58A**, 1.
Santos, E.: 1985, in G. Tarozzi and A. van der Merwe (eds); see book list.
Scalera, G. C.: 1983, *Lett. Nuovo Cimento* **38**, 16; 1984, *Lett. Nuovo Cimento* **40**, 353.
Schrödinger, E.: 1935, *Proc. Cambridge Phil. Soc.* **31**, 555.
Selleri, F.: 1978, *Found. Phys.* **8**, 103; 1980, *Lett. Nuovo Cimento* **27**, 1.
Selleri, F.: 1984, *Lett. Nuovo Cimento* **39**, 252.
Selleri, F.: 1988, in A. van der Merwe *et al.* (eds); see book list.
Shimony, A.: 1978, *Int. Philos. Quart.* **18**, 3.
Stapp, H. P.: 1977, *Nuovo Cimento* **40B**, 191; 1985, *Am. J. Phys.* **15**, 306; 1988, in A. van der Merwe *et al.* (eds); see book list.
Vigier, J.-P.: 1979, *Lett. Nuovo Cimento* **24**, 258; 1985, in G. Tarozzi and A. van der Merwe (eds); see book list.
Wan, K. K., and McLean, R. G.: 1983, *Phys. Lett.* **94A**, 198; 1984, *Phys. Lett.* **102A**, 163.
Wan, K. K., and Timson, D. R. E.: 1985, *Phys. Lett.* **111A**, 165.
Wightman, A. S.: 1976, *Proc. Symposia Pure Math.* **28**, 147.
Wigner, E. P.: 1970, *Am. J. Phys.* **38**, 1005.
Wolfe, H. C.: 1936, *Phys. Rev.* **49**, 274.
Wu, C. S., and Shaknov, I.: 1950, *Phys. Rev.* **77**, 136.
Yanase, M. M.: 1971, in B. d'Espagnat (ed.); see book list.
Zeilinger, A.: 1986, *Phys. Lett.* **118A**, 1.

Chapter 6

The EPR Paradox in the Real World

6.1. Criticisms of Einstein Locality

In the previous chapter we defined 'Einstein locality' as a set of three assumptions: (1) EPR reality criterion, (2) time arrow, (3) and separability. The arguments which have been raised against this conception will now be reviewed. Some of them we shall discard by showing that they are equivalent to rejecting *a priori*, one way or another, the very idea of local realism. Now, it should be clear from the discussion of the previous chapter that the rejection of local realism *is* a possible solution of the paradox and that the great question we are facing is precisely how to choose between local realism and some predictions of the existing quantum theory. Criticisms of Einstein locality consisting essentially of a denial of local realism can then be taken as statements of full faith in the existing theory and of a readiness to abandon any conception incompatible with it. Such statements are respectable, but of no help in scientifically solving the paradox.

Nevertheless, some points of criticism are left that must be taken seriously and that can be answered only by giving a probabilistic generalization of Einstein locality:

1. A first argument against the EPR reality criterion objects that the latter defines as real *only* a property that can be predicted with certainty: Since only in the case of an eigenstate of some Hermitian operator can one make predictions with certainty, it follows that applications of the reality criterion to the quantum domain are limited to the case of eigenstates. Now, why, it may be asked, should a physical system described by a state vector that is a superposition of eigenstates be less real?

This criticism attacks only *the limited validity* of the EPR reality criterion, without undermining its validity where it applies. Einstein, Podolsky, and Rosen were aware of this limitation and wrote: ". . . this criterion, while far from exhausting all possible ways of recognizing a physical reality, at least provides us with one such way." We may add that this criterion has in itself a tremendous predictive power and that physics stands to benefit from exploiting it fully, even if its limited applicability is recognized. Anyway, it will be shown in the present chapter that a generalized reality criterion can be given for which all physical situations (e.g., the quantum mechanical superpositions) are treated in the same manner.

2. There is a weak point in the EPR reality criterion where it presupposes predictability *with certainty* of the physical quantity under consideration: How can one ever be absolutely certain of anything in a concrete experimental situation? In 1935, the predictability with certainty appeared to be a consequence of quantum theory for certain types of state vectors, but the situation is different today, following the Wigner—Araki—Yanase theorem (Yanase, 1971), which proved the necessity of imperfect measurement, even in principle, if quantum theory has to be considered correct for all physical interactions, including measurements.

Also this point of criticism — fully justified when directed against the 1935 definition of Einstein locality — can be overcome by advancing a probabilistic generalization of Einstein locality in a manner to be explained in the next section.

3. The EPR idea of separability insists on the fact that two quantum objects *spatially* separated should be incapable of influencing one another. Some people object that the quantum formalism does not represent a reality in ordinary spacetime: the Schrödinger equation for n particles represents a wave propagating in a $3n$-dimensional space (configuration space). In order to obtain a nonphysical space, it suffices to take $n > 1$; it could then happen that separation in ordinary space is not enough to ensure separation in configuration space. Thus, the physical predictions of the theory could lead to nonlocal effects. Related to this consideration is the idea that the state vector of two spin-$\frac{1}{2}$ particles is an element of an abstract spin space, and that two objects separated in ordinary space could be contiguous in spin space. From this point of view, Einstein, Podolsky, and Rosen look like naive realists who do not understand the subtleties of the quantum formalism.

The proper response to the foregoing is, of course, that the mathematical structure of quantum mechanics is very useful and accurate in making empirical predictions, but that it does not need to be *true in nature*. Those who believe differently are rather naive platonists, unable to appreciate the complexity of the relationship between the mathematical formalism and the physical world. Mathematics is man-made, and so are configuration space and spin space. Four-dimensional spacetime, on the other hand, exists in some sense independently of humans. Separability in ordinary space should therefore be viewed as the most important fact, particularly because, as is well known, all interactions decrease rapidly with distance.

4. Related to the previous point is Bohr's complementarity. Bohr's answer to the original EPR paper does not question the correctness of the EPR reasoning once all its implicit and explicit premises are accepted (Bohr, 1935). Rather, Bohr maintains that the quantum mechanical formalism cannot be adapted to a philosophical point of view like that of Einstein. Bohr

argues in favor of 'a final renunciation of the classical ideal of causality and a radical revision of our attitude towards the problem of physical reality', while stressing that his notion of complementarity is a 'new feature of natural philosophy' which implies a 'radical revision of our attitude as regards physical reality'. Bohr's line of reasoning is the following.

All measurements performed on atomic and subatomic systems must necessarily be prepared, carried out, and, when results are obtained, expressed in classical terms. This is so because the physicist lives in a macroscopic world where classical laws and, even more important, classical concepts (space, time, causality, etc.) hold and have become the unavoidable means for expressing all human experience. The experimental physicist will therefore naturally *try* to express his experimental results in classical terms, that is, he will try to express the regularities of the microscopic world as causal processes in space and time. This goal is however impossible to achieve, and the physicist finds himself suddenly exposed to an irrational element, because the existence of the quantum of action h implies a finite mutual *interaction* between the measured object and the measuring instrument. Thus arise effects on the observed system which cannot even be eliminated logically.

In other words, the existence of h implies a perturbation of the measured object that is completely unpredictable and thus, in a way, irrational. These facts not only set a limit to the *extent* of the information obtainable by measurements, but they also set a limit to the *meaning* which can be attributed to such information. Bohr's opinion is that an independent reality cannot be attributed to atomic phenomena. The word "phenomenon" should *exclusively* be used to refer to the observations obtained under specified circumstances, including a description of the entire experimental apparatus. From this point of view, physics must deal exclusively with acts of observation, all reference to an unobserved elementary reality being banished from a truly scientific reasoning. Obviously, then, no paradox exists when one considers two correlated systems. From Bohr's perspective, the EPR assumption about the elements of reality appears immediately useless: One is free to make it, but it means only that there is an element of reality associated *with an act of measurement concretely performed*, for there is no other reality that can be considered. In particular, the EPR conclusion that position and momentum correspond to two simultaneously existing elements of reality appears totally unjustified (Bohr writes that the conclusion contains 'an essential ambiguity'), because one can never perform simultaneous measurements of position and momentum; if there is no concrete measurement, there is also nothing real to which an element of reality can be attributed.

If this is Bohr's point of view, one should stress that it attempts to forbid a causal completion of quantum theory (that is, a causal description in space and time of individual atomic events, a description necessarily more detailed than that provided by quantum mechanics, which makes only statistical predictions). Bohr's complementarity thus has the same aim as von Neumann's theorem (von Neumann, 1955). Accordingly, the existence of causal space-time models, doing precisely what von Neumann's theorem tried to forbid (Selleri, 1985), not only makes this theorem obsolete but shows at the same time how arbitrary Bohr's conclusion was. This point was discussed in Chapter 2.

5. It is often stated that the EPR reality criterion, by associating elements of reality with observables that cannot be measured simultaneously, essentially assigns to both observables well-defined values ('counterfactuality'). This point of view arises from a positivistic misunderstanding of the EPR realistic position. What we mean is perhaps best illustrated with a classical example. One can predict with certainty that water brought below 0 °C will solidify to form ice; the EPR reality criterion allows us to associate an element of reality λ_1 with this property of water. A second property of water is that it becomes vapor about 100 °C, and also this can be predicted with certainty; the EPR reality criterion then associates a second element of reality λ_2 with this new property of water. The simultaneous existence of the elements of reality λ_1 and λ_2, which reflect nothing but well-understood features of the interaction potential between water molecules, as today is well known, does not of course imply that we should be able *at the same time* to make water freeze and boil. The situation with quantum measurements could be analogous.

6. The most formidable challenge to Einstein locality is said to come from experimental evidence. Röhrlich expressed the opinion (Röhrlich, 1983) that "Local hidden-variables theory is dead. It received its *coup de grace* by two precision experiments carried out . . . in Paris." As a matter of fact, though, all the experiments performed with atomic cascades in order to test Bell's inequality have been analyzed with the help of some additional hypotheses (Selleri, 1988b). A logical refutation of these hypotheses is therefore all that is required to restore full agreement between Einstein locality and quantum predictions as far as the existing empirical evidence is concerned (Selleri and Zeilinger, 1988). However, further relevant experiments that have heretofore not been performed now become especially important. The situation is thus fully open to debate, and opposite claims reflect more than anything else old-fashioned ideas biased by ideological choices which obfuscate logical thinking.

The choice between Einstein locality and quantum theory remains however to be made at the experimental level. It would suffice to carry out the usual photon correlation experiments with nearly ideal detectors in order to decide finally between these alternatives. But this remains unfortunately a remote possibility. Other promising paths of investigation are those involving photon experiments with polarizers and retardation plates (De Caro, 1989), $K^0\overline{K}^0$ correlation experiments (Selleri, 1983), and the coincident detection of correlated atom pairs (Lo and Shimony, 1981). The above criticisms of

Einstein locality are far from providing really strong arguments against the realistic and rationalistic world view of Albert Einstein. Nevertheless, they do require a generalization of the ideas of reality and separability. This is so because the EPR reality criterion can be used only in those very rare (nonexisting?) cases in which one can make *predictions with certainty*, and because it can be applied only to the situations described as *eigenstates* within quantum theory but not to the general case of superpositions.

The point above illuminates the path one should choose in order to find a generalized reality criterion. In the general case of quantum mechanical superpositions, *only probabilities can be predicted*; our new reality must therefore be linked to predictable probabilities. That such a generalization can actually be achieved will be shown in Section 6.2. Here we shall next only briefly discuss the different definitions of probability — one of those fundamental concepts with which scientists and philosophers are still struggling without obtaining a unification of the different points of view:

> Ideas involving the theory of probability play a decisive part in modern physics. Yet we still lack a satisfactory, consistent definition of probability; or, what amounts to much the same, we still lack a satisfactory axiomatic system for the calculus of probability.

These words, contained in the 1980 edition of Karl Popper's *Logic of Scientific Discovery* (Popper, 1980), underline well-known difficulties in the formulation of probabilistic problems in physics and in science generally.

The classical definition of probability, due to Laplace, expresses the numerical value of a probability as the ratio between the number of favorable cases and the number of (equally) possible cases. Many objections have been raised against this definition. Frequently mentioned is the fact that 'equally possible' is just another way of saying 'equally probable', so that the definition is circular. Attempted responses to the critical status of the notion of probability have led in our century to several new formulations of the theory of probability; these are usually classified as follows:

(1) The *subjective interpretation*, in which a central role is played by the 'degree of rational belief' which one may assign to a given statement about the future development of a system.
(2) The *objective interpretation*, which treats every numerical probability assignment as a statement about the relative frequency with which the event considered occurs within a sequence of similar events.
(3) What can be called the *agnostic position*, which is based on the idea that mathematicians have clearly defined probability as a special case of measure theory, and that, as a consequence, scientists need no longer agonize about the question of what probability really is.

In none of these positions is probability defined *for a single system* in objective terms.

Popper's propensity formulation of quantum theory (Popper, 1980) points in the direction of introducing probabilities for individual systems. Popper argues *against* the view "that a singular probability cannot be objective and that an objective theory of probability must be a *purely* statistical or frequency theory". He sees a problem to be solved in the difference between Schrödinger's and Born's formulations of quantum wave theory. For Born, the squares of the wave moduli merely express the probabilities (relative frequencies of occurrence) of different results in quantum experiments, such as position measurements and momentum measurements. Schrödinger, by contrast, stresses the objectivity and reality of the waves (in configuration space) and the fact that the wave moduli provide a weight or strength or measure of the probabilities. In this approach there is an obvious connection with statistical distributions, but the main idea is that of introducing objectively real properties for the *single* system. Also Popper's propensity formulation of probabilities in quantum physics tries to bridge the gap between objective properties of a single system and objective probabilities in the statistical ensemble of many repetitions of the system under consideration. It allows the single system to have objective tendencies or propensities of varying strengths and shows that, if these are interpreted as tendencies to realize particular properties, they lead to corresponding statistical frequencies in a virtual (or in a real) statistical ensemble. *It should be stressed that this introduction of objective propensities for the single system finds its conceptual support precisely in its relationship with the statistical frequencies*, which have a well established status of *objective* properties. This outcome could hardly be different, particularly in the quantum mechanical context, given the statistical nature of quantum theory.

The individual definition of probability is thus, one sees, still to be established on independent grounds. This conclusion is important, because we will see that the definition of 'probability' plays a crucial role in the study of the EPR paradox, the Clauser and Horne 1974 formulation of 'objective local theories' being acceptable only with an individual definition of the probabilistic conception.

6.2. Probabilistic Einstein Locality

As we saw in the previous section, the original formulations of the EPR paradox, based on the idea of deterministic and separables elements of reality ('local hidden variables'), cannot be considered fully satisfactory today. The problem is that determinism very probably does not apply to the quantum domain, a fact brought to full light also by the clarification of the important distinction between causality and determinism. Our new definition of Einstein locality rests on the following considerations (Selleri, 1988a).

Let a set S of physical objects of the same type (e.g., photons) be given;

denoting these objects by $\alpha_1, \alpha_2, \ldots, \alpha_N$, we write

$$S = \{\alpha_1, \alpha_2, \ldots, \alpha_N\}. \tag{6.1}$$

Suppose an instrument I is given that is capable of measuring a dichotomic physical quantity A on the systems composing S, and let the possible results be A_1 and A_2. We then assume:

If it is possible to predict the existence of a subset

$$S' = \{\alpha_{i_1}, \alpha_{i_2}, \ldots, \alpha_{i_\ell}\}, \tag{6.2}$$

of S without disturbing the objects composing S′ and S, and if it is also possible to predict correctly that future measurements of A on S′ will give the results A_1 and A_2 *with respective probabilities* p_1 *and* p_2, *then it is said that a physical property* λ' *belongs to S′ which fixes the probabilities*:

$$p_1 = p_1(A; \lambda'), \qquad p_2 = p_2(A; \lambda'). \tag{6.3}$$

This statement will be called the *generalized reality criterion* (GRC); it constitutes a natural generalization of the famous criterion for (deterministic) reality put forward by Einstein, Podolsky, and Rosen in 1935.

The most important use of the GRC occurs in connection with EPR experiments, where each of the systems α_i composing S is physically correlated with a (separated) system β_i. Let the set of these new systems be denoted by

$$T = \{\beta_1, \beta_2, \ldots, \beta_N\}. \tag{6.4}$$

(For instance, the particles α_i and β_i could have been produced simultaneously by the decay of an unstable system M_i, $i = 1, 2, \ldots, N$.) Suppose that at some time the space parts of the quantum mechanical wave packets describing an α system and the corresponding β system are separated by a very large distance. This requirement will be viewed as a sufficient condition for the validity of the generalized separability principle (GSP), which is formulated as follows:

> *Measurements performed on the set T (β systems) cannot generate the physical properties required by the GRC belonging to any subset S′ of S (α systems).*

Of course, in the GSP the roles of the sets S and T can also be exchanged. See Figure 6.1 for relation between sets S, T and E.

In an EPR experiment one will detect S' by measuring a (dichotomic) observable B on the β particles. If $T' \subset T$ is the set of β particles for which the result $B = B_1$ has been found, it can often be predicted that another observable A will be found to have the values A_1 and A_2, with respective probabilities p_1 and p_2, in the subset S' of α particles correlated with the β particles belonging to T'.

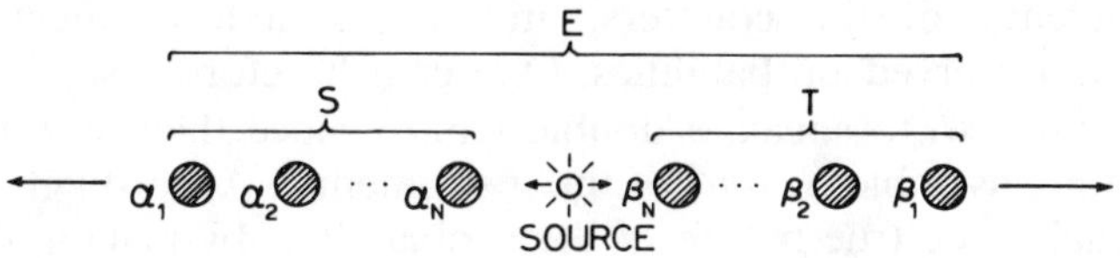

Fig. 6.1. S is the set of the α particles; T is the set of the β particles; E is the set of the (α, β) pairs.

Application of the GSP guarantees that the physical property λ' is not created at a distance from the measurements performed on T. This property λ' is a part of the physical reality of S' that is not detected *directly* in the usual experiments where eigenvalues and probabilities are measured. It is, therefore, of the same general nature as the so-called 'hidden variables', even though it is obviously more general than these, since it is attributed to statistical ensembles, rather than to individual systems. At present the existence of λ' is conjectural, but, if this conjecture were correct, it should obviously become possible to detect its existence with suitable experiments.

In a typical EPR experiment there are two observers, O_α who performs measurements of the observable A on the set S of the α particles, and O_β who measures B on the set T of the β particles. Assuming that O_β performs the first set of measurements (in the laboratory frame) and that T' is the subset of T for which $B = B_1$ (constant) has been found, then O_β himself can predict the existence of the subset S' (composed of the α particles individually correlated with the β particles contained in T') for which O_α later will find the results $A = A_1$ and $A = A_2$ with respective probabilities p_1 and p_2. Notice that p_1 and p_2 are in general different from the corresponding probabilities pertaining to the whole ensemble S. Therefore, if O_α later does find the probabilities predicted by O_β, it can be concluded that there is something in the physical reality of S' which somehow generates p_1 and p_2 if and when A is measured by O_α with the instrument I. It can also be concluded that these probabilities are *not* generated by *I independently of the properties of the objects composing* S', since neither the instrument I nor the observer O_α possesses information on the composition of S'. It is only when the experiment has been completed that O_β will tell O_α which subset of the indices $\{1, 2, \ldots, N\}$ should be used in composing the set S' for which the probabilities p_1 and p_2 are predicted. Under these conditions it seems natural to attribute the predicted emergence of p_1 and p_2 (which are detected as frequencies) at least partly to the physical properties of S'. It is on this basis that the physical property λ' is attributed to S'. The value of λ' fixes p_1 and p_2, as can be seen from (6.3).

In general, the probabilities, which become concrete when the observable A is actually measured, could *also* depend on the instrument I that is being used. This happens in many real experiments, where the precision of the analyzers, the efficiency of the counters, and so on indeed affect the determination of the observed probabilities. One can therefore assume that these probabilities have, so to speak, a double nature, since they reflect the physical reality of the ensemble S' *and* of the instrument I. Accordingly, the notation (6.3) highlights the true nature of the probabilities by making them dependent both on the physical property λ' of S' and on the measured observable A. (For simplicity, we include in the symbol A the properties of the instrument I, local vacuum fluctuation effects, and so on.)

The foregoing discussion shows why the probabilities p_1 and p_2 were not attributed directly to S' as real entities, whereas the property λ' was introduced. Had we defined the probabilities directly as real, then in practice none of the results to be obtained would have changed, but our approach would have been exposed to the criticism that we have attributed to S', as real, something that can emerge only through the concrete intervention of a measuring apparatus.

The exact nature of λ' is not of interest here. This property will generally result from the existence of a large number of 'elements of reality' that cooperate to generate a situation for S' in which the probabilities are p_1 and p_2 and not different.

An important point to stress is that the property λ' will be attributed to S' even if no actual measurement is carried out on S', *provided that previous and exhaustive experience with repetitions of the sets S, S′, T has shown that the probabilities* p_1 *and* p_2 *are correctly predicted*. By this nothing more is implied than the standard scientific procedure. If a large body of empirical evidence, assembled by repeated observations, proves that p_1 and p_2 invariably show up in S' — and thus, by the GRC, that the property λ' belongs invariably to S' — then one can conclude that λ' is real, also if no future measurement on the given set S' is made. In this manner one eliminates only the possibility that the action of the instrument I on the α particles creates a property λ' of these particles *retroactively in time*. Retroaction has, on the other hand, been proposed as a solution of the EPR paradox by some physicists (Rayski, 1988).

We can go a step further and by virtue of the GSP completely disentangle the eixstence of λ' from all types of measurement. Consider the measurement of B made by O_β on the set T of β particles. The results $B = B_1$ and $B = B_2$ split T into two subsets T' (all cases $B = B_1$) and T'' (all cases $B = B_2$). Recall that T' is related to S' in the way stated above and that λ' is a physical property of S'. Naturally, a completely symmetrical reasoning could be made for T'', the related set of α particles S'', its predictable probabilities, and its physical property λ''. The splitting of T into T' and T'' takes place if B is measured. It is this splitting that allows one to *identify* the sets S' and S'' to which physical properties λ' and λ'' can be attributed. But these physical properties cannot be created at a distance by the measurements on T, such influencing being excluded by the GSP. *Consequently, an unknown but concrete splitting of S into S′ (with the property* λ'*) and S″ (with the property*

λ'') *exists even if no measurement at all is made on the β particles.* Merely concluding that S' and S'' (with their respective properties λ' and λ'') exist will however suffice for our purposes, as will be shown later, even if it is not possible to identify those α particles that compose S' and S''.

The considerations put forth in this section constitute our form of local realism, which is obviously far weaker and more general than the one originally proposed by EPR. Our GRC does not suffer from the limitations of the EPR reality criterion, against which the objection has for instance been raised that one cannot see 'why a superposition should be less real than an eigenstate'. The present approach, by contrast, can be used equally starting from either eigenstates or from superpositions, since probabilities can be predicted for all types of quantum mechanical wave functions. Furthermore, our argument is not limited to the world of quanta; it is applicable to all physical domains, also independently of the use of any particular theory for the prediction of probabilities. The latter can often be known simply from previous experimental investigations.

It will next be shown that GRC + GSP lead to the conclusion that variable probabilities *must* be associated with different subsets of S. This result will be obtained by proving that statistical ensembles like S and S' cannot be homogeneous.

Consider a large set E of correlated pairs $\alpha_i + \beta_i$ $(i = 1, 2, \ldots, N)$. Suppose that a dichotomic observable $A(a) = \pm 1$ is measured on the set S of α objects and that either the observable $B(b) = \pm 1$ or the observable $B(b') = \pm 1$ are measured on the set T of β objects. The measurement of $B(b)$ divides E into two parts: $E_+(b)$ is the subset for which $B(b) = +1$ has been found, while $E_-(b)$ is the subset for which $B(b) = -1$ instead obtains. Assume that:

$$\begin{aligned}
&\omega(a_+ | b_+) \text{ is the probability of finding } A(a) = +1 \text{ in } E_+(b),\\
&\omega(a_- | b_+) \text{ is the probability of finding } A(a) = -1 \text{ in } E_+(b),\\
&\omega(a_+ | b_-) \text{ is the probability of finding } A(a) = +1 \text{ in } E_-(b),\\
&\omega(a_- | b_-) \text{ is the probability of finding } A(a) = -1 \text{ in } E_-(b).
\end{aligned} \tag{6.5}$$

If these probabilities can be predicted correctly, then one can attribute them as real to $E_\pm(b)$ by virtue of the GRC in the previously specified sense. They are of course real properties of some *unknown* subensemble also if no measurement of $B(b)$ has been performed. In fact, if this were not the case, one should admit that the measurement of $B(b)$ on the β systems creates these probabilities for the α systems, in violation of the GSP. Therefore, one deduces from GRC + GSP that, *even if no measurement of $B(b)$ is made*, there exists at least one unknown subensemble[1] $E_+(b)$ of E with two properties:

(i) Its population $N_+(b)$ divided by the total population N of E equals the *a priori* probability $q(b_+)$ of measuring $B(b)$ on the β systems and finding $+1$.

(ii) It has a physical property λ_+, leading to the necessary emergence of the probability $\omega(a_+ | b_+)$ when $A(a)$ is measured on the α objects of $E_+(b)$, for the result $A(a) = +1$.

In connection with the point (ii) above, it should be stressed that an important property of the probabilities $\omega(a_\pm | b_\pm)$ implied by the GSP is that they depend on b only in the sense that the ensembles $E_\pm(b)$ to which they belong do depend on b. Otherwise, given the physical separation between the objects α and β which guarantees the validity of the GSP, they depend exclusively on the observable $A(a)$ of the α objects to which they are relative. In other words, one could say that the measurements of $B(b)$ on the β objects give rise to a b-dependent splitting of E in $E_+(b)$ and $E_-(b)$. In each one of these subsets a probability is given for finding $A(a) = +1$ which is *physically independent* of b. The previous considerations are important for the deduction of the result (6.6) below.

Other probabilities can be introduced for a different splitting of E into $E_+(b')$ and $E_-(b')$ arising from the measurement of $B(b')$ on the β objects. For example, one assumes that $\omega(a_+ | b'_+)$ is the probability of finding $A(a) = +1$ in $E_+(b')$. One can again apply the GRC and declare $\omega(a_+ | b'_+)$ to be a consequence of a real property λ'_+ of $E_+(b')$. We have thus introduced three probabilities:

$\omega(a_+ | b_+)$, consequence of a real property of $E_+(b)$,

$\omega(a_+ | b_-)$, consequence of a real property of $E_-(b)$,

$\omega(a_+ | b'_+)$, consequence of a real property of $E_+(b')$.

They all refer to the result $A(a) = +1$. As conditional probabilities they do depend on b or on b', that is, on observables defined on the β objects, but, we repeat, only in the sense that the ensembles $E_+(b)$, $E_-(b)$, $E_+(b')$ to which they are attributed depend on b or on b'. Since

$$E = E_+(b) \cup E_-(b),$$

it follows that $E_+(b')$, which is also a subset of E, is necessarily composed of a fraction γ of elements of $E_+(b)$ and of a fraction $1 - \gamma$ of elements of $E_-(b)$, with $0 \leqslant \gamma \leqslant 1$. If the ensembles were homogeneous, in the sense that every subset of each of them had a probability for $A(a) = +1$ exactly equal to the one for the larger ensemble, then one should necessarily have [see Figure 6.2]

$$\omega(a_+ | b'_+) = \gamma\omega(a_+ | b_+) + (1 - \gamma)\omega(a_+ | b_-). \tag{6.6}$$

Equation (6.6) gives $\omega(a_+ | b'_+)$ as a weighted average of two other probabilities. But this means that $\omega(a_+ | b'_+)$ must be internal to the interval of positive numbers $\omega(a_+ | b_+)$ and $\omega(a_+ | b_-)$:

$$\omega(a_+ | b_-) \leqslant \omega(a_+ | b'_+) \leqslant \omega(a_+ | b_+) \tag{6.7}$$

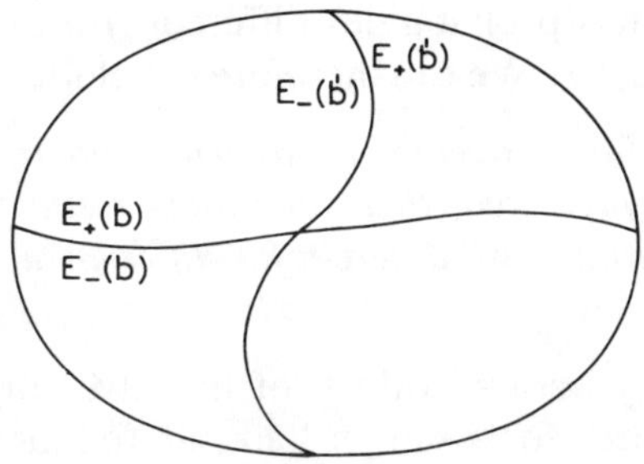

Fig. 6.2. Two different splittings of the set E of (α, β) pairs due to measurements of different observables $[B(b)$ and $B(b')]$ on the β particles. $E_+(b')$ is the union of a part of $E_+(b)$ and a part of $E_-(b)$.

if, for instance, $\omega(a_+|b_-) \leqslant \omega(a_+|b_+)$. The latter condition cannot however be generally satisfied. Even if it were true by chance, a repetition of the reasoning with the arguments b' and b interchanged would lead to a condition which could in general not be satisfied, the new condition being, of course, that $\omega(a_+|b_+)$ should lie inside the interval defined by $\omega(a_+|b'_+)$ and $\omega(a_+|b'_-)$. Now there is ample empirical and theoretical evidence in the quantum domain for the existence of situations in which the observed and/or predicted probabilities do not satisfy the previous conditions. As a numerical example, take

$$\omega(a_+|b_+) = 0.6, \omega(a_+|b_-) = 0.4, \omega(a_+|b'_+) = 0.8.$$

Obviously, in this case $\omega(a_+|b'_+)$ does not lie within the interval defined by $\omega(a_+|b_-)$ and $\omega(a_+|b_+)$. That this example is relevant to the quantum mechanical situation can be seen as follows. Consider the case in which the two quantum objects α and β discussed above are two photons in a positive parity, zero angular momentum state given by

$$|0+\rangle = \frac{\{|x_\alpha\rangle|x_\beta\rangle + |y_\alpha\rangle|y_\beta\rangle\}}{\sqrt{2}}, \tag{6.8}$$

where the letters x and y denote states of linear polarization along the x and y axis, respectively. As is well known, the conditional probabilities predicted by the state vector (6.8) are

$$\begin{aligned} \omega(a_+|b_+) &= \cos^2(a-b), \\ \omega(a_+|b_-) &= \sin^2(a-b), \\ \omega(a_+|b'_+) &= \cos^2(a-b'). \end{aligned} \tag{6.9}$$

It is very easy to choose values of the arguments $(a - b)$ and $(a - b')$ such that the inequalities (6.7) are not satisfied. This evidence undermines the basis of the reasoning leading to (6.6). Therefore, *the ensembles* $E_+(b)$ *and* $E_+(b')$ *cannot be homogeneous.* This lack of homogeneity means, for instance, that a subensemble of $E_+(b)$ has probabilities different from those which can be attributed to the whole of $E_+(b)$. We can therefore conclude:

> *If probabilities are consequences of objectively real physical properties of the ensembles for which they are predicted, then they must result from averages of objective probabilities in general different for different subsets composing these ensembles.*

Notice that the existing quantum theory knows nothing of this structure of probabilities. The probabilities (6.9) are, for example, considered devoid of any more detailed description, and the corresponding ensembles are described as homogeneous. We have therefore discovered a new and more general sense in which quantum theory has to be considered incomplete. In the old sense, incompleteness meant that a perfectly deterministic description of quantum phenomena should be considered possible. As is well known, several theorems of von Neumann's type were proposed in order to shield the theory from possible deterministic completions. No theorem has ever been proposed against the type of incompleteness discussed above.

6.3. New Proof of Bell's Inequality

In the previous section it was concluded that quantum mechanical ensembles cannot be homogeneous if our general formulation of probabilistic Einstein locality holds in nature. It will next be shown that this formulation leads necessarily to the validity of Bell's inequality. That ensembles are inhomogeneous means that probabilities cannot have constant values.

It becomes thus necessary to introduce (variable) probabilities for subsets. We call

$p_i(a_+)$ the probability of $A(a) = +1$ for subset E_i,

$q_i(b_+)$ the probability of $B(b) = +1$ for subset E_i,

where $i = 1, 2, \ldots, n$, n being the total number of subsets. Of course these subsets must be thought of as given *a priori*, that is physically determined in the source and not by acts of measurement performed on the pairs. They must satisfy

$$E = E_1 \cup E_2 \cup \cdots \cup E_n. \tag{6.10}$$

These probabilities are local: for example $p_i(a_+)$ does not depend on b, that is, on the observable measured on the (separated) systems β which form physical pairs with the systems α. This form of locality will be assumed from now on. It can obviously be deduced from the GSP, and is more general than the usual locality assumed in connection with the deterministic approach. All the ensemble probabilities can now be expressed by means of weighted averages of subset probabilities. One has, for instance,

$$p(a_+) = \sum_{i=1}^{n} p_i(a_+)\left(\frac{N_i}{N}\right), \; q(b_+) = \sum_{i=1}^{n} q_i(b_+)\left(\frac{N_i}{N}\right), \tag{6.11}$$

for the probabilities to obtain $A(a) = +1$ and $B(b) = +1$, respectively, in the full set E, where N_i is the population of the subset E_i. Also conditional probabilities can be written in this way. For example,

$$\begin{aligned} \omega(a_+|\,b_+) &= \sum_{i=1}^{n} p_i(a_+)\left[\frac{N_i'(b)}{N_+(b)}\right], \\ \omega(a_+|\,b_-) &= \sum_{i=1}^{n} p_i(a_+)\left[\frac{N_i''(b)}{N_-(b)}\right], \end{aligned} \tag{6.12}$$

where $N_i'(b)$ and $N_i''(b)$ are the numbers of pairs of the set E_i which belong also to $E_+(b)$ and $E_-(b)$, respectively. Obviously

$$N_i'(b) + N_i''(b) = N_i, \tag{6.13}$$

since every pair of E_i must belong either to $E_+(b)$ or to $E_-(b)$.

The joint probabilities of finding $A(a) = \pm 1$ and $B(b) = \pm 1$ are given by the fundamental theorem of probabilities and can be written

$$\begin{aligned} \Omega(a_+, b_+) &= \omega(a_+|\,b_+)q(b_+) = \sum_{i=1}^{n} p_i(a_+)\left[\frac{N_i'(b)}{N}\right], \\ \Omega(a_-, b_+) &= \omega(a_-|\,b_+)q(b_+) = \sum_{i=1}^{n} p_i(a_-)\left[\frac{N_i'(b)}{N}\right], \\ \Omega(a_+, b_-) &= \omega(a_+|\,b_-)q(b_-) = \sum_{i=1}^{n} p_i(a_+)\left[\frac{N_i''(b)}{N}\right], \\ \Omega(a_-, b_-) &= \omega(a_-|\,b_-)q(b_-) = \sum_{i=1}^{n} p_i(a_-)\left[\frac{N_i''(b)}{N}\right]. \end{aligned} \tag{6.14}$$

The joint probabilities are useful for the calculation of the correlation function, given by

$$\begin{aligned} P(a, b) &= \Omega(a_+, b_+) - \Omega(a_-, b_+) - \Omega(a_+, b_-) + \Omega(a_-, b_-) \\ &= \sum_{i=1}^{n} \pi_i(a)\left[\frac{N_i'(b)}{N}\right] - \sum_{i=1}^{n} \pi_i(a)\left[\frac{N_i''(b)}{N}\right], \end{aligned} \tag{6.15}$$

where

$$\pi_i(a) = p_i(a_+) - p_i(a_-). \tag{6.16}$$

Since $\pi_i(a)$ is given by the difference of two probabilities, one must have

$$|\pi_i(a)| \leqslant 1. \tag{6.17}$$

It is now a simple matter to show that

$$|P(a, b) - P(a', b)| \leqslant \sum_{i=1}^{n} |\pi_i(a) - \pi_i(a')| \times$$

$$\times \left\{\left[\frac{N'_i(b)}{N}\right] + \left[\frac{N''_i(b)}{N}\right]\right\},$$

and

$$|P(a, b') + P(a', b')| \leqslant \sum_{i=1}^{n} |\pi_i(a) + \pi_i(a')| \times$$

$$\times \left\{\left[\frac{N'_i(b)}{N}\right] + \left[\frac{N''_i(b)}{N}\right]\right\}.$$

By using (6.13) one can easily obtain

$$|P(a, b) - P(a', b)| + |P(a, b') + P(a', b')| \leqslant$$

$$\leqslant \sum_{i=1}^{n} \left(\frac{N_i}{N}\right) \{|\pi_i(a) - \pi_i(a')| + |\pi_i(a) + \pi_i(a')|\} \leqslant$$

$$\leqslant \sum_{i=1}^{n} \left(\frac{N_i}{N}\right) 2 \leqslant 2, \tag{6.18}$$

that is, Bell's inequality (Bell, 1965).

It has been possible to show that the same approach used in this section allows one to obtain all the other inequalities for linear combinations of correlation functions and of joint probabilities: This matter will be discussed in Section 6.6.

Probabilistic realism as defined in the previous section means that probabilities that can be predicted correctly before measurements are performed must be based on concrete physical properties of those ensembles for which they are predicted. Probabilistic separability means that probabilities must remain the same independently of the measurements made on a different distant (and thus physically separated) ensemble. The present scheme is based on very general and very reasonable assumptions. It rests on such a weak form of realism that one wonders if the physicists of the Copenhagen and Göttingen schools could perhaps have accepted it. Nevertheless, Bell's inequality could still be deduced. Since this inequality is known to be violated by the quantum mechanical predictions, it follows that not even the present weak form of local realism is compatible with the existing quantum theory.

6.4. Probabilities for Pairs of Correlated Systems

In 1974 Clauser and Horne defined the 'objective local theories' in the following way.

Let an EPR-type experiment be given with independent pairs of correlated objects emitted by a source, with object α (object β) flying to the right (to the left), where the analyzer 1 and the detector 1 (the analyzer 2 and the detector 2) are placed. The 'state' of the two objects is denoted by λ which is given, over a statistical ensemble of similar pairs, with a positive-definite and normalized density function $\rho(\lambda)$. The variable λ is taken to vary in a 'space of states' Γ. If the two analyzers have each an adjustable parameter, $a(b)$ being the value of the parameter of apparatus 1 (of apparatus 2), Clauser and Horne make the following assumption:

Neither the density $\rho(\lambda)$ nor the space Γ depend on the parameters a and b. (CH1)

Let the probabilities of a count being triggered at apparatus 1 and 2 be $p(a, \lambda)$ and $q(b, \lambda)$, respectively, and let $P(a, b, \lambda)$ be the probability that both counts are triggered. In terms of the quantities just defined, the corresponding ensemble probabilities are

$$p(a) = \int_\Gamma d\lambda\, \rho(\lambda) p(a, \lambda),$$
$$q(b) = \int_\Gamma d\lambda\, \rho(\lambda) q(b, \lambda), \tag{6.19}$$

$$D(a, b) = \int_\Gamma d\lambda\, \rho(\lambda) D(a, b, \lambda). \tag{6.20}$$

Clauser and Horne focused their attention on a class of theories in which

$$D(a, b, \lambda) = p(a, \lambda) q(b, \lambda) \tag{CH2}$$

and called this 'the factorizability condition'.

The assumptions (CH1) and (CH2) define the 'objective local theories' of Clauser and Horne.

Now, this formulation of local realism is often taken to be the most general one, covering a priori all conceivable situations which can be described in terms of (probabilistic) local realism. It will however be shown in the present section that this is not true, and that general formulations of local realism lead to the violation of the condition (CH1) if probabilities are defined in terms of frequencies. The meaning of 'probability' is crucial in this context. Clauser and Horne seem to oscillate between an individual definition and a statistical one. We shall show that their assumptions can be justified only with the former definition, which is however unsatisfactory in other respects, since no known formulation of probability calculus adopts a probability referred to the individual object.

Our conclusions will therefore fully support reservations and criticisms of the Clauser–Horne paper made by different authors (Suppes and Zanotti, 1976; Selleri and Tarozzi, 1980).

It will next be shown that the first probabilistic formulation of local realism (Wigner, 1970) does not satisfy the assumption (CH1). Later (Section 6.5) it will also be shown that our general formulation of local realism leads to violations of (CH1). The common reason for these disagreements rests on different definition of the very idea of 'probability'. In spite of this disagreement, the inequalities previously deduced from the Clauser–Horne formulation of probabilistic local realism remain unaffected, as will be shown in Section 6.6.

Wigner made two basic assumptions. The first one was that the results of all conceivable measurements are simultaneously prefixed, even in the case of incompatible observables. This realistic standpoint does not contradict Heisenberg's relations, because the latter can be taken simply to mean that a concrete measurement made on a given object modifies the prefixed values of other observables of that object not compatible with the measured one. Before the action of the instrument it is however possible that the results of all conceivable measurements be predetermined.

The second assumption was locality. A measurement made on $\alpha[\beta]$ does not modify the prefixed values $B(b_1), B(b_2), \ldots, B(b_n)$ $[A(a_1), A(a_2), \ldots, A(a_m)]$ of the observables of $\beta[\alpha]$. Therefore, if, for example, $A(a_1)$ is measured on an α object, the values of $B(b_1), B(b_2), \ldots, B(b_n)$ are not modified by that measurement, but those of $A(a_2), \ldots, A(a_m)$, in general, are.

We are obviously dealing with a local realistic and deterministic approach, since the result of every possible measurement is predetermined by some concrete properties of the measured object ('hidden variables'). This does not mean that an active role of the apparatus is excluded, it only means that the interaction between object and apparatus is steered to a prefixed outcome ('result of measurement') by the hidden variables of the object.

If we consider again the dichotomic observables

$$\begin{aligned} &A(a_1), A(a_2), \ldots, A(a_m) \quad \text{for } \alpha; \\ &B(b_1), B(b_2), \ldots, B(b_n) \quad \text{for } \beta; \end{aligned} \tag{6.21}$$

the previous assumptions imply that the set E of the $N(\alpha, \beta)$ pairs splits into 2^{m+n} subsets, with well-defined populations, in each of which the prefixed outcomes of the $m + n$ possible measurements are constant.

Defining the vectors V with $m + n$ components equal to ± 1:

$$V = [A(a_1), A(a_2), \ldots, A(a_m), B(b_1), B(b_2), \ldots, B(b_n)], \tag{6.22}$$

we notice that there are 2^{m+n} different V vectors, one for every subset.

One can then use V to denote a particular subset of E and write

$$E(V) = E[A(a_1), A(a_2), \ldots, A(a_m), B(b_1), B(b_2), \ldots, B(b_n)] \tag{6.23}$$

for the subset and

$$N(V) = N[A(a_1), A(a_2), \ldots, A(a_m), B(b_1), B(b_2), \ldots, B(b_n)] \tag{6.24}$$

for its population.

Obviously, one has

$$\sum_{V \in W} N(V) = N, \tag{6.25}$$

where

$$W = \{V\} \tag{6.26}$$

is the set of all possible vectors of type (6.22), which are of course 2^{m+n} in number.

The joint probability of having $A(a_\mu) = +1$ and $B(b_\nu) = +1$ is given by

$$\Omega(a_\mu+, b_\nu+) = \sum_{V \in W} \frac{N(V)}{N} \cdot \frac{A(a_\mu) + 1}{2} \cdot \frac{B(b_\nu) + 1}{2}, \tag{6.27}$$

since the last two factors multiply by 1 (by 0) the contributions coming from V vectors having $A(a_\mu) = B(b_\nu) = 1$ [$A(a_\mu) = -1$ and/or $B(b_\nu) = -1$].

In a deterministic model all probabilities are either 0 or 1, and one can write

$$p(a_\mu+, V) = \frac{A(a_\mu) + 1}{2}, \qquad q(b_\nu+, V) = \frac{B(b_\nu) + 1}{2} \tag{6.28}$$

if $p(a_\mu+, V)$ and $q(b_\nu+, V)$ are the probabilities for having $A(a_\mu) = +1$ and $B(b_\nu) = +1$, respectively. These probabilities can be thought to depend on V, even though they actually depend only on one component of the vector V.

Inserting (6.28) in (6.27), one obtains

$$\Omega(a_\mu+, b_\nu+) = \sum_{V \in W} \frac{N(V)}{N} p(a_\mu+, V) q(b_\nu+, V). \tag{6.29}$$

This result is similar to the Clauser—Horne formula (CH2), if V is substituted for λ. There are, however, two important differences:

(i) The 'density' $N(V)/N$ depends on the chosen observables, as shown in Equation (6.24).

(ii) The set W of possible vectors V depends on their total number $m+n$.

These two conclusions are in disagreement with the assumption (CH1). Thus already Wigner's 1970 formulation of local realism was incompatible with a full generality of the Clauser—Horne conditions. This is hardly surprising, given the different definitions of probability used by Wigner (probability as frequency in an ensemble of similar objects) and by Clauser and Horne (probability referred to the individual quantum object in the microstate λ and/or to the single pair of correlated quantum systems).

One can show in all generality that an averaging over one variable leads to expressions that do not need to satisfy the CH conditions for the remaining variables (Selleri, 1988a). Let $\{\lambda\}$ be the set of variables specifying the state of $\alpha + \beta$ and λ_0 just one of these variables. Denote by $\{\lambda'\}$ the variables that remain after λ_0 has been deleted from the original set. If $\rho(\lambda', \lambda_0)$ is the density function and if the CH conditions hold for $\{\lambda\} = \{\lambda', \lambda_0\}$, then one can write

$$\Omega(a_+, b_+) = \int \mathrm{d}\lambda_0 \int \mathrm{d}\lambda' \, \rho(\lambda', \lambda_0) p(a_+, \lambda', \lambda_0) q(b_+, \lambda', \lambda_0). \tag{6.30}$$

Now suppose the CH conditions remain valid also after the averaging over the variable λ_0 has been executed. One should then have something of the type

$$\rho(\lambda')\bar{p}(a_+, \lambda')\bar{q}(b_+, \lambda') = \int \mathrm{d}\lambda_0 \, \rho(\lambda', \lambda_0) p(a_+, \lambda', \lambda_0) q(b_+, \lambda', \lambda_0). \tag{6.31}$$

That this outcome is not possible in general, one can see on dividing the relation (6.31) by its partial derivative with respect to the argument a:

$$\frac{\bar{p}(a_+, \lambda')}{\dfrac{\partial}{\partial a} \bar{p}(a_+, \lambda')} = \frac{\displaystyle\int \mathrm{d}\lambda_0 \, \rho(\lambda', \lambda_0) p(a_+, \lambda', \lambda_0) q(b_+, \lambda', \lambda_0)}{\displaystyle\int \mathrm{d}\lambda_0 \, \rho(\lambda', \lambda_0) \left[\frac{\partial}{\partial a} p(a_+, \lambda', \lambda_0) \right] q(b_+, \lambda', \lambda_0)}. \tag{6.32}$$

Indeed, while the left-hand side of (6.32) depends only on a (and not on b), its right-hand sidc will *in general* depend on both a and b. Consequently, only those very particular cases in which the right-hand side of (6.4) does not depend on b can maintain the CH conditions after the averaging over λ_0.

One can conclude that any validity of CH(1) and CH(2) that might originally exist would in most cases get lost as soon as an averaging procedure is carried out. This fact casts serious doubts on the validity of the two CH conditions *in the world*.

Of course, an optimist could maintain that the world is very simple and that we know almost everything about it. The next variable λ which is introduced for describing an atomic system would provide a perfect reproduction of the physical reality of such a system. In the case of correlated EPR pairs, it could then be a reasonable *ansatz* to postulate (CH1) + (CH2) for fixed λ.

However, a more skeptical person could assume that the depth of 10^{-15} cm or so which we have reached so far is very remote from 'the bottom of the world'. Moreover, such a bottom might very well not exist at all, and every atomic system could be 'inexhaustible in depth', as some thinkers have claimed. In either case, our skeptic would be correct in concluding that any description of reality given today must necessarily be based on the presupposition that our observed results are physical averages over all those variables that describe the as yet unexplored levels of reality. But then the assumption of (CH1) + (CH2) for EPR pairs would be totally unjustified, as we have seen.

We shall next show that even in the macroscopic domain there are situations in which correlated systems do not satisfy condition (CH1).

The example concerns a statistical ensemble of identical twins (Liddy, 1983). One can imagine that every pair of twins is split up by placing the identical individuals in two different towns. We introduce three variables for the description of these twins:

> a is the *weight.* A dichotomic observable $A(a)$ is defined to be $+1$ for a heavy person and -1 for a light person, the point of division beign taken at 50 kg (we are considering growing youngsters).
>
> b is the *height.* A dichotomic observable $A(b)$ is defined to be $+1$ for a tall person and -1 for a short person, the cut being made at 160 cm.
>
> λ is the age in years.

Obviously one may regard $A(a)$ and $A(b)$ as discontinuous functions of λ:

$$A(a) = A(a, \lambda), A(b) = A(b, \lambda), \tag{6.33}$$

since in general it happens that a given person at some age jumps from light to heavy and/or from short to tall if these properties have the dichotomic definition given above.

Next we introduce probabilities. Let $p(a_{\pm}, \lambda)$ be the probability that a given twin is heavy (light) at age λ, and let $p(b_{\pm}, \lambda)$ be the probability that a given twin is tall (short) at age λ. One must have

$$p(a_{+}, \lambda) + p(a_{-}, \lambda) = 1 = p(b_{+}, \lambda) + p(b_{-}, \lambda). \tag{6.34}$$

Only two independent probabilities are therefore left. We consider, of course, our probabilistic theory of twins to be a *statistical theory*, of the same type as quantum theory (and of its eventual generalizations through the introduction of one or a few new parameters). Therefore, *all the previous probabilities are defined in statistical terms, that is, they are to be viewed as frequencies of twins of age λ being heavy/light/tall/short.*

Coming now to the EPR-type situation, we assume every pair of identical twins to be composed of the individuals α and β. Measurements of the two observables $A_\alpha(a)$ and $A_\alpha(b)$, defined as above, are made on α in a certain town, while measurements of the similar observables $A_\beta(a)$ and $A_\beta(b)$ are made on β in a different town.

Also the probabilities must be introduced separately for α and β. One may, for instance, define $p_\alpha(a_+, \lambda)$ to be the probability that α will be heavy at age λ, and so on. In this way, eight λ-dependent probabilities can be introduced. But only four of them are linearly independent, since relations of the type (6.34) must hold.

Joint probabilities, too, can be defined. For example, $\Omega(a_+, b_-, \lambda)$ may signify the probability that α will be found heavy and β short at the common age λ. Obviously, 16 probabilities of this type exist:

$$\Omega(x_\pm, y_\pm, \lambda), \qquad x, y = a, b.$$

There are, however, several relations which either equate to zero some of these probabilities or connect them to probabilities for a single twin (α or β). We are interested only in eight of these relations; the first four read:

$$\begin{aligned} \Omega(a_+, a_-, \lambda) &= 0, \\ \Omega(a_-, a_+, \lambda) &= 0, \\ \Omega(b_+, b_-, \lambda) &= 0, \\ \Omega(b_-, b_+, \lambda) &= 0. \end{aligned} \tag{6.35}$$

Equations (6.35) are consequences of the assumed identity of the twins making up a set. For instance, one twin cannot be tall and his identical partner short at the same age. The other four relations of interest are

$$\begin{aligned} \Omega(a_+, a_+, \lambda) &= p_\alpha(a_+, \lambda) = p_\beta(a_+, \lambda), \\ \Omega(a_-, a_-, \lambda) &= p_\alpha(a_-, \lambda) = p_\beta(a_-, \lambda), \\ \Omega(b_+, b_+, \lambda) &= p_\alpha(b_+, \lambda) = p_\beta(b_+, \lambda), \\ \Omega(b_-, b_-, \lambda) &= p_\alpha(b_-, \lambda) = p_\beta(b_-, \lambda). \end{aligned} \tag{6.36}$$

These equalities are also consequences of the identity of twins. The first Equation (6.36), for example, says that the probability of finding both twins heavy is the same as the probability that either one of them will be heavy — an obviously valid conclusion.

Importantly, it follows from (6.36) that, for instance,

$$\Omega(a_+, a_+, \lambda) \neq p_\alpha(a_+, \lambda) p_\beta(a_+, \lambda), \tag{6.37}$$

an equality sign replacing the inequality sign only when the probabilities are all either zero or unity. The meaning of (6.37) is clear: *the CH factorizability assumption does not hold in the present, rigorously local example of identical twins.*

Conceivably three escape routes from this conclusion exist. One possibility would be to say that we must delve deeper and include further physical variables in the set of λ's. Perhaps the inclusion of parameters describing personal histories, genetic endowments, and so on, would recover factorizability for our separable twins. The argument given above, Equation (6.32), shows however that such a solution cannot always be valid, unless all the existing physical variables are included in the theoretical scheme, which requirement may well be impossible to satisfy, at least at the present time.

A second possibility consists in adopting the view that the CH-factorizability assumption should be accepted as a completely new definition of *objective probabilities for individual systems.*

We scarcely need to recall here that the objective definition of probability known today is that of statistical probability, and that an extension of this notion to individual systems should not be done by *ansatz*, in the vein of Clauser and Horne, but by a careful derivation from solid physical principles.

The third way to save factorizability is to observe that (6.37) does not obtain if all probabilities are all either 0 or 1. Formally this solution is certainly valid, but the catch is that probabilities equal to 1 or to 0 represent *certainties* and that the whole probabilistic scheme then becomes a deterministic one. The situation we face is that determinism very probably does not apply to the quantum domain.

Therefore it is quite possible today to defend causality, but almost impossible to believe in the universal validity of determinism. From this point of view, it is finally evident that Laplacian determinism is dead, that Brownian motion is an *objectively casual* phenomenon, and that *objective casuality is part of causality.*

None of the foregoing attempts to save CH factorizability is *a priori* fully convincing. After all, factorizable probabilities represent statistically independent phenomena, and one cannot see why Einstein locality should be *equivalent* to statistical independence.

6.5. A New Factorizability Condition

In a previous section we have been led to the conclusion that, e.g., $E(b+)$ and $E(b-)$, and therefore E, cannot be homogeneous ensembles as far as the probabilities of finding $A(a) = \pm 1$ are concerned. A symmetrical conclusion holds for the probabilities of $B(b) = \pm 1$ in the ensemble E.

In the present section, we shall exploit systematically these proprieties of E by introducing the subensembles

$$\sigma_1(a), \sigma_2(a), \ldots, \sigma_r(a), \tag{6.38}$$

homogeneous for $A(a) = \pm 1$, and

$$\tau_1(b), \tau_2(b), \ldots, \tau_s(b), \tag{6.39}$$

homogeneous for $B(b) = \pm 1$. Obviously:

$$\begin{aligned} \sigma_1(a) \cup \sigma_2(a) \cup \cdots \cup \sigma_r(a) &= E, \\ \tau_1(b) \cup \tau_2(b) \cup \cdots \cup \tau_s(b) &= E. \end{aligned} \tag{6.40}$$

The notation reflects the fact that a subset that is homogeneous for an observable (e.g., $A(a)$) is not expected in general to be homogeneous for a different observable (e.g. $B(b)$).

One can also introduce subensembles which are homogeneous for two or more observables.

Considering $A(a)$ and $B(b)$, the homogeneous subensembles are

$$E_k(a, b) = \sigma_i(a) \cap \tau_j(b), \tag{6.41}$$

where the single index k has been chosen, for simplicity, to correspond (in a one-to-one way) to the pair of indices i, j.

The previous approach can easily be generalized to an arbitrary number of observables of the first and the second object. Considering m dichotomic observables for α and n dichotomic observables for β, as in (6.21), the following homogeneous subensembles of E can be introduced:

$$\begin{aligned} &\sigma_1(a_1), \sigma_2(a_1), \ldots, \sigma_{r_1}(a_1), \text{ homogeneous for } A(a_1), \\ &\sigma_1(a_2), \sigma_2(a_2), \ldots, \sigma_{r_2}(a_2), \text{ homogeneous for } A(a_2), \\ &\vdots \\ &\sigma_1(a_m), \sigma_2(a_m), \ldots, \sigma_{r_m}(a_m), \text{ homogeneous for } A(a_m), \\ &\tau_1(b_1), \tau_2(b_2), \ldots, \tau_{s_1}(b_1), \text{ homogeneous for } B(b_1), \\ &\tau_1(b_2), \tau_2(b_2), \ldots, \tau_{s_2}(b_2), \text{ homogeneous for } B(b_2), \\ &\vdots \\ &\tau_1(b_n), \tau_2(b_n), \ldots, \tau_{s_n}(b_n), \text{ homogeneous for } B(b_n). \end{aligned} \tag{6.42}$$

Obviously, the union of the ensembles of every line in (6.42) gives always E, just as in (6.40).

By means of suitable intersections, one can introduce smaller subensembles in which all the observables considered have constant probability.

One can write

$$\begin{aligned} E_k(a_1, \ldots, a_m, b_1, \ldots, b_n) = \sigma_{i_1}(a_1) \cap \sigma_{i_2}(a_2) \cap \cdots \cap \sigma_{i_m}(a_m) \cap \\ \cap \tau_{j_1}(b_1) \cap \tau_{j_2}(b_2) \cap \cdots \cap \tau_{j_n}(b_n) \end{aligned} \tag{6.43}$$

for a typical subensemble homogeneous for all the probabilities of the ± 1 results of the $m + n$ observables considered. In (6.43) the single index k has been chosen, for simplicity, to correspond in a one-to-one way to the set of indices

$$\{i_1, i_2, \ldots, i_m, j_1, j_2, \ldots, j_n\}. \tag{6.44}$$

The number $t(m, n)$ of such sets can easily be calculated if one notices that i_1 can be chosen in r_1 different ways, ..., i_m in r_m different ways, j_1 in s_1 different ways, ..., j_n in s_n different ways, all choices being independent. Therefore

$$t(m, n) = r_1 r_2 \ldots r_m s_1 s_2 \ldots s_n. \tag{6.45}$$

Notice, however, that $t(m, n)$ is in general expected to change if some arguments $a_1, \ldots, b_n$ are modified, as is also clear from the notation (6.42), where r_1, for example, has the same index as a_1, precisely because a change of a_1 can give rise to a modification of r_1.

Therefore the set I of indices k will depend on the arguments of the observables:

$$I = I(a_1, \ldots, a_m, b_1, \ldots, b_n). \tag{6.46}$$

The notation can be simplified if one introduces a 'vector' V, similar to (6.22), having $m + n$ components and given by

$$V \equiv \{a_1, \ldots, a_m, b_1, \ldots, b_n\}, \tag{6.47}$$

since the homogeneous subensembles and their populations can be written

$$E_k(V) = E_k(a_1, \ldots, a_m, b_1, \ldots, b_n), \tag{6.48}$$

and

$$N_k(V) = N_k(a_1, \ldots, a_m, b_1, \ldots, b_n). \tag{6.49}$$

The basic probabilities, which are constant for all pairs of a given subensemble $E_k(V)$, can be written

$$\begin{aligned} &p_k(a_1 \pm), \quad \text{probabilities for} \quad A(a_1) = \pm 1,\\ &p_k(a_2 \pm), \quad \text{probabilities for} \quad A(a_2) = \pm 1,\\ &\vdots\\ &p_k(a_m \pm), \quad \text{probabilities for} \quad A(a_m) = \pm 1, \end{aligned} \tag{6.50}$$

and

$$\begin{aligned} &q_k(b_1 \pm), \quad \text{probabilities for} \quad B(b_1) = \pm 1,\\ &q_k(b_2 \pm), \quad \text{probabilities for} \quad B(b_2) = \pm 1,\\ &\vdots\\ &q_k(b_n \pm), \quad \text{probabilities for} \quad B(b_n) = \pm 1. \end{aligned} \tag{6.51}$$

These probabilities can be used for calculating the single-object probabilities for the whole ensemble E, as follows:

$$p(a_\mu \pm) = \sum_{k \in I} \frac{N_k(V)}{N} p_k(a_\mu \pm); \; q(b_\nu \pm) = \sum_{k \in I} \frac{N_k(V)}{N} q_k(b_\nu \pm), \quad (6.52)$$

where $\mu = 1, \ldots, m$ and $\nu = 1, \ldots, n$.

Also joint probabilities can be calculated in the same way, provided one recognizes that two observables of the same object are often represented by noncommuting operators in quantum theory and cannot be measured simultaneously. Operators relative to two observables like $A(a_\mu)$ and $B(b_\nu)$ instead always commute and the previous obstacle does not exist.

In a previous section we showed that for a given dichotomic observable $B(b_\nu)$ there are two subensembles $E(b_\nu+)$ and $E(b_\nu-)$, such that

$$E(b_\nu+) \cup E(b_\nu-) = E, \quad (6.53)$$

having populations $N(b_\nu+)$ and $N(b_\nu-)$, respectively, with

$$N(b_\nu+) + N(b_\nu-) = N, \quad (6.54)$$

such that the *apriori* probability that a measurement of $B(b_\nu)$ on E gives the value ± 1 is given by

$$q(b_\nu \pm) = \frac{N(b_\nu \pm)}{N}. \quad (6.55)$$

If we call $N_k(b_\nu \pm)$ the population of $E_k(V) \cap E(b_\nu \pm)$, we have the conditional probabilities

$$\begin{aligned} \omega(a_\mu \pm \mid b_\nu+) &= \sum_{k \in I} \frac{N_k(b_\nu+)}{N(b_\nu+)} p_k(a_\mu \pm), \\ \omega(a_\mu \pm \mid b_\nu-) &= \sum_{k \in I} \frac{N_k(b_\nu-)}{N(b_\nu-)} p_k(a_\mu \pm). \end{aligned} \quad (6.56)$$

Therefore, the joint probabilities become

$$\begin{aligned} \Omega(a_\mu \pm, b_\nu+) &= \omega(a_\mu \pm \mid b_\nu+) q(b_\nu+) = \sum_{k \in I} \frac{N_k}{N} p_k(a_\mu \pm) q_k(b_\nu+), \\ \Omega(a_\mu \pm, b_\nu-) &= \omega(a_\mu \pm \mid b_\nu-) q(b_\nu-) = \sum_{k \in I} \frac{N_k}{N} p_k(a_\mu \pm) q_k(b_\nu-), \end{aligned} \quad (6.57)$$

since the a-priori probabilities are equal to the corresponding frequencies

$$q(b_\nu \pm) = \frac{N(b_\nu \pm)}{N}, \; q_k(b_\nu \pm) = \frac{N_k(b_\nu \pm)}{N_k}. \quad (6.58)$$

If one writes

$$\frac{N_k(V)}{N} \equiv \rho_k(V), \tag{6.59}$$

one must obviously have

$$\sum_{k \in I} \rho_k(V) = 1, \tag{6.60}$$

so that $\rho_k(V)$ can be considered a 'weight', in general different for different subsets $E_k(V)$.

The joint probabilities (6.57) can now be written

$$\Omega(a_\mu \pm, b_\nu \pm) = \sum_{k \in I} \rho_k(V) p_k(a_\mu \pm) q_k(b_\nu \pm), \tag{6.61}$$

which looks like the Clauser—Horne factorizability formula, the role of the hidden variable λ being played by the index k. There are however very important differences, because now the probability density (6.59) depends on the parameters of the observables under consideration; and so do the set of indices I [equivalent to the set Γ of values of λ in the CH theory].

In spite of this, when the sum over k is performed, all dependence on the arguments $a_1, \ldots, a_m, b_1, \ldots, b_n$ disappears, with the exception of a_μ and b_ν, as indicated on the left-hand side of (6.61).

We conclude that the (CH1) condition, which we reproduced in Section 6.4, does not hold in a probabilistic theory based on local realism.

The conclusions drawn in the present section in the general case are the same as those of Section 6.2, in which Wigner's probabilistic approach (based on a deterministic background) was discussed. Also, the reason for this similarity is clear: we used a statistical definition of probability (probability as frequency in the statistical ensemble) while Clauser and Horne insisted on an individual definition (probability for one or two correlated objects in the state λ).

Commenting on their choice of a hidden-variable density function $\rho(\lambda)$ that does not depend on a and b, Clauser and Horne pointed out that in so doing they were discarding such objective and local possibilities as these:

(1) Systems of some type originate at the source which, after being reflected by the analyzers, return to the source and modify the probability of production of different values of λ in a manner dependent upon a and b.
(2) Systems originate at the analyzers and impinge upon the source, thus modifying the density $\rho(\lambda)$ in an a- and b-dependent way.
(3) Systems originate from the interaction of the background light cones of both analyzers and the source. They propagate into the spatial region of the whole apparatus and simultaneously modify the emis-

sions from the source and the experimenters' selections of the parameters a and b.

These far-fetched local generations of dependences of $\rho(\lambda)$ on the parameters a and b chosen in a given experiment have nothing to do with our conclusion that the density depends on the chosen set of parameters. The latter conclusion is in general a necessary consequence of probabilistic local realism and is due to the existence of homogeneous subensembles for all dichotomic observables. In order to find subensembles homogeneous for all the observables considered, one must necessarily perform intersections. It is natural that a subset obtained by intersection depends on the number and on the quality of the sets that are used for generating it.

6.6. All the Inequalities of Einstein Locality

In 1978—1979 Roy and Singh deduced from local realism inequalities more general than Bell's inequality and providing physical restrictions on the correlation functions $P(a_\mu, b_\nu)$ that cannot be obtained from Bell's inequality (Roy and Singh, 1978 and 1979). Before writing down some of these inequalities, it is important to stress that, given a linear combination of the correlation functions,

$$\sum_{\mu=1}^{m} \sum_{\nu=1}^{n} c_{\mu\nu} P(a_\mu, b_\nu),$$

the $m \cdot n$ coefficients $c_{\mu\nu}$ define an $m \times n$ matrix which can be taken to represent the original linear combination.

The first Roy—Singh inequality is

$$\sum_{\mu=1}^{4} \sum_{\nu=1}^{5} c^1_{\mu\nu} P(a_\mu, b_\nu) \leqslant 6, \tag{6.62}$$

where

$$c^1_{\mu\nu} = \begin{pmatrix} 1 & 1 & 1 & 0 & 1 \\ 1 & 1 & -1 & 1 & 0 \\ 1 & 1 & 0 & -1 & -1 \\ 1 & -1 & 0 & 0 & 0 \end{pmatrix}; \tag{6.63}$$

a second one is

$$\sum_{\mu=1}^{4} \sum_{\nu=1}^{7} c^2_{\mu\nu} P(a_\mu, b_\nu) \leqslant 8, \tag{6.64}$$

where

$$c^2_{\mu\nu} = \begin{pmatrix} 1 & 1 & 1 & 1 & 0 & 0 & 0 \\ 1 & -1 & 0 & 0 & 1 & 1 & 0 \\ 1 & 0 & -1 & 0 & -1 & 0 & 1 \\ 1 & 0 & 0 & -1 & 0 & -1 & -1 \end{pmatrix}; \tag{6.65}$$

and a third one

$$\sum_{\mu=1}^{6} \sum_{\nu=1}^{8} c^3_{\mu\nu} P(a_\mu, b_\nu) \leqslant 16, \tag{6.66}$$

where

$$c^3_{\mu\nu} = \begin{pmatrix} 1 & 1 & -1 & -1 & 1 & 1 & 1 & 1 \\ 1 & 1 & 1 & 1 & -1 & -1 & 1 & 1 \\ 1 & 1 & 1 & 1 & 1 & 1 & -1 & -1 \\ 1 & -1 & -1 & 1 & 1 & -1 & 1 & -1 \\ 1 & -1 & 1 & -1 & -1 & 1 & 1 & -1 \\ 1 & -1 & 1 & -1 & 1 & -1 & -1 & 1 \end{pmatrix}. \tag{6.67}$$

It is easy to prove that inequalities (6.62)—(6.67) provide restrictions on $P(a_\mu, b_\nu)$ not implied by Bell's inequality.

Take for example, $P(a_4, b_2) = P(a_2, b_3) = P(a_3, b_4) = P(a_3, b_5) = 0$ and the remaining $P(a_\mu, b_\nu)$ occurring in (6.62) all equal to 2/3; then all Bell's inequalities involving these $P(a_\mu, b_\nu)$ are obeyed; but the inequality (6.62) is violated, as one can immediately check. One can easily find similar examples for (6.64) and (6.66).

Inequalities that are consequences of local realism and are at the same time more general than the original one obtained by Bell were first obtained by Pearle (1970), who did however not show that independent physical restrictions were implied. Also d'Espagnat obtained similar results (d'Espagnat, 1975).

Given an arbitrary linear combination of correlation functions, it can be shown that a corresponding inequality must be satisfied if local realism holds in nature (Garuccio and Selleri, 1980). Also a very large, probably infinite, subset of these inequalities provide physical restrictions not contained in Bell's inequality (Garuccio, 1988). This means that there are correlation functions satisfying Bell's inequality for all possible choices of the arguments a_μ and b_ν which are nevertheless excluded by Einstein locality.

In the present section all these inequalities will be put on a more solid footing by deducing them again from the satisfactory probabilistic formulation of local realism advanced in the previous sections.

Given the dichotomic observables $A(a_\mu)$ ($\mu = 1, 2, \ldots, m$) for the α objects and $B(b_\nu)$ ($\nu = 1, 2, \ldots, n$) for the β objects, consider the linear combination of correlation functions

$$\sum_{\mu\nu} c_{\mu\nu} P(a_\mu, b_\nu), \tag{6.68}$$

where the $c_{\mu\nu}$ are mn real coefficients. It will next be shown that Einstein locality implies

$$\left| \sum_{\mu\nu} c_{\mu\nu} P(a_\mu, b_\nu) \right| \leqslant M_0, \tag{6.69}$$

where

$$M_0 = \max \left\{ \sum_{\mu\nu} c_{\mu\nu} \xi_\mu \eta_\nu \right\}, \tag{6.70}$$

the right-hand side being calculated for that choice of the sign factors ξ_μ ($= \pm 1$) and η_ν ($= \pm 1$) which maximizes it.

In order to proceed with the proof, it is useful to remember (6.57),

$$\Omega(a_\mu \pm, b_\nu \pm) = \sum_{k \in I} \frac{N_k}{N} p_k(a_\mu \pm) q_k(b_\nu \pm), \tag{6.71}$$

and the definition of correlation function

$$P(a_\mu, b_\nu) = \Omega(a_\mu +, b_\nu +) - \Omega(a_\mu +, b_\nu -) - \\ - \Omega(a_\mu -, b_\nu +) + \Omega(a_\mu -, b_\nu -),$$

which give the result

$$P(a_\mu, b_\nu) = \sum_{k \in I} \frac{N_k}{N} \tilde{p}_k(a_\mu) \tilde{q}_k(b_\nu), \tag{6.72}$$

where

$$\tilde{p}_k(a_\mu) = p_k(a_\mu +) - p_k(a_\mu -), \tag{6.73}$$

$$\tilde{q}_k(b_\nu) = q_k(b_\nu +) - q_k(b_\nu -). \tag{6.74}$$

Remembering that

$$p_k(a_\mu +) + p_k(a_\mu -) = 1, \tag{6.75}$$

$$q_k(b_\nu +) + q_k(b_\nu -) = 1, \tag{6.76}$$

it follows that

$$-1 \leqslant \tilde{p}_k(a_\mu) \leqslant 1, \tag{6.77}$$

$$-1 \leqslant \tilde{q}_k(b_\nu) \leqslant 1. \tag{6.78}$$

One can then write

$$\begin{aligned}\sum_{\mu\nu} c_{\mu\nu} P(a_\mu, b_\nu) &= \sum_{\mu\nu} c_{\mu\nu} \sum_{k \in I} \frac{N_k}{N} \tilde{p}_k(a_\mu)\tilde{q}_k(b_\nu) \\ &\leqslant \sum_{k \in I} \frac{N_k}{N} M_k,\end{aligned} \tag{6.79}$$

where

$$M_k = \max \left\{ \sum_{\mu\nu} c_{\mu\nu}\tilde{p}_k(a_\mu)\tilde{q}_k(b_\nu) \right\}. \tag{6.80}$$

Now, the right-hand side of (6.80) is linear in the quantities $\tilde{p}_k(a_\mu)$ and in the independent quantities $\tilde{q}_k(b_\nu)$. From this observation, it follows that the maximum is to be found on the boundary of these quantities, that is for

$$|\tilde{p}_k(a_\mu)| = 1, \, |\tilde{q}_k(b_\nu)| = 1. \tag{6.81}$$

Furthermore, the maximum depends only on the coefficients $c_{\mu\nu}$, and not at all on the particular arguments entering in $\tilde{p}_k$ and in $\tilde{q}_k$, so that one can perform the substitutions

$$\tilde{p}_k(a_\mu) \to \xi_\mu = \pm 1, \tag{6.82}$$

$$\tilde{q}_k(b_\nu) \to \eta_\nu = \pm 1. \tag{6.83}$$

One thus obtains

$$M_k = \max_{\xi, \eta} \left\{ \sum_{\mu\nu} c_{\mu\nu}\xi_\mu \eta_\nu \right\}. \tag{6.84}$$

The latter result is independent of k so that, when inserted in (6.79), it leads to

$$\sum_{\mu\nu} c_{\mu\nu} P(a_\mu, b_\nu) \leqslant M_0, \; M_0 = \max_{\xi, \eta} \left\{ \sum_{\mu\nu} c_{\mu\nu}\xi_\mu \eta_\nu \right\}, \tag{6.85}$$

as a consequence of (6.59)—(6.60). A parallel reasoning leads to the conclusion that $-M_0$ is the minimum value of the left-hand side of (6.79), so that one indeed obtains the result (6.69)—(6.70).

We may add the following comments:

(1) The upper limit given by the right-hand side of (6.70) furnishes the most stringent inequality deducible from Einstein locality, since it is easy to check that a theoretical model exists, based on Einstein locality, which assigns to the left-hand side of (6.69) a value precisely equal to the limiting value.

(2) If the $c_{\mu\nu}$ have factorizable signs ($c_{\mu\nu} = |c_{\mu\nu}| s_\mu t_\nu$ with $s_\mu = \pm 1$ and $t_\nu = \pm 1$) the corresponding inequality (6.69) is trivial because

$$M_0 = \sum_{\mu, \nu} |c_{\mu\nu}|, \tag{6.86}$$

the maximum of the right-hand side of (6.69) being obtained for $\xi_\mu = s_\mu$ and $\eta_\nu = t_\nu$. Such an inequality must be obeyed by all theories, local or nonlocal, since the correlation functions, by definition, satisfy $|P(a_\mu, b_\nu)| \leqslant 1$.

(3) Comparing the bilinear form

$$F = \sum_{\mu, \nu} c_{\mu\nu} \xi_\mu \eta_\nu, \tag{6.87}$$

used in order to calculate M_0 (see (6.70)), with the expression

$$L = \sum_{\mu, \nu} c_{\mu\nu} P(a_\mu, b_\nu),$$

one notes a one-to-one correspondence between experimental parameters a_μ and signs ξ_μ and between b_ν and η_ν. As a consequence, if in L a given argument a_μ or b_ν occurs only once, the modulus alone of the corresponding $c_{\mu\nu}$ appears in M_0. Indeed, let ξ_1 be the once-only occurring sign in F. It follows that

$$\begin{aligned} M_0 &= \max_{\xi, \eta} \left\{ c_{1\ell} \xi_1 \eta_\ell + \sum_{\mu=2}^{m} \sum_{\nu=1}^{n} c_{\mu\nu} \xi_\mu \eta_\nu \right\} \\ &= |c_{1\ell}| + \max_{\xi, \eta} \left\{ \sum_{\mu=2}^{m} \sum_{\nu=1}^{n} c_{\mu\nu} \xi_\mu \eta_\nu \right\}, \end{aligned}$$

since one can always choose ξ_1 in such a way that

$$c_{1\ell} \xi_1 \eta_\ell = |c_{1\ell}|.$$

Given the fact that $|c_{1\ell} P(a_1, b_\ell)| \leqslant |c_{1\ell}|$ is always true, by definition of the correlation function, the inequality

$$\left| \sum_{\mu=1}^{m} \sum_{\nu=1}^{n} c_{\mu\nu} P(a_\mu, b_\nu) \right| \leqslant \max \left\{ \sum_{\mu=1}^{m} \sum_{\nu=1}^{n} c_{\mu\nu} \xi_\mu \eta_\nu \right\}$$

can immediately be reduced to the simpler one

$$\left| \sum_{\mu=2}^{m} \sum_{\nu=1}^{n} c_{\mu\nu} P(a_\mu, b_\nu) \right| \leqslant \max \left\{ \sum_{\mu=2}^{m} \sum_{\nu=1}^{n} c_{\mu\nu} \xi_\mu \eta_\nu \right\}.$$

Therefore, in the irreducible inequality, every argument a_μ and every argument b_ν appears more than once.

(4) If L can be written as $L_1 + L_2$, in such a way that no argument a_μ or b_ν is common to correlation functions occurring in L_1 and L_2, then the inequality deducible from Einstein locality can be reduced to two more elementary inequalities. In fact, the correspondence between L and F ensures that also F can be split into two parts F_1 and F_2 having no argument ξ_μ or η_ν in common. Consequently

$$M_0 = \max_{\xi, \eta} \{F\} = \max_{\xi, \eta} \{F_1\} + \max_{\xi, \eta} \{F_2\},$$

and the inequality $L \leqslant M_0$ is the sum of two reduced inequalities $L \leqslant M_{01}$ and $L_2 \leqslant M_{02}$.

(5) By using the results of the previous four points as systematic guides for the construction of inequalities, it is not difficult to show that only trivial or reducible inequalities exist in the case of linear combinations of two, three, or five correlation functions. Meaningful is, on the other hand, the case of four correlation functions. Indeed, take $m = n = 2$ in (6.69):

$$\left| \sum_{\mu=1}^{2} \sum_{\nu=1}^{2} c_{\mu\nu} P(a_\mu, b_\nu) \right| < \max_{\xi, \eta} \left\{ \sum_{\mu=1}^{2} \sum_{\nu=1}^{2} c_{\mu\nu} \xi_\mu \eta_\nu \right\} \tag{6.88}$$

and put

$$c_{\mu\nu} = |c_{\mu\nu}| \sigma_{\mu\nu},$$

where

$$\sigma_{\mu\nu} = \text{sign}\{c_{\mu\nu}\}.$$

The upper limit M_0 of (6.88) is then

$$M_0 = \max_{\xi, \eta} \left\{ \sum_{\mu=1}^{2} \sum_{\nu=1}^{2} |c_{\mu\nu}| \sigma_{\mu\nu} \xi_\mu \eta_\nu \right\}.$$

Consider next the four equations

$$\xi_1 \eta_1 = \sigma_{11}, \xi_1 \eta_2 = \sigma_{12}, \xi_2 \eta_1 = \sigma_{21}, \xi_2 \eta_2 = \sigma_{22}, \tag{6.89}$$

which would assign, if satisfied simultaneously, the maximum conceivable value to M_0. It is always possible to satisfy any three of these equations. From the first three, for example, one gets

$$\eta_1 = \sigma_{11}\xi_1,\ \eta_2 = \sigma_{12}\xi_1,\ \xi_2 = \sigma_{21}\sigma_{11}\xi_1.$$

It therefore suffices to assign ξ_1 arbitrarily and to adopt the values of ξ_2, η_1, η_2 that ensue from the foregoing relations in order to obtain the sum of the moduli of the first three $c_{\mu\nu}$ on the right-hand side of (6.88).

From the fourth relation (6.89) one has however

$$\sigma_{22} = \xi_2\eta_2 = (\sigma_{21}\sigma_{11}\xi_1)(\sigma_{12}\xi_1) = \sigma_{11}\sigma_{12}\sigma_{21},$$

which can also be written as

$$\sigma_{12}\sigma_{12}\sigma_{21}\sigma_{22} = 1.$$

Therefore, if the product of the four signs is positive, an inequality of the type (6.88) is trivial because

$$M_0 = \sum_{\mu,\nu} |c_{\mu\nu}|.$$

If, on the other hand,

$$\sigma_{11}\sigma_{12}\sigma_{21}\sigma_{22} = -1,$$

then only three of Equations (6.89) can be satisfied. This means that one of the $|c_{\mu\nu}|$ on the right-hand side of (6.88) will appear with the minus sign. We must obviously choose the smallest $|c_{\mu\nu}|$ to have the minus sign, so that

$$\left|\sum_{\mu=1}^{2}\sum_{\nu=1}^{2} c_{\mu\nu}P(a_\mu, b_\nu)\right| < \sum_{\mu=1}^{2}\sum_{\nu=1}^{2} |c_{\mu\nu}| - 2\min_{\mu',\nu'}(|c_{\mu'\nu'}|). \tag{6.90}$$

The simplest example of this type corresponds to

$$c_{11} = -c_{12} = c_{21} = c_{22} = 1,$$

for which (6.90) becomes

$$|P(a_1, b_1) - P(a_1, b_2) + P(a_2, b_1) + P(a_2, b_2)| \leqslant 2, \tag{6.91}$$

that is, Bell's inequality.

(6) Bell's inequality (6.91) is characterized by the fact that

$$M_0 = \frac{1}{2}\sum_{\mu,\nu} |c_{\mu\nu}|,$$

as one can immediately check by noticing that the moduli of the coefficients of the correlation functions in (6.91) add up to four.

We shall call an inequality of the type (6.69) for which

$$M_0 < \frac{1}{2}\sum_{\mu,\nu} |c_{\mu\nu}| \tag{6.92}$$

a *superinequality.* There are many examples of superinequalities (SI), as was shown by Garuccio (1988). Of great interest is the fact that, for every SI, there are choices of correlation functions that violate the SI without violating any conceivable Bell's inequality (of the type (6.91)).

To understand this important point, consider the inequality (6.69) with m values of μ and n of ν and the mn-dimensional space of the $P(a_\mu, b_\nu)$. The physical accessible region of this space is the hypercube defined by $-1 \leqslant P(a_\mu, b_\nu) \leqslant 1$ for all the correlation functions. Any vertex of this hypercube is a point in which all the correlation functions are ± 1. Among these vertices there is one, which we shall call vertex V_0, for which

$$P(a_\mu, b_\nu) = \sigma_{\mu\nu} = \text{sign}(c_{\mu\nu}).$$

This (often nonlocal) choice obviously leads to

$$\sum_{\mu,\nu} c_{\mu\nu} P(a_\mu, b_\nu) = \sum_{\mu,\nu} |c_{\mu\nu}|.$$

Consider next the origin O of the previous mn-dimensional space, and, on the segment OV_0, take a point Z defined by

$$P(a_\mu, b_\nu) = 0.5\sigma_{\mu\nu} \tag{6.93}$$

for all μ, ν. This choice obviously results in

$$\sum_{\mu,\nu} c_{\mu\nu} P(a_\mu, b_\nu) = 0.5 \sum_{\mu,\nu} |c_{\mu\nu}|, \tag{6.94}$$

which is a violation of the corresponding SI. But the choice (6.93), excluded, as we have just seen, by Einstein locality, cannot violate any inequality of Bell's type, since $\sigma_{\mu\nu} = \pm 1$, and the addition of four correlation functions satisfying (6.93) can give at most 2, which does not contradict (6.91)! *The conclusion is that the limitations implied by Einstein locality are not all contained in Bell's inequality but are independently contained in the 'super-inequalities' as well.*

The case of three different values of μ and of three or more for ν was analyzed by Garg, who proved for this restricted ensemble of inequalities that the necessary and sufficient condition for the validity of Einstein locality is that every possible Bell's inequality be satisfied (Garg, 1983).

Interesting results are obtained in the case of four different values of μ and four of ν. In this case there are 16 correlation functions and the number of different 4×4 matrices with coefficients $c_{\mu\nu}$ having integer values in the range $[-2, +2]$ is 5^{15}. Garuccio analyzed with a computer only 13 500 000 matrices of this set (0.009% of the total) and found 1050 superinequalities. Therefore the entire previous set of 4×4 inequalities should contain about 10^7 superinequalities.

Given a 4×4 matrix one can easily show, by permuting or exchanging the signs of rows and/or columns, that it is possible to obtain 255 other equivalent matrices, i.e., matrices with the same Einstein-locality maximum M_0. Therefore one can conclude that the number of 4×4 *independent* linear combinations, leading to superinequalities, with coefficients in the range $[-2, +2]$ is nearly 45 000.

As examples, we write in the following three of these superinequalities. The first one is

$$\sum_{\mu\nu}^{4,4} c_{\mu\nu}^{(1)} P(a_\mu, b_\nu) \leqslant 11, \tag{6.95}$$

where

$$c_{\mu\nu}^{(1)} = \begin{pmatrix} 2 & 2 & 2 & 2 \\ 2 & 2 & 1 & -2 \\ -2 & 2 & 0 & 0 \\ 1 & 1 & -2 & 0 \end{pmatrix}; \tag{6.96}$$

the second one reads

$$\sum_{\mu\nu}^{4,4} c_{\mu\nu}^{(2)} P(a_\mu, b_\nu) \leqslant 10, \tag{6.97}$$

where

$$c_{\mu\nu}^{(2)} = \begin{pmatrix} 2 & 2 & 2 & 0 \\ 2 & 1 & -1 & -2 \\ 2 & 1 & -1 & 2 \\ 2 & -2 & 0 & 0 \end{pmatrix}; \tag{6.98}$$

the third one is

$$\sum_{\mu\nu}^{4,4} c_{\mu\nu}^{(3)} P(a_\mu, b_\nu) \leqslant 6, \tag{6.99}$$

where

$$c_{\mu\nu}^{(3)} = \begin{pmatrix} 1 & 1 & 1 & -1 \\ -1 & 1 & 1 & -1 \\ 0 & 2 & -1 & 1 \\ 0 & 0 & 1 & 1 \end{pmatrix}. \tag{6.100}$$

This last inequality was discovered by Kemperman in 1984 (private communication by D. Mermin).

In order to clarify the content of (6.93), we analyze in detail the inequality (6.95). The maximum possible value of the left-hand side is obviously

$$\overline{M}' = \sum_{\mu\nu}^{4,\,4} |c_{\mu\nu}^{(1)}| = 23$$

and is obtained for the following choice of $P(a_\mu, b_\nu)$:

$$\{\overline{P}(a_\mu, b_\nu)\} = \begin{pmatrix} 1 & 1 & 1 & 1 \\ 1 & 1 & 1 & -1 \\ -1 & 1 & 0 & 0 \\ 1 & 1 & -1 & 0 \end{pmatrix}; \tag{6.101}$$

(we use a matrix representation for the vector $\{\overline{P}(a_\mu, b_\nu)\}$ in the 4×4 space). Starting from (6.101), it is possible to define a new set of correlation functions

$$\{\overline{P}'(a_\mu, b_\nu)\} = \begin{pmatrix} 0.5 & 0.5 & 0.5 & 0.5 \\ 0.5 & 0.5 & 0.5 & -0.5 \\ -0.5 & 0.5 & 0 & 0 \\ 0.5 & 0.5 & -0.5 & 0 \end{pmatrix}. \tag{6.102}$$

Since all the $\overline{P}'(a_\mu, b_\nu)$ are within ± 0.5, all Bell's inequalities with the $\overline{P}'(a_\mu, b_\nu)$ are satisfied. Therefore, using the Bell inequality only, one could conclude that the set (6.102) describes a physical system compatible with Einstein locality. This is however not true, since the set (6.102) introduced in inequality (6.95) leads to its violation, the left-hand side becoming 11.5. Therefore the set of correlation functions (6.102) cannot be obtained from a local theory.

Moreover, a hypersphere of center $\{\overline{P}'(a_\mu, b_\nu)\}$ and radius $R = 0.5/2$ exists such that all sets of correlation functions inside this circle satisfy Bell's inequality but violate inequality (6.95).

It is possible to verify that the region in 4×4 space of $P(a_\mu, b_\nu)$ with these features is not unique; indeed the Kemperman inequality (6.100), for example, is violated for the following two sets of correlation functions:

$$\{P'(a_\mu, b_\nu)\} = 0.5 \begin{pmatrix} 1 & 1 & 1 & -1 \\ -1 & 1 & 1 & -1 \\ 0 & 1 & -1 & 1 \\ 0 & 0 & 1 & 1 \end{pmatrix} \tag{6.103}$$

and

$$\{P''(a_\mu, b_\nu)\} = \begin{pmatrix} 1 & 0.6 & 0.6 & -0.6 \\ 0.2 & 0.6 & 0.6 & -0.6 \\ 0.6 & 0.2 & 0.2 & -0.2 \\ 0.6 & 0.2 & 1 & -0.2 \end{pmatrix}, \tag{6.104}$$

as one can easily check.

New and more general inequalities have been deduced from local realism in 1989 (Lepore, 1989) for arbitrary linear combinations of *joint probabilities*. The physical content of some of these new inequalities is not deducible from any inequality of the type (6.69), while every inequality of the latter type can be considered as a special case of inequality for joint probabilities, given the fact that every correlation function is a linear combination of four joint probabilities.

Let us consider again the general case of m values $a_1, a_2, \ldots, a_m$ of the instrumental parameter of the first measuring apparatus and n values $b_1, b_2, \ldots, b_n$ of the instrumental parameter of the second measuring apparatus, as in (6.21). Let

$$\Omega(a_\mu h; b_\nu k) \tag{6.105}$$

be the joint probability of measuring $A(a_\mu)$ and obtaining h and of measuring $B(b_\nu)$ and obtaining k ($h, k = \pm 1$). One can then consider the linear combination of joint probabilities

$$J = \sum_{\substack{h,k\\ \mu,\nu}} c_{\mu\nu}^{hk} \Omega(a_\mu h; b_\nu k), \tag{6.106}$$

where the $c_{\mu\nu}^{hk}$ are 4 mn arbitrary real coefficients. The same procedure used for deducing (6.69) can be used also in the present case, and one obtains

$$m_0 \leqslant J \leqslant M_0. \tag{6.107}$$

These inequalities hold true for all conceivable local realistic theories if

$$\begin{aligned} m_0 &= \min_{\xi,\eta} \{F(\xi_1, \ldots, \xi_m, \eta_1, \ldots, \eta_n)\}, \\ M_0 &= \max_{\xi,\eta} \{F(\xi_1, \ldots, \xi_m, \eta_1, \ldots, \eta_n)\}, \end{aligned} \tag{6.108}$$

where the minimum and maximum must be calculated by considering only the values 0 and 1 of the variables ξ_μ, η_ν. In (6.108) the definition used was

$$F(\xi_1, \ldots, \eta_n) = \sum_{\mu\nu} c_{\mu\nu}^{++} \xi_\mu \eta_\nu + \sum_{\mu\nu} c_{\mu\nu}^{+-} \xi_\mu (1 - \eta_\nu) + \\ + \sum_{\mu\nu} c_{\mu\nu}^{-+} (1 - \xi_\mu) \eta_\nu + \sum_{\mu\nu} c_{\mu\nu}^{--} (1 - \xi_\mu)(1 - \eta_\nu). \quad (6.109)$$

In order to show as simply as possible that the set of inequalities (6.107) is not equivalent to the set (6.69) and implies more stringent restrictions, we use the following model (Garg and Mermin, 1982): The joint probabilities are assumed to be

$$\Omega(a_\mu h; b_\nu k) = \tfrac{1}{4}[1 - c(h + k) + A_{\mu\nu} h \cdot k], \quad (6.110)$$

with

$$0 < c \leqslant \tfrac{1}{3} \quad (6.111)$$

and

$$A_{11} = A_{22} = 1; A_{\mu\nu} = -\tfrac{1}{3} \quad \text{if } (\mu, \nu) \neq (1, 1), (2, 2). \quad (6.112)$$

($\mu = 1, 2, 3$; $\nu = 1, 2, 3$). From the definition of the correlation function, one has immediately

$$P(a_\mu, b_\nu) = A_{\mu\nu},$$

and it is then possible to show that (6.112) satisfies all the possible inequalities of the set (6.69) since the model (6.110)—(6.112) can describe a *local* physical situation, *if* $c = 0$ (Lepore, 1989). As a matter of fact, the nonlocal features of the Garg—Mermin model are hidden in the value (6.111) of c, as we shall soon see.

Considering now a particular case of (6.106)—(6.107), with three parameters a_1, a_2, a_3 of the first apparatus and three parameters b_1, b_2, b_3 of the second apparatus, one can write the following linear combination of joint probabilities:

$$J_0 = \Omega(a_3+, b_3+) - \Omega(a_2-, b_2-) + \Omega(a_1-, b_1+) + \Omega(a_2+, b_2+) + \\ + \Omega(a_3-, b_2-) - \Omega(a_3-, b_1+) + 2\Omega(a_1+, b_2+) + \\ + \Omega(a_2-, b_3-) - \Omega(a_1+, b_3-).$$

By calculating the minimum value of the associated function

$$F(\xi_1, \ldots, \eta_n)$$

one gets $\min\{F\} = 0$. Einstein locality therefore implies

$$J_0 \geqslant 0. \quad (6.113)$$

If now one substitutes (6.110) in (6.113), one obtains

$$c \leqslant 0.$$

Positive values of c are thus seen to be forbidden by Einstein locality. Due to (6.111), the Garg—Mermin model is nonlocal, in spite of the fact that it satisfies all the inequalities of the set (6.69). This also means that there are physical restrictions implied by inequality (6.113), and then by the set of inequalities (6.107) to which (6.113) belongs, that no inequality of the set (6.69) can contain.

6.7. Tests of the EPR Paradox in Particle Physics

The EPR paradox, originally discovered in nonrelativistic quantum mechanics, is fully applicable also in the relativistic domain. The simplest way to realize this is perhaps to consider the rare but allowed decay

$$\pi^0 \rightarrow e^+ + e^-,$$

which accounts for a fraction of about 2×10^{-7} of π^0 decays. After the application of C and P conservation to the decay process, one can show that, in the center of mass system of the electron—positron pair, the only possible spin state is the singlet state. As a consequence, the EPR paradox exists also for this relativistic process where, for example, the two final particles have velocities very near to that of light.

Other, more practical, examples are discussed below.

A significant recent trend has been the search for examples of the EPR paradox from hitherto unexplored domains in order to subject the incompatibility between quantum theory and local realism to a more critical scrutiny. A few such interesting cases are found in elementary particle physics. Törnqvist pointed out that the decay of the $J^P = 0^-$ meson η_c, of mass 2980 MeV/c^2, into a Λ hyperon plus $\overline{\Lambda}$ antihyperon pair (with the subsequent decays $\Lambda \rightarrow p + \pi^-$ and $\overline{\Lambda} \rightarrow \bar{p} + \pi^+$) was one such candidate (Törnqvist, 1981). If $\hat{a}$ and $\hat{b}$ are unit vectors in the directions of the emitted π^- and π^+ mesons in the p and $\bar{p}$ rest frames, respectively, then the suitably normalized decay rate summed over p and $\bar{p}$ spins is, according to quantum theory, given by

$$r_0(a, b) = 1 + \alpha_\Lambda^2 \hat{a} \cdot \hat{b}, \tag{6.114}$$

where $\alpha_\Lambda = -0.642 \pm 0.013$ is the Λ-decay asymmetry parameter. Törnqvist showed that the above prediction disagreed up to 10% from a limit deducible from certain types of local realist models. However, the generality of this result for all possible local realist models remains to be probed.

Another, more promising, example is the decay of a $J^{PC} = 1^{--}$ vector meson into a pair of neutral pseudoscalar mesons. Of this type is the decay of the $\Phi(1020)$ vector meson into a K^0—$\bar{K}^0$ pair. This decay takes place through strong interactions, and the 'particle' is essentially a resonant state. Nevertheless, its quantum numbers are sharply defined, and one can study the EPR paradox by starting from the kaon—antikaon pair produced in the final state, as shown by various authors (d'Espagnat, 1976; Ross, 1980; Six, 1982; Selleri, 1983). Actually, Lee and Yang were the first to consider this process as interesting from the point of view of the EPR paradox (Inglis, 1961). Here we seek to review the present status of this example.

Invoking charge conjugation invariance of the strong interaction, the wave function of the K^0—$\bar{K}^0$ pair at the time of its production ($t = 0$) from the decay of the $J^{PC} = 1^{--}$ state is given by

$$|\Psi_0\rangle = \frac{[|K^0\rangle_a|\bar{K}^0\rangle_b - |\bar{K}^0\rangle_a|K^0\rangle_b]}{\sqrt{2}}, \tag{6.115}$$

where a (b) refers to the left (right) hemisphere. The subsequent time development of the K^0—$\bar{K}^0$ pair is described in terms of the eigenstates of the effective Hamiltonian, which includes weak interactions. In the situation under consideration, the weak interactions induce decays of both K^0 and $\bar{K}^0$ and also give rise to K^0—$\bar{K}^0$ transitions (A pure $|K^0\rangle$ ($|\bar{K}^0\rangle$) state, evolving in time, becomes a superposition of $|K^0\rangle$, $|\bar{K}^0\rangle$, and decay products. This gives rise to K^0—$\bar{K}^0$ oscillations). The effective Hamiltonian is written as $H = M - i\Gamma/2$, where M and Γ are the Hermitian mass and decay matrices, respectively. The eigenstates of this H are $|K_L\rangle$ and $|K_S\rangle$, with the eigenvalues $\alpha_L = m_L - i\gamma_L/2$ and $\alpha_S = m_S - i\gamma_S/2$, respectively, where m_L (m_S) and γ_L (γ_S) are, respectively, the mass and the decay width of $|K_L\rangle$ ($|K_S\rangle$); $m_L - m_S = 0.53 \times 10^{10}\ \hbar\ s^{-1}$ and $\gamma_S = 582\gamma_L = 1.12 \times 10^{10}\ \hbar\ s^{-1}$. We assume CP invariance; the implications of CP violation in this type of example will be discussed later. Note that

$$|K_L\rangle = \frac{[|K^0\rangle + |\bar{K}^0\rangle]}{\sqrt{2}},\ |K_S\rangle = \frac{[|K^0\rangle - |\bar{K}^0\rangle]}{\sqrt{2}}. \tag{6.116}$$

They time-evolve as

$$U(t,0)|K_L\rangle = |K_L\rangle\exp(-i\alpha_L t) + |\phi_L(t)\rangle, \tag{6.117}$$

with a corresponding equation for $|K_S\rangle$. Here $|\phi_L(t)\rangle$ ($|\phi_S(t)\rangle$) represents the decay products from $|K_L\rangle$ ($|K_S\rangle$); $|\phi_L\rangle$ ($|\phi_S\rangle$) is taken as orthogonal to the state $|K_L\rangle$ ($|K_S\rangle$). CP invariance requires that $\langle K_L|K_S\rangle = 0$. In terms of the states $|K_L\rangle$ and $|K_S\rangle$, the wave function $|\Psi_0\rangle$, given by Equation (6.115), can be written as

$$|\Psi_0\rangle = \frac{[|K_S\rangle_a|K_L\rangle_b - |K_L\rangle_a|K_S\rangle_b]}{\sqrt{2}}. \tag{6.118}$$

The time evolution of the nonseparable form of the two-particle wave function $|\Psi_0\rangle$ correlates the oscillations between the $|K^0\rangle$ and $|\bar{K}^0\rangle$ states, such that it carries the essence of nonlocal correlation, reminiscent of the EPR-type situation. If the left (right) kaon is observed to be a K^0 (strangeness $S = +1$) at a particular instant, then the right (left) kaon can be predicted, with certainty, to be observed as a $\bar{K}^0$ ($S = -1$) at that same instant. Alternatively, if the left (right) kaon decays in the K_S mode (CP = +1), then the right (left) kaon is bound to decay as a K_L (CP = −1) at some future instant. It is to be noted that there is a subtle distinction between the K^0—$\bar{K}^0$ and K_L—K_S correlations; while the former holds only for equal proper times, the latter is a time-independent consequence of the non-separable form of the wave function (Selleri, 1983).

An experimental test of this EPR-type situation, based on the measurement of the joint probability $\Omega_{--}(t_a, t_b)$, has been suggested (Six, 1982). $\Omega_{--}(t_a, t_b)$ refers to a double $\bar{K}^0$ observation (i.e., on two sides), at times t_a and t_b on the left and right, respectively. The quantum mechanical prediction is given by

$$\Omega_{--}(t_a, t_b) = |\langle \bar{K}^0_a \bar{K}^0_b | \Psi(t_a, t_b)\rangle|^2,$$

where $|\Psi(t_a, t_b)\rangle$ is the state evolved from $|\Psi_0\rangle$ at $t = 0$:

$$|\Psi(t_a, t_b)\rangle = \frac{1}{\sqrt{2}} \{|K_S\rangle_a |K_L\rangle_b \exp[-i(\alpha_S t_a + \alpha_L t_b)] - \\ - |K_L\rangle_a |K_S\rangle_b \exp[-i(\alpha_L t_a + \alpha_S t_b)]\}, \tag{6.119}$$

whence

$$\Omega_{--}(t_a, t_b) = \frac{1}{8} \{\exp[-\gamma_S t_a - \gamma_L t_b] + \exp[-\gamma_L t_a - \gamma_S t_b] - \\ - 2\exp[-\gamma(t_a + t_b)] \cos \Delta m(t_a - t_b)\}, \tag{6.120}$$

where $\gamma = (\gamma_L + \gamma_S)/2$ and $\Delta m = m_L - m_S$.

An upper bound on $\Omega_{--}(t_a, t_b)$ for the K^0—$\bar{K}^0$ system, using a general argument based on the notion of local realism, has been obtained (Selleri, 1983):

$$\Omega_{--}(t_a, t_b) \leqslant \frac{1}{8} \{\exp[-(\gamma_S t_a + \gamma_L t_b)] + \exp[-(\gamma_L t_a + \gamma_S t_b)]\}. \tag{6.121}$$

This local realistic upper bound differs from the quantum mechanical prediction (6.120) by the absence of the interference term. Quantum mechanics therefore leads to a prediction that violates (6.121) whenever the interference term is positive, that is, whenever $\cos \Delta m(t_a - t_b) < 0$. The maximum possible discrepancy is calculated to be about 12% for $\gamma_S(t_b - t_a) \simeq 5$.

Time-integrated joint probabilities have recently been examined (Datta and Home, 1986) for the case of the B^0—$\bar{B}^0$ system. This system is almost identical to the K^0—$\bar{K}^0$ system, the only difference being that $\gamma_L = \gamma_S$ ($= \gamma \sim 10^{12}\hbar\ s^{-1}$) for the eigenstates of the B^0—$\bar{B}^0$ system, which are analogous to the $|K_L\rangle$, $|K_S\rangle$ states. It is produced in the decay of a spin-1 $\Gamma(4S)$ vector meson.

Considerable controversy has been generated by a curious *gedanken* example of the EPR paradox based on the rare but well-known processes which take place in violation of CP invariance (Datta *et al.*, 1987).

The example involves a pair of correlated neutral pseudo-scalar mesons (M^0—$\bar{M}^0$) originating from the decay of a $J^{PC} = 1^{--}$ vector meson; typical instances are the decays of the $\Phi(1020)$ resonance into K^0—$\bar{K}^0$ and $\Gamma(4S)$ into B^0—$\bar{B}^0$. The exponentially decaying states, with associated masses and lifetimes, are $|M_L\rangle$ and $|M_S\rangle$ (which are certain linear combinations of $|M^0\rangle$ and $|\bar{M}^0\rangle$); in the standard formalism (outlined above) these eigenstates of the effective Hamiltonian are used to describe the quantum mechanical time-evolution of the system. In the presence of CP nonconservation, $|M_L\rangle$ and $|M_S\rangle$ are nonorthogonal. This nonorthogonality of the physically relevant states (unique characteristic of the quantum mechanical treatment of CP nonconservation) gives a new twist to the EPR paradox.

The two-particle wave function at the time of production ($t = 0$) of the M^0—$\bar{M}^0$ pair is given by

$$|\Psi_0\rangle = \frac{[|M_S M_L\rangle - |M_L M_S\rangle]}{\sqrt{2}}, \tag{6.122}$$

and the first (second) member of each pair refers to the left (right) hemisphere. The time-evolved wave function can be written in the form

$$|\Psi(t)\rangle = C_1|M_L\phi_S\rangle + C_2|M_S\phi_L\rangle + C_3|\chi\rangle, \tag{6.123}$$

where C_1, C_2, C_3 are time-dependent constants, and $|\chi\rangle \sim |M_S M_L\rangle - |M_L M_S\rangle$ represents the undecayed piece with $\langle\chi|\chi\rangle = 1$. $|\phi_L\rangle$ ($|\phi_S\rangle$) corresponds to the decay products on the right from $|M_L\rangle$ ($|M_S\rangle$). The important difference between (6.123) and the EPR-type correlations in other standard examples lies in the fact that $|M_L\rangle$ and $|M_S\rangle$ are nonorthogonal eigenstates of the effective Hamiltonian $H = M - i\Gamma/2$, where M and Γ are noncommuting due to CP noninvariance. In writing (6.123), we have not considered those components of the wave function that contain decay products on the left, as they are not relevant for our subsequent discussion, which is focused on the flux of, say, undecayed $|\bar{M}^0\rangle$ on the left.

It can be easily seen from (6.123) that the above flux on the left would involve a contribution due to the overlap between the decay product states $|\phi_L\rangle$ and $|\phi_S\rangle$ on the right, which is nonvanishing in the presence of CP noninvariance. Note that $\langle\phi_L(t)|\phi_S(t)\rangle$ is proportional to $\langle M_L|M_S\rangle$; contribution to this nonorthogonality comes essentially from the common decay products of M_L and M_S in the presence of CP violation. That the statistical property of the particles on one side has some formal dependence pertaining to the overlap between the physical states of the particles on the other side is the key feature of the example suggested by DHR. Whether this overlap can be physically tempered (even in principle) by suitable 'measurements' is the controversial issue raised by this example.

In the absence of CP violation, the mutually orthogonal $|\phi_L\rangle$ and $|\phi_S\rangle$ states can be unambiguously distinguished. However, in the presence of a small but nonvanishing CP-violating interaction, if one can partially discriminate between $|\phi_L\rangle$ and $|\phi_S\rangle$ on one side by exploiting the differences in their physical attributes, there arises a possibility to affect the above overlap, thereby leading to a curious nonlocal effect concerning the flux of undecayed kaons on the other side. Such a scheme, of course, envisages measurements partially destroying the coherence of the original pure state and leading to statistical mixtures of nonorthogonal states. A concept of this type of nonorthodox measurement ('partial collapse'), though unconventional, is not inadmissible and can be dealt with, in principle, by suitable generalizations of the standard quantum theory of measurement, as shown by various authors (Yuen, 1982; Bloch and Burba, 1974; Kraus, 1974; Davies, 1976). As a matter of fact, there are various examples of realistic measurements, such as approximate or nonideal measurements, which cannot be described by the standard scheme based on orthogonal projections alone.

A simple example of nonorthogonal measurement is provided by the Stern—Gerlach experiment for spin-$\frac{1}{2}$ atoms, in which the magnetic field is assumed to be very weak and the counters are placed so close together that each of the two separated beams has a finite probability of being registered in both counters. 'Partial distinction' between nonorthogonal states in a sense involves 'unsharp' or 'imprecise' simultaneous measurement of noncommuting observables, a concept whose tenability has been analyzed (Wootters and Zurek, 1979; Busch, 1986; Mittelstaedt *et al.*, 1987; Greenberger and Yasin, 1988).

In their original treatment, DHR had first assumed the collapse to a mixed state including nonorthogonal components to be 'total', and then the error involved (due to overlap between the probability distributions of the invariant masses of the decay products) was estimated. DHR argued that in principle the error could be made small compared with a suitably defined measure of the nonlocal effect. The scheme for estimating this error suffered however from a certain ambiguity, depending upon the choice of a parameter. This ambiguity can be avoided by directly incorporating the notion of 'partial collapse' and by taking into account probability conservation through a formal density-matrix treatment (Datta *et al.*, 1988).

Of course, if one chooses to confine one's attention only to standard quantum measurements involving, for example, the invariant masses of the individual and mutually orthogonal decay products of $|\phi_L\rangle$ and $|\phi_S\rangle$, then there will be no nonlocal effect at the statistical level (Squires and Siegwart,

1988; Clifton and Redhead, 1988; Finkelstein and Stapp, 1987). However, the new feature in this case is whether one can envisage nonorthodox or generalized measurements, in the sense of selection processes, to tinker with the overlap between the two final states $|\phi_L\rangle$ and $|\phi_S\rangle$, an overlap which contributes to the flux of undecayed kaons on the other side. One such possibility, albeit at the *gedanken* level, is to exploit the difference in the lifetimes of the states $|M_L\rangle$ and $|M_S\rangle$ to select out partially the decay products corresponding to, say, the $|\phi_L\rangle$ state. This may be done, for example, by using an apparatus on one side which registers only the decay products originating within the specified time interval (say, ΔT) around the time of the order of the lifetime of $|M_L\rangle$. These decay products will come predominantly from $|M_L\rangle$. There will also be a small but nonvanishing probability (which can be made as small as desired at the level of a thought experiment by assuming the lifetimes of $|M_L\rangle$ and $|M_S\rangle$ to be widely different) of the decay products from $|M_S\rangle$ being registered during ΔT. But the relevant point is whether the overlap between the common decay products from $|M_S\rangle$ (which originated well before ΔT) and those from $|M_L\rangle$ during ΔT can get affected by this process.

It has been argued (Hall, 1987; Ghirardi *et al.*, 1988) that the type of wave-function collapse envisaged in the DHR example necessarily corresponds to a 'measurement' which simultaneously affects the two separated subsystems; this would ensure that no action at a distance is involved here.

6.8. On the Possibility of New Experiments

New ideas for experiments on the EPR paradox in fundamental particle physics have been discussed in the previous section. Here we shall instead present some important possible improvements of the experiments performed with pairs of atomic photons.

The first idea (Falciglia *et al.*, 1983) is based on the introduction of the source of atomic-photon pairs into a 10^2—10^3 G magnetic field. For a (0, 1, 0) cascade, the intermediate $J = 1$ atomic level splits into three levels with $m_J = 0, \pm 1$, the energy separation being $\mu_B |\mathbf{B}|$ (μ_B is the Bohr magneton, **B** the magnetic field). As is well known, the $m_J = 0$ state does not contribute if the photons propagate in directions parallel/antiparallel to that of **B**. There remain two types of atomic transitions for the emission of the photon pair:

(1) $(J = 0, m_J = 0) \to (J = 1, m_J = +1) \to (J = 0, m_J = 0)$. This leads to the emission of a photon pair in the polarization state $|R_\alpha\rangle|R_\beta\rangle$.
(2) $(J = 0, m_J = 0) \to (J = 1, m_J = -1) \to (J = 0, m_J = 0)$. This leads to the emission of a photon pair in the polarization state $|L_\alpha\rangle|L_\beta\rangle$.

Therefore, if the photons are detected along $\pm\mathbf{B}$, and if the energy separation is large enough to allow one to detect, at least in principle, the energy difference between process (1) and process (2) above, one expects the (paradoxical) state vectors (5.78) to reduce themselves to a (nonparadoxical) mixture of state vectors $|R_\alpha\rangle|R_\beta\rangle$ and $|L_\alpha\rangle|L_\beta\rangle$. The correlation function should make a sudden jump when the magnetic field is switched on, since the vectors (5.78) violate, and the previous mixture satisfies, Bell's inequality. Therefore, the 'reduction of the wave-packet', a typical quantum mechanical feature, finds in the case of the EPR paradox a striking possibility of experimental control.

The zero angular momentum states (5.78) satisfy a peculiar invariance condition: If on the trajectories of the two photons are placed quarter-wave plates *with parallel optical axes*, it is easy to show that the two vectors (5.78) are respectively transformed into the vectors (5.79). But the latter are mathematically identical to the former, and the action of the plate is thus seen to have no physical implication. This means that the photon polarization correlations remain exactly the same, at least for ideal plates. This prediction can be checked experimentally in an EPR experiment by inserting a quarter-wave plate into the trajectory of each photon before it interacts with the analyzer. The coincidence counting rates should remain the same for all relative orientations of the analyzer axes.

Two experiments of the previous type have been performed, and surprisingly both gave results in agreement with the strong inequalities. In the first one (Clauser, 1976), the retardation induced by the plates had a measured drift of $\pm 8.5°$ between different calibrations of the plates. The probability for transmission plus detection of both photons for the (1, 1, 0) cascade used by Clauser is predicted to be

$$D(a, b) = \tfrac{1}{4}[\varepsilon_+^1\varepsilon_+^2 - \varepsilon_-^1\varepsilon_-^2 F_2(\theta)\cos(\xi_1 - \xi_2)\cos 2(a-b)]\eta_1\eta_2, \qquad (6.124)$$

where θ is the half-angle subtended by the primary lenses, $F_2(\theta)$ is a numerical parameter measuring the polarization decorrelation due to lack of photon collinearity; ξ_1 and ξ_2 are retardations induced by the quarter-wave plates (in the ideal case $\xi_1 = \xi_2 = \pi/2$); the remaining symbols are the same as in (5.80). Clauser tried to check Freedman's inequality $\delta \leqslant 0$, with δ defined in (5.60), and found experimentally $\delta_{\text{ex.}} = -0.015 \pm 0.025$. He claimed that the previous result was in agreement with the quantum theoretical prediction $\delta = 0.002$. The previous value is however incompatible with the quoted values of the retardations. Even if one assumed a very pessimistic value of $\xi_1 - \xi_2 = 17°$, one would get, from Clauser's experimental data, $\delta = 0.024$, which deviates from the experimental result by 1.5 standard deviations.

In the Stirling experiment (Duncan, 1985), some data were collected with polarizers having fixed and parallel axes; but the investigators used a varying relative angle ϕ of the quarter-wave plates (one plate had a fast axis rotated by $\pi/4$ with respect to the common direction of the polarizer axes and kept fixed, the other one was rotated during the experiment). The quantum formula for the double transmission and detection probability for this experiment is

$$D(\phi) = \tfrac{1}{4}[\varepsilon_+^1\varepsilon_+^2 + \varepsilon_-^1\varepsilon_-^2 F\cos(\xi_1 - \xi_2)\cos 2\phi]\eta_1\eta_2, \tag{6.125}$$

where the geometrical factor $F = 0.996$. Experimentally it was found $\delta = -0.055 \pm 0.016$. The quantum predictions can reproduce such a value of δ, which does *not* disagree with Freedman's inequality, only by assuming a highly patological value of $\xi_1 - \xi_2 \simeq 41°$. A more reasonable value of $\xi_1 - \xi_2$ of about 7°, which one can deduce from the plate retardations as a function of photon wavelength, would lead to a prediction for δ at variance with its measured value.

With these comments I do not wish to replace the experimentalists in the analysis of the collected data but only to call attention to the fact that the insertion of quarter-wave plates in EPR experiments has been attempted twice and that both times: (i) the predicted invariance of the relative counting rate was not observed; (ii) Freedman's (strong) inequality was satisfied; (iii) very strange effects had to be invoked in order to explain the data.

The calculation of the quantum mechanical predictions for an EPR experiment with two plates, of arbitrary retardation and arbitrary orientation, and with two polarizers has recently been performed (De Caro, 1989). Several exciting particular cases emerge from this general calculation, e.g., the possibility to violate the strong inequalities simply by varying the plate retardations.

Several experiments have been proposed in order to overcome the distinction between weak and strong inequalities or, in other words, in order to have a concrete physical situation in which the disagreement between local realism ('Einstein locality') and the quantum mechanical predictions is fully realized and need not be superimposed by means of arbitrary additional assumptions. An interesting proposal of this type (Lo and Shimony, 1981) suggested measuring spin correlations of two Na atoms coming from the dissociation of a Na_2 molecule in a singlet state. For the detection of Na atoms, better instruments are available than for optical photons (see however the critical remarks: Santos, 1984). Another suggested idea (Drummond, 1983) considered the cooperative emission of photons from excited atoms and showed that, in principle, one can obtain quantum mechanical correlations of two wave packets, each containing N photons that violate Bell's inequality. The efficiency of a photodetector detecting at least one of the N photons is obviously much larger than the single-photon detection probability (for N high enough), and one can thus approach the high detection region, where the distinction between strong and weak inequalities declines. A partly similar situation was studied by Reid and Walls (1984).

The standard experimental arrangement used in the observations of the Hanbury—Brown—Twiss effect was considered (Chubarov and Nikolayev, 1985). Radiation from a source is split by a semitransparent mirror into two beams and registered by two photon-counting detectors D_1 and D_2. Two polarizers are inserted in front of the detectors, as in standard EPR experiments. These authors showed that the coincidence counting rate has to obey inequalities of the strong type (deduced with the help of the usual additional assumptions) for different orientations of the polarizers. If the quantum states so analyzed are photonic states with sub-Poissonian statistics, one can instead have cases where the quantum mechanical predictions lead to violations of the strong inequalities.

Extension of the previous idea (Ou *et al.*, 1987) showed that violations of the strong inequalities can be obtained even for the nearly Poissonian, but highly correlated, signal of photon pairs that are created in the process of spontaneous parametric down conversion.

A proposed experimental study of the EPR paradox by using nuclear beta decay has been advanced (Skalsey, 1986): Beta-decay electrons, in cascade with the conversion electrons, are longitudinally polarized due to parity violation in the weak decay. Therefore the detection of the electron direction is equivalent to a spin measurement.

Study of the EPR paradox by using Rydberg atoms has also been proposed (Oliver and Stroud, 1987). Nearly 100% efficient state-selective ionizers can be used as detectors in Rydberg-atom experiments, and this could provide a decisive advantage over the usual EPR experiments with pairs of atomic photons. In this idea, the photon polarization is replaced by the two-level-atom Bloch vector. An interesting method for producing pairs of Rydberg atoms with the right quantum mechanical correlation ('state vector of the second type') is also proposed by these authors.

6.9. Variable Probabilities

The notion of 'variable probabilities' as a possible solution of the EPR paradox is provided by the results of the experiments performed with atomic photon pairs, which indicate that the inequalities of the strong type (deduced from local realism and from additional assumptions) are violated. This result seems very likely to be correct, even though there is a protracted debate on the role of rescattering in the atomic source; see Sanz and Sanchez-Gomez (1987).

It is perfectly logical to adopt the viewpoint that it is not local realism but the additional assumptions which should be blamed for the violation of the strong inequalities. To follow this outlook, one must then study local realist models in which the logical negation of the additional assumptions is explicitly taken into account. The relevant models should then satisfy the following conditions:

(1) If a pair of photons emerge from two regions of space where two

polarizers can be located, the probability of their joint detection from two photomultipliers depends on the presence and/or on the orientations of the polarizers ($\overline{\text{CHSH}}$ property).

(2) Considering a photon in the state λ, the probability of its detection with a polarizer interposed on its path can be larger than the detection probability with the polarizer removed ($\overline{\text{CH}}$ property).

(3) For a photon in the state λ, the sum of the detection probabilities in the 'ordinary' and in the 'extraordinary' beams emerging from a two-way polarizer depends on the polarizer's orientation ($\overline{\text{GR}}$ property).

A comprehensive overview of this line of research can be found in the review articles by Pascazio (1988) and by Ferrero *et al.* (1988).

From this point of view, it follows that the issue of quantum theory versus local realism can be settled only if weak inequalities (deduced from local realism alone, without any auxiliary assumption) can be experimentally tested. This appears not to be possible in the near future in the context of experiments involving pairs of photons. This situation can be better in the case of some proposed particle-physics experiments and in the case of experiments using pairs of atoms where the detectors have higher efficiency.

The interesting point to be stressed here is that it is possible to do meaningful studies, even for the case of low efficiency detectors, by replacing the usual additional assumptions (called CHSH, CH, and GR) by more physically appropriate conditions. It is very unlikely that the large disagreement between quantum theory and local realism for the ideal high efficiency detectors would reduce to perfect agreement for the low efficiency detectors.

In this connection it should be interesting to study the use of symmetrical functions for describing the detection processes of the two photons in view of the demonstration by Caser (1984) that the quantum mechanical predictions for such a case cannot agree with the factorizable probabilities of Clauser and Horne.

Along these lines of thought, there remains a lot of work to be done beyond the proof, reviewed in the following, that such models can indeed reproduce exactly the quantum mechanical predictions for low-efficiency detectors. What is still missing is a physical idea about the true nature of photon pairs and of their interactions (local or nonlocal) with polarizers and detectors. An attempt of this type was made by Garuccio and Selleri (1984), but their predictions turn out to be incompatible with experiments.

Following an idea of Selleri and Zeilinger (1988), it will next be shown that local and deterministic models exist which are able to reproduce exactly the quantum mechanical predictions for low-efficiency detectors. It is rather obvious that if deterministic models of this type exist, also truly probabilistic local models should exist with the same basic physical property — indistinguishability from the quantum mechanical predictions for low enough efficiency of the detectors used. In fact, the class of probabilistic local models is much wider than that of deterministic local models, which it includes.

Selleri and Zeilinger (SZ in the following) generalized in two important ways the deterministic approach first proposed by Wigner (1970):

(i) Additional variables which determine whether a specific photon will trigger the detector or not were included.
(ii) Four different polarizer directions, two on each side, were considered, while Wigner had studied a total of three.

In the SZ model, each individual photon pair (α, β) is described by the set of ten variables:

$$(s, s', \sigma, \sigma', \delta; t, t', \tau, \tau', \varepsilon), \tag{6.126}$$

with the first five variables pertaining to photon α and the second five to photon β. Each of the ten variables can only be zero or unity, and in this consists the deterministic nature of the model. In fact, all the variables (6.126) could be considered probabilities, and probabilities that can assume only the values 0 or 1 are certainties (negative and positive certainties, respectively).

The variables (6.126) determine the interactions that a given photon, α or β, will have with polarizers and detectors. One assumes that:

$s = 1$ ($s' = 1$) determines that photon α will traverse its polarizer oriented along direction a (a');
$\sigma = 1$ ($\sigma' = 1$) determines that photon α will be registered by its detector after having passed the polarizer oriented along direction a (a');
$\delta = 1$ determines that photon α will be registered by its detector if no polarizer is in its beam path;
$t = 1$ ($t' = 1$) determines that photon β will traverse its polarizer oriented along direction b (b');
$\tau = 1$ ($\tau' = 1$) determines that photon β will be registered by its detector after having passed the polarizer oriented along direction b (b');
$\varepsilon = 1$ determines that photon β will be registered by its detector if no polarizer is in its beam path;
$s, s', \sigma, \sigma', \delta = 0$ ($t, t', \tau, \tau', \varepsilon = 0$) determines that photon α (photon β) will not pass its polarizer or will not be registered by the detector.

We note that in this way the fate of each individual photon, whether it will traverse its polarizer or not and whether it will be registered by its detector or not, is strictly determined by the whole set of variables the photon pair carries. Also, it is inherently assumed in our model that, if the photon does not encounter a polarizer on its flight from the source to the detector, the variable δ (ε) is active at the detector but it is switched off — and the variable σ (τ) is switched on — upon interaction with the polarizer. One could equally well have assumed only one variable describing whether a photon is detected or not, which then could change upon interaction with the polarizer.

All N_0 photon pairs emitted by the source may therefore be grouped into subsets according to the specific variables they carry. If one defines by $n(s, s', \sigma, \sigma', \delta; t, t', \tau, \tau', \varepsilon)$ the population of that subset of photon pairs that carry the specific set $(s, s', \sigma, \sigma', \delta; t, t', \tau, \tau', \varepsilon)$ of variables, one obtains the normalization condition

$$\sum n(s, s', \sigma, \sigma', \delta; t, t', \tau, \tau', \varepsilon) = N_0. \tag{6.127}$$

Due to the ten different variables present in the model, it may at first sight appear that one has to deal with $2^{10} = 1024$ different subsets. Yet, it is easy to see that subsets for which, say, $s = 0$ and simultaneously $\sigma = 1$ are not physically meaningful, since this would imply detection with certainty of a photon which has not traversed the polarizer. The same considerations apply to the other three polarizer orientations, which reduces the number of different subsets by a factor of $(3/4)^4$ to 324.

In order to define the experimental observables in terms of sums over the populations of the subsets of the model, one abbreviates

$$n(\ldots) = n(s, s', \sigma, \sigma', \delta; t, t', \tau, \tau', \varepsilon) \tag{6.128}$$

whenever it is understood that the sum is to be taken over the whole ensemble. The various joint probabilities may therefore be written as:

$$\begin{aligned}
D(\infty, \infty) &= \frac{1}{N_0} \sum n(\ldots)\delta\varepsilon, \\
D(a', \infty) &= \frac{1}{N_0} \sum n(\ldots)s'\sigma'\varepsilon, \\
D(\infty, b) &= \frac{1}{N_0} \sum n(\ldots)\delta t\tau, \\
D(a, b) &= \frac{1}{N_0} \sum n(\ldots)s\sigma t\tau, \\
D(a, b') &= \frac{1}{N_0} \sum n(\ldots)s\sigma t'\tau', \\
D(a', b) &= \frac{1}{N_0} \sum n(\ldots)s'\sigma' t\tau, \\
D(a', b') &= \frac{1}{N_0} \sum n(\ldots)s'\sigma' t'\tau',
\end{aligned} \tag{6.129}$$

where, for example, $D(a, b)$ is the joint probability of detecting photon α with its polarizer oriented along direction a and photon β with its polarizer oriented along direction b. As has become customary, the symbol ∞ stands for the situation where the corresponding polarizer is removed from the beam.

The individual detection probabilities are given by

$$p(a') = \frac{1}{N_0} \Sigma\, n(\ldots)s'\sigma',$$
$$q(b) = \frac{1}{N_0} \Sigma\, n(\ldots)t\tau. \tag{6.130}$$

One can define the quantity Γ as follows:

$$\Gamma = D(a, b) - D(a, b') + D(a', b) + D(a', b')$$
$$= \frac{1}{N_0} \Sigma\, n(\ldots)[s\sigma(t\tau - t'\tau') + s'\sigma'(t\tau + t'\tau')]. \tag{6.131}$$

The model can now be tested against two different types of inequalities. The first type includes the *weak inequality*, which provides a less stringent limitation on the experimental results than another one, the *strong inequality*. It can be shown that the weak inequality is identical with Bell's original inequality, while the strong inequality reduces to Bell's inequality only when the detector efficiencies tend to unity.

In order to demonstrate first that the model satisfies the weak inequality, one uses the relationship

$$-XY \leqslant xy - xy' + x'y + x'y' - x'Y - Xy \leqslant 0, \tag{6.132}$$

valid if $0 \leqslant x, x' \leqslant X$ and $0 \leqslant y, y' \leqslant Y$. Setting now

$$x = s\sigma, x' = s'\sigma',$$
$$y = t\tau, y' = t'\tau', \tag{6.133}$$

and thus $X, Y = 1$, one obtains

$$-1 \leqslant s\sigma(t\tau - t'\tau') + s'\sigma'(t\tau + t'\tau') - s'\sigma' - t\tau \leqslant 0, \tag{6.134}$$

which inequality is valid for each individual photon pair. Multiplication with the population density $n(\ldots)$, summation over all populations, and normalization by N_0 yield

$$-1 \leqslant \Gamma - p(a') - q(b) \leqslant 0. \tag{6.135}$$

This is exactly the weak inequality. Therefore, as expected, the local model is in agreement with Bell's inequality and thus necessarily in disagreement with quantum mechanics.

Turning now to the strong inequality, one can first show that it is not violated if the no-enhancement hypothesis is assumed to be valid. With the present notation, the no-enhancement hypothesis can be written as

$$\begin{aligned} &0 \leqslant s\sigma,\, s'\sigma' \leqslant \delta, \\ &0 \leqslant t\tau,\, t'\tau' \leqslant \varepsilon. \end{aligned} \tag{6.136}$$

Since all variables are dichotomic, these latter relations imply that a photon which is registered by the detector with the polarizer in place would always be detected if the polarizer were removed, but not necessarily *vice versa*. Using again the relationship of Equation (6.132) with the definitions (6.133) and, hence, $X = \delta$ and $Y = \varepsilon$, one obtains

$$-\delta\varepsilon \leqslant s\sigma(t\tau - t'\tau') + s'\sigma'(t\tau + t'\tau') - s'\sigma'\varepsilon - \delta t\tau \leqslant 0, \tag{6.137}$$

which inequality is again valid for each individual photon pair. Multiplication by the population density $n(\ldots)$, summation over all populations, and normalization by N_0 now yield

$$-D(\infty, \infty) \leqslant \Gamma - D(a', \infty) - D(\infty, b) \leqslant 0. \tag{6.138}$$

Here is so obtained exactly the strong inequality which, again as expected, is not violated by the model provided the no-enhancement hypothesis is assumed to be valid.

One can give a simple example where the model violates the strong inequality. For that purpose one assumes that for all photon pairs $s\sigma = s'\sigma' = t\tau = t'\tau' = 1$. Therefore $\Gamma = 2$, and the right-hand side of the inequality (6.138) becomes

$$\Gamma \leqslant \frac{1}{N_0} \sum n(\ldots)(\varepsilon + \delta). \tag{6.139}$$

Clearly, this inequality is violated if ε or δ vanish even for a single photon only. Admittedly, this case is not a very physical one, both because of the assumptions made and because it exhibits average enhancement, i.e., the total count rate with polarizers present is larger than that without polarizers. Yet, the SZ model, having a large number of adjustable parameters that are the populations of the individual subensembles, is certainly rich enough to yield physically more reasonable cases which still violate the strong inequality. This will be shown next.

One can formulate some natural requirements on the counter efficiencies. If $\eta_1(a)$ is the efficiency of counter 1 to detect photons after they have traversed the polarizer oriented along direction a and η_1 the efficiency of counter 1 without polarizer present (and likewise for the other detector), it is natural to require that

$$\begin{aligned} &\eta_1(a) = \eta_1(a') = \eta_1, \\ &\eta_2(b) = \eta_2(b') = \eta_2. \end{aligned} \tag{6.140}$$

These conditions also guarantee that there is no average enhancement, since the transmitivities of the polarizers are never larger than unity. In terms of the SZ model, the counter efficiencies have to be written as ensemble averages. They are

$$\eta_1(a) = \frac{\sum n(\ldots)s\sigma}{\sum n(\ldots)s}, \; \eta_1 = \frac{\sum n(\ldots)\delta}{\sum n(\ldots)}, \tag{6.141}$$

and likewise for the other directions and the other photon. Also, it is reasonable to require the detection probabilities to be statistically independent on the average, i.e.,

$$\eta_1\eta_2 = \eta_{12}, \tag{6.142}$$

where η_{12} is the probability for coincidence counts, in the SZ model again a coincidence count rate averaged over the whole ensemble.

One can explicitly construct a specific example that fulfills these conditions but violates the strong inequality. To simplify our notation, note that in the inequalities the variables determining the photon behavior with polarizers present appear in products with the corresponding detector variable. One can therefore define the new variables

$$\begin{aligned} S &\equiv s\sigma, \; S' \equiv s'\sigma', \\ T &\equiv t\tau, \; T' \equiv t'\tau'. \end{aligned} \tag{6.143}$$

This implies that, for example, $S = 1$ determines that photon 1 will traverse its polarizer oriented in direction a and that it will be registered by its detector. The populations now are functions of six variables only:

$$N(S, S', \delta; T, T', \varepsilon), \tag{6.144}$$

with some of these populations being sums over some of the populations n discussed above. Clearly there are only 64 different populations N. In order to show that the populations (6.144) can reproduce the quantum mechanical predictions exactly, one assumes, as is certainly possible, the following structure of the populations:

$$N(S, S', \delta; T, T', \varepsilon) = N_0 F(S, S'; T, T')G(\delta, \varepsilon), \tag{6.145}$$

where all 16 quantities F and all four quantities G are nonnegative and fulfil the normalization conditions

$$\sum_{S, S', T, T'} F(S, S'; T, T') = 1, \tag{6.146}$$

$$\sum_{\delta, \varepsilon} G(\delta, \varepsilon) = 1. \tag{6.147}$$

For an explicit example, it is sufficient to assume that only the following nine quantities F are positive and have the values

$$
\begin{aligned}
F_1 &= F(1,0;1,0) = \frac{1}{4}\left(1+\frac{1}{\sqrt{2}}\right)\eta_1\eta_2,\\
F_2 &= F(1,0;0,1) = \frac{1}{4}\left(1-\frac{1}{\sqrt{2}}\right)\eta_1\eta_2,\\
F_3 &= F(0,1;1,0) = \frac{1}{4}\left(1+\frac{1}{\sqrt{2}}\right)\eta_1\eta_2,\\
F_4 &= F(0,1;0,1) = \frac{1}{4}\left(1+\frac{1}{\sqrt{2}}\right)\eta_1\eta_2,\\
F_5 &= F(1,0;0,0) = \frac{1}{2}\,\eta_1 - \frac{1}{2}\,\eta_1\eta_2,\\
F_6 &= F(0,1;0,0) = \frac{1}{2}\,\eta_1 - \frac{1}{2}\left(1+\frac{1}{\sqrt{2}}\right)\eta_1\eta_2,\\
F_7 &= F(0,0;1,0) = \frac{1}{2}\,\eta_1 - \frac{1}{2}\left(1+\frac{1}{\sqrt{2}}\right)\eta_1\eta_2,\\
F_8 &= F(0,0;0,1) = \frac{1}{2}\,\eta_2 - \frac{1}{2}\,\eta_1\eta_2,\\
F_9 &= F(0,0;0,0) = (1-\eta_1)(1-\eta_2) + \frac{\eta_1\eta_2}{2\sqrt{2}}.
\end{aligned}
\tag{6.148}
$$

$F_1, \ldots, F_4$ can give rise to double or single detections; $F_5, \ldots, F_8$ can give rise to single detection only; and F_9 can never produce any detection. Likewise one assumes

$$
\begin{aligned}
G_1 &= G(1,1) = \eta_1\eta_2,\\
G_2 &= G(1,0) = \eta_1 - \eta_1\eta_2,\\
G_3 &= G(0,1) = \eta_2 - \eta_1\eta_2,\\
G_4 &= G(0,0) = (1-\eta_1)(1-\eta_2).
\end{aligned}
\tag{6.149}
$$

It is easy to see that the populations (6.145) to (6.149) are able to reproduce the quantum mechanical predictions for the case of maximum violation of the strong inequality. The reader may easily check, for example, that:

$$
\begin{aligned}
p(a) &= (F_1 + F_2 + F_5)\left(\sum G_i\right) = \frac{1}{2}\,\eta_1, \\
D(\infty, \infty) &= \left(\sum F_i\right) G_1 = \eta_1 \eta_2, \\
D(a', \infty) &= (F_3 + F_4 + F_6)(G_1 + G_3) = \frac{1}{2}\,\eta_1 \eta_2, \qquad (6.150) \\
D(a, b) &= F_1 \left(\sum G_i\right) = \frac{1}{4}\left(1 + \frac{1}{\sqrt{2}}\right)\eta_1 \eta_2, \\
D(a, b') &= F_2 \left(\sum G_i\right) = \frac{1}{4}\left(1 - \frac{1}{\sqrt{2}}\right)\eta_1 \eta_2,
\end{aligned}
$$

and so on, the sums being extended over all possible values. These predictions obviously refer to the case of ideal polarizers, but it is a simple matter to modify (6.145)—(6.149) in such a way as to obtain the fully realistic case. It is also evident that the populations may be chosen so as to violate the strong inequality in cases of far-from-maximum violation, thus agreeing with experiment for all angular orientations of the polarizers.

Note

1. Actually there must be several equivalent subensembles of this type, since in a truly probabilistic situation it cannot be fixed *a priori* which β particles will produce the result $B(b) = +1$ upon measurement. This complication is irrelevant for our purposes.

References

Books

d'Espagnat, B. (ed.): 1971, *Foundations of Quantum Mechanics* (Italian Physical Society, Proceedings of the International School of Physics 'Enrico Fermi', Course IL), Academic Press, New York.

Hartkämper, A., and Neumann, H. (eds): 1974, *Foundations of Quantum Mechanics and Linear Spaces*, Springer-Verlag, Berlin.

Kostro, L. (ed.): 1988, *Problems in Quantum Physics*, World Scientific, Singapore.

Popper, K.: 1980, *The Logic of Scientific Discovery*, Hutchinson, London.

Selleri, F. (ed.): 1988, *Quantum Mechanics versus Local Realism: The Einstein, Podolsky, and Rosen Paradox*, Plenum, New York.

Suppes, P. (ed.): 1980, *Studies in the Foundations of Quantum Mechanics*, Philosophy of Science Association, East Lansing.
von Neumann, J.: 1955, *The Mathematical Foundations of Quantum Mechanics*, Princeton University Press, Princeton.
van der Merwe, A., Selleri, F., and Tarozzi, G.: 1988, *Microphysical Reality and Quantum Formalism*, Kluwer Academic Publishers, Dordrecht.

Articles

Bell, J. S.: 1965, *Physics* **1**, 195.
Bloch, I., and Burba, D. A.: 1974, *Phys. Rev. D* **10**, 3206.
Bohr, N.: 1935, *Phys. Rev.* **48**, 696.
Busch, P.: 1986, *Phys. Rev. D* **33**, 2253.
Caser, S.: 1984, *Phys. Lett. A* **102**, 152.
Clauser, J. F.: 1976, *Nuovo Cimento B* **33**, 740.
Clauser, J. F., and Horne, M. A.: 1974, *Phys. Rev. D* **10**, 526.
Clifton, R. K., and Redhead, M. L. G.: 1988, *Phys. Lett. A* **126**, 295.
Chubarov, M. S., and Nikolayev, E. P.: 1985, *Phys. Lett.* **110A**, 199.
Datta, A., and Home, D.: 1986, *Phys. Lett. A* **119**, 3.
Datta, A., Home, D., and Raychaudhuri, A.: 1987, *Phys. Lett. A* **123**, 4.
Datta, A., Home, D., and Raychaudhuri, A.: 1988, *Phys. Lett. A* **130**, 187.
Davies, E. B.: 1976, *Quantum Theory of Open Systems*, Academic Press, New York.
De Caro, L.: 1989, *Found. Phys. Lett.* **2**, 151
d'Espagnat, B.: 1975, *Phys. Rev. D* **11**, 1454; 1976, *Conceptual Foundations of Quantum Mechanics*, Benjamin, London.
Drummond, P. D.: 1983, *Phys. Rev. Lett.* **50**, 1407.
Duncan, A.: 1985, private communication.
Einstein, A., Podolsky, B., and Rosen, N.: 1935, *Phys. Rev.* **47**, 777.
Falciglia, F., *et al.*: 1983, *Lett. Nuovo Cimento* **37**, 66.
Ferrero, M., Marshall, T. W., and Santos, E.: 1988, in F. Selleri (ed.); see book list.
Finkelstein, J., and Stapp, H. P.: 1987, *Phys. Lett. A* **126**, 295.
Garuccio, A.: 1988, in F. Selleri (ed.); see book list.
Garuccio, A., and Selleri, F.: 1980, *Found. Phys.* **10**, 209; 1984, *Phys. Lett.* **103A**, 99.
Garg, A.: 1983, *Phys. Rev. D* **28**, 785.
Garg, A., and Mermin, N. D.: 1982, *Phys. Rev. Lett.* **49**, 901; *Found. Phys.* **14**, 1.
Ghirardi, G. C., *et al.*: 1988, *Europhysics Lett.* **6**, 95.
Greenberger, D. M., and Yasin, A.: 1988, *Phys. Lett. A* **128**, 391.
Hall, M. J. W.: 1987, *Phys. Lett. A* **125**, 89.
Inglis, D. R.: 1961, *Rev. Mod. Phys.* **33**, 1.
Kraus, K.: 1974, in A. Hartkämper and H. Neumann (eds); see book list.
Lepore, V. L.: 1989, *Found Phys. Lett.* **2**, 15.
Liddy, D. E.: 1983, *J. Phys. A* **16**, 2703.
Lo, T. K., and Shimony, A.: 1981, *Phys. Rev. A* **23**, 3003.
Marshall, T. W.: 1983, *Phys. Lett.* **99A**, 163.
Mittelstaedt, P., *et al.*: 1987, *Found. Phys.* **17**, 891.
Oliver, B. J., and Stroud, C. R., Jr: 1987, *J. Opt. Soc. Am. B* **4**, 1426.
Ou, Z. Y., Hong, C. K., and Mandel, L.: 1987, *Phys. Lett.* **122A**, 11.
Pascazio, S.: 1988, in F. Selleri (ed.); see book list.
Pascazio, S., and Reignier, J.: 1987, *Phys. Lett.* **126A**, 163.
Pearle, P. M.: 1970, *Phys. Rev.* **2**, 1418.
Rayski, J.: 1988, in F. Selleri (ed.); see book list.
Reid, M. D., and Walls, D. F.: 1984, *Phys. Rev. Lett.* **53**, 955.

Roy, S. M., and Singh, V.: 1978, *J. Phys. A* **11**, 167; 1979, *J. Phys. A* **12**, 1003.
Ross, M.: 1980, University of Helsinki preprint.
Röhrlich, F.: 1983, *Science* **221**, 1251.
Santos, E.: 1984, *Phys. Rev. A* **30**, 2128; 1988, in F. Selleri (ed.); see book list.
Sanz, A. L., and Sanchez-Gomez, J. L.: 1987, *Europhysics Lett.* **3**, 519.
Selleri, F.: 1983, *Nuovo Cimento Lett.* **36**, 521; 1985, *Rivista storia d. scienza* **2**, 545; 1988a, in A. van der Merwe *et al.* (eds); see book list. 1988b, in L. Kostro (ed.); see book list.
Selleri, F., and Tarozzi, G.: 1980, *Nuovo Cimento Lett.* **29**, 533.
Selleri, F., and Zeilinger, A.: 1988, *Found. Phys.* **18**, 1141.
Six, J.: 1982, *Phys. Lett. B* **114**, 200.
Skalsey, M.: 1986, University of Michigan preprint.
Squires, E., and Siegwart, D.: 1988, *Phys. Lett. A* **126**, 295.
Suppes, P., and Zanotti, M.: 1976, in *Logic and Probability in Quantum Mechanics*, Reidel, Dordrecht.
Törnqvist, N.: 1981, *Found. Phys.* **11**, 171; 1986, *Phys. Lett. A* **117**, 1.
Wigner, E. P.: 1970, *Am. J. Phys.* **38**, 1005.
Wootters, W. K., and Zurek, W. H.: 1979, *Phys. Rev.* **D19**, 473.
Yanase, M. M.: 1971, in B. d'Espagnat (ed.); see book list.
Yuen, H. P.: 1982, *Phys. Lett. A* **91**, 101.

Chapter 7

Perspectives of Physical Realism

Quantum mechanics is a philosophically committed theory, in the strong sense that its inner mathematical structure leads to empirical predictions which contradict in an observable manner some simple consequences of very general physical ideas — so general that they can be considered philosophical ideas! This leads one naturally to consider the role of society, culture, and prejudice in the birth of the new theory. Furthermore, this opens up the possibility of conceiving experimental situations in which one can choose empirically among different expectations arising from different points of view. It is argued that the case for realism presents itself interestingly strong.

7.1. Objectivity of Scientific Knowledge

As we saw in the first chapter, the antirealistic position of Bohr and Heisenberg was shared by Born, Jordan, and Pauli. This general point of view was discussed and developed by several philosophers of science, for instance by Reichenbach who wrote:

You say that while you are in your office your house stands unchanged in its place. How do you know? . . . The trouble is that unless you can find a better answer to that question than is supplied by the argument of common sense, you will not be able to solve the problem of whether light and matter consist of particles or waves. (Reichenbach, 1960.)

Several popular books were written on these subjects, and a large educated public was informed about the new trend in fundamental physics. Rather extreme inferences were drawn; for instance:

To put the conclusion crudely — the stuff of the world is mind-stuff. . . . The mind-stuff of the world is, of course, something more general than our individual conscious minds; but we may think of its nature as not altogether foreign to the feelings in our consciousness. (Eddington, 1928, p. 47.)

The objective position of the Copenhagen—Göttingen school, with its insistence on the acts of observation as basic components of quantum theory, had important aspects in common with logical empiricism (positivism). No wonder then that the opposers of the newly born paradigm expressed themselves forcefully against these kinds of ideas. Thus Planck wrote:

However much Positivism may regard itself as preceding without presuppositions, it is nonetheless committed to a fundamental premise if it is not to degenerate into an unintelligible solipsism. This premise is that every physical measurement can be reproduced . . . But this simply means that the factor which is decisive for the result of the measurement lies beyond the observer. (Planck, 1949, p. 171.)

Similarly Einstein declared:

I see Mach's greatness in his incorruptible skepticism and independence; in my younger years, however, Mach's epistemological position also influenced me very greatly, a position which today appears to me to be essentially untenable. (Schilpp, 1970, p. 21.)

Even stronger was Schrödinger's position. For him, positivistic conceptions

. . . call to mind that sense of misgiving, that cold clutch of dreary emptiness which comes over everybody, I expect, when they first encounter the description given by Kirchhoff and Mach of the task of physics (or of science generally) . . . (Quoted by Scott, 1967, p. 114.)

All this could perhaps lead to the conclusion that these disagreements were of a purely philosophical nature, that is to say that they were not relevant for the practical applications of quantum mechanics. This conclusion would be reinforced by the attitudes concerning the paradoxes of quantum theory that existed before 1960: Einstein, de Broglie, and Schrödinger saw paradoxes, while Bohr, Heisenberg, and Pauli denied that any inconsistencies existed. As we have seen in detail in the cases of de Broglie's paradox and of the EPR paradox in the original version, these different conclusions depend exclusively on conceptual differences about physical reality, since the reasonings on which the contradictory conclusions were based appeared logically flawless.

The situation changed dramatically with the discovery of the Bell inequality. This deepening of the EPR paradox led to the possibility of discriminating experimentally between quantum mechanics and Einstein's idea of the separable elements of reality. As we saw in detail elsewhere, it is possible to conclude that, if quantum mechanics is complete, then there arises an incompatibility between its unlimited validity and the existence of separable elements of reality. If, on the other hand, quantum mechanics is not complete, one can deduce Bell's inequality as an immediate consequence of the EPR reasoning. Also in this case, therefore, one obtains an incompatibility between the existence of separable elements of reality and quantum mechanics. But this simply means that completeness is a false problem and that one has in all cases incompatibility between quantum mechanics and separable reality! This is the choice facing modern physics, and there cannot be any doubt that, if quantum mechanics is correct also for EPR-type correlations, then one must give up either the reality of atomic systems or their separability.

It is from experimental grounds that scientific knowledge mostly arises, and this is a statement on which all physicists agree. Another point of agreement is that, if one is given a logically correct reasoning, then the only remaining debate can be around the correctness of the postulated premises or about the importance of the consequences.

Experiments and logic are therefore the pillars on which the unity of physics is supported. This is the minimal methodological apparatus that everybody accepts. Disagreements arise as soon as one asks the question whether or not there is something else that should be of interest. Logical empiricism is based on the fundamental assumption that nothing else matters, but we saw that very distinguished physicists (Planck, Einstein, Schrödinger, Ehrenfest, and de Broglie) disagreed sharply with such a point of view. Furthermore, it is clear that these great masters are representative of a rather widespread attitude in modern physics which considers it possible to obtain significant information through physical research on the objective (that is observation-independent) properties of atomic and subatomic entities. As a consequence, if correct reasoning along with a reliance on experiments is the common basis for all physical research, out of which the unity of physics arises, still there is room for wide disagreements on different grounds. It is in particular possible to disagree with the philosophical content of quantum theory and, more generally, to develop a critical attitude (positively critical, of course) toward other modern scientific disciplines.

An important question we must answer is one that concerns the very possibility to criticize, to not accept in its entirety a successful scientific theory such as quantum mechanics, relativity, or statistical thermodynamics. Is it not natural to think that the great success of these theories guarantees their scientific rigor; or do we perhaps have some doubt that they represent the highest achievements of the human mind? If we have misgivings about these achievements, if we criticize them, do we not, just because of our attitude, become antiscientific, thus begin to tread a slippery devolutionary path?

Actually, the fact that critique is not meaningless is demonstrated by the frequency with which scientists have, sometimes vehemently, criticized each other's theories. For instance, the quantum mechanics of the Copenhagen and Göttingen schools were considered unacceptable by Planck, Einstein, Schrödinger, and de Broglie; and even Dirac himself had since 1970 declared this theory to be inadequate, predicting that it would be improved upon eventually (Dirac, 1975).

On the other hand, modern epistemology leaves a well-defined space for possible imperfections, defects, and errors in scientific theories. Suffice it here to quote Popper, for whom in a sense every theory is a free creation of the human intellect which can only be falsified by experiments, never proven true. The general stand he takes with regard to scientific theories is that "... even the best theories are always our own inventions. They are beset by errors" (Popper, 1985b, p. 396). It is from this perspective that a champion of scientific rationality like Popper allows himself to consider modern physics as 'a bastion of subjectivistic philosophy'.

It is only natural that every scientific theory should be profoundly affected by arbitrary choices introduced in it (whether consciously or not) by its creator. Einstein has stated repeatedly that *physics is a creation of man in his effort to comprehend the real world.*

I do not believe it possible to obtain a correct description of physics unless one takes a clue from these Einsteinian concepts. It might be objected that one thus creates a contradictory situation, in which a scientific theory is at one and the same time arbitrary and objective; and in a way this is so, but in a noble sense. A theory is indeed a very complex construction which synthesizes and organizes empirical evidences and data along with cultural and philosophical lines of thought. The outcome cannot help being a closely interlaced fabric of the general prejudices of the researchers who realize the synthesis and of the novel acquisitions of knowledge pertaining to the real world.

Naturally there are excellent reasons for believing that even the fundamental theories of modern physics result, in turn, from the interlacing of arbitrary elements and objective elements, even though it is sometimes difficult to distinguish the one type from the other. However, it is always possible to put forward hypotheses, which are sometimes very plausible. As far as the objective contents of theories are concerned, the following examples could be considered:

1. The Milky Way is a regular galaxy viewed from its inside.
2. Matter has a molecular and atomic structure.
3. Every atom contains a positively charged nucleus, which is of the order 10^5 times smaller than the atom.
4. There exist subatomic systems (p, n, π, Λ, Σ, etc.).
5. A few hundred million years ago, the continents used to be all tied together.
6. All living terrestrials originate from a common evolutionary process.
7. Mass and energy are mutually interconvertible.

It is simply inconceivable that future scientists would disavow the statements above, even though we might quite possibly be overlooking many of the details of some of them; that is, there is still a lot of work to be done in *depth* for nearly every single statement, yet this cannot possibly serve as a reason for discarding any of them.

It is more difficult to identify the historically determined components of contemporary science. If one could unequivocally recognize them as such, it would clearly be important to replace them with statements closer to objective reality. One may however guess that the relativity principle, the acausal nature of quantum mechanics, and the nonseparability of quantum systems, could be modified wholly or in part.

Thus, every scientific theory appears to result from a tight interweaving of two types of contents: the irreversible content, which is essentially a true reflection (albeit an inaccurate one, inasmuch as it is schematic and 'out of focus') of the properties of the material world, and the arbitrary content, which gets injected into the theory either to suit the preferences of its creators or in response to strong cultural and social pressure, which in some historical periods, for instance during the Weimar Republic, have affected and distorted scientific thinking.

Both scientism and historicism lie outside the present description, due to their one-sidedness.

The *scientism* typical of certain neutralistic visions of science, which is common among scientists and epistemologists of positivist persuasion, states essentially that science cannot be criticized except for possible lack of rigor in the application of mathematical or empirical methods. This view, if valid, by itself would guarantee the uniqueness and truth of existing science. But from the same standpoint it is difficult to understand the great unsettled quarrels that took place among famous scientists (for instance between Einstein and Bohr); these are accordingly simply ignored or painfully ascribed to the senility of the unfavored contender.

The *historicism* typical of rightist inclinations (such as Spengler's or Croce's) or leftist inclinations (such as the Frankfurt brand of Marxism) maintains, on the contrary, that the contents of science are, deep down, all historically determined and that cultural fads and/or the production rates in factories are ultimately the processes fixing the nature of science in any given epoch. But from this perspective one has a hard time understanding why certain concepts of science such as that of electron can survive five or six different paradigms in different epochs and different societies.

One is therefore well advised to avoid the false quandary scientism/historicism and simply to accept the fact that a scientific theory can work very well even if heavily biased through multifarious philosophical choices. One must take the various theories at their face value, and we believe that the description introduced above might be a step in the right direction.

In the following section we shall discuss the discipline of mathematics in order to show that a critical understanding is possible even of this most rational of all sciences.

7.2. Mathematics and Reality

Many people consider the use of mathematics in a physical theory to be the surest sign of that theory's solid scientific standing. We may agree with this assessment, in the sense that, if mathematics were removed from culture, our society would, in all probability, collapse into a primitive state. Having said this, we none the less deem it appropriate to try to sketch a critical evaluation of the role of mathematics.

The successes of modern theoretical physics are quite astounding if one recalls numerical predictions exact to twelve figures (the magnetic moment of the electron) or the great conceptual conquests (first at the theoretical and then at the experimental level) such as the reciprocal transformation of matter into energy according to $E = mc^2$ or the prediction of antiparticles. All this would seem to testify forcefully that mathematics acts as a direct cognitive instrument, even abstracting from any physical understanding of microscopic phenomena in causal and spacetime terms. It almost appears that this formal instrument enjoys a direct and privileged relationship with material reality, as if the latter were trying to follow a rational, preordained design.

Confidence in the mathematical approach to our knowledge of reality has grown considerably in the present century. The mathematical rigor characteristic of theoretical physics has been elevated by the neo-positivists to a status where it serves as the universal criterion for what is scientific and what is not, thereby reducing biology, astronomy and geology to the rank of pseudo-sciences, almost as though their exceptional achievements should be forgotten or at least suspended until a mathematical formulation of their propositions could be realized.

The tendency to understand reality in intuitive terms was fought by Heisenberg, who favored 'a retreat into the mathematical scheme' (Heisenberg, 1959, p. 179). A large part of contemporary theoretical physics has fallen in with this line. In particle physics, for example, it is maintained that the properties guaranteeing the existence of a certain type of particle are its quantum numbers (mass, half-life, charge, spin, isotopic spin, strangeness, baryonic number, and so on). It is also maintained that such a characterization is possible only within a framework based on mathematically formulated theoretical relationships, not within a description employing intuitive or spacetime terms.

Another consideration that seems to favor a mathematization of science involves the well-known fact that the mathematical formalism of a theory generally outlives the physico-philosophical vision that underlies and generated it. Thus, Maxwell's equations and Schrödinger's equation still form an essential part of our scientific heritage, while, by contrast, the physical models that Maxwell and Schrödinger took as their starting points have fallen by the wayside.

The problem of mathematics was posed as follows by Albert Einstein in 1921:

Here arises a puzzle that has disturbed scientists of all periods. How it is possible that mathematics, a product of human thought that is independent of experience, fits so excellently the objects of physical reality? Can human reason without experience discover by thinking properties of real things? (Quoted by Frank, 1957, p. 85.)

Einstein himself, as is well known, was capable of using mathematics to make some of the most far-reaching conquests of our science. We may, for example, recall that when the two giant theories of classical physics, Newtonian mechanics and Maxwellian electrodynamics, were unified within the superior framework of special relativity (which allowed the resolution of the conflict between the invariance of Newton's equations and the non-invariance of Maxwell's equations for transformations between inertial systems), new and unforeseen consequences descended on the world of physics. Foremost among these was perhaps the $E = mc^2$ relation, expressing the reciprocal transformability of matter (measured by mass) and motion (measured by energy), a consequence which has profoundly affected our present world, in both positive and negative ways, and which vividly demonstrates how practical the implications of a theoretical discovery can be.

A further illustration of the creative potential of mathematics is found in the role it played in bringing about the unification of relativity and quantum mechanics through Dirac's equation, and in this equation's prediction of antiparticles, to be looked at soon.

Thus mathematics has undoubtedly led to crucial discoveries. Nevertheless, one should not forget that, particularly at the start of this century, developments took place which gave a substantial shock to the naive confidence on the part of some in a direct and positive relationship between mathematics and reality. We may call to mind:

(1) The discovery of the fact that no logical contradiction flows from the denial of the axiom of parallel lines (Euclid's fifth postulate) and the resulting advent of non-Euclidean geometries. It was discovered that an infinite number of geometries exist, all of them perfectly rational. And since there is only one real world, any illusion of a *one-to-one* correspondence between mathematics and reality came to grief.

(2) The discovery, within set theory, of 'antinomies', or nonresolvable paradoxes, which showed how one could deduce, in an absolutely rigorous fashion, real and proper contradictions, starting out from statements or principles whose self-evident nature could not, it seemed, be challenged (one may think of Russell's paradox). Out of this was born the tendency no longer to seek the justification for a mathematical axiom in empirical evidence, and thus there opened up a real and proper fracture between mathematics and reality.

(3) The discovery of Gödel's theorem, which consisted in the statement that one could not introduce in mathematics an axiomatic system that was simultaneously *consistent* and *complete*; to be precise, if one wants the system of axioms to be consistent, one must of necessity give up completeness. [A system of axioms is *consistent* or noncontradictory if one can never

deduce from it both a particular statement and its negation. A system of axioms is *complete* if one can always deduce from it a given (meaningful) proposition or its negation.]

At the general level, the fact that divergent currents exist regarding the nature of mathematics was verified at the Königsberg Congress (1930) for German speakers and at the London Congress (1931) for Russian speakers. At the first of these meetings there were three invited plenary lectures: that of Carnap (the *logicistic* foundation of mathematics), that of Heyting (the *intuitionistic* foundation), and that of von Neumann (the *formalistic* foundation). In London, on the other hand, it was the mechanicists and the dialecticians who were in opposition to one another.

Centrone wrote:

> The decomposition of the sole objective universe, the true field of experimentation for the Platonists, following on the collapse of Russell's project of building a 'Great Logic,' opened the way to a dispersion of the universes of discourse and to their extreme adaptability in an operational key. (Centrone, 1978.)

Looking at this matter carefully, however, the crisis in the relation between mathematics and reality began with Kepler, i.e., with the very birth of modern theoretical physics, at least from a logical point of view, if not from that of a full realization of the existence of this crisis. Even more than Galileo and Newton, Kepler was inspired by the idea that the universe had been constructed by following a perfectly rational plan.

For him, *everything that exists* was created according to a preestablished mathematical scheme, the goal of science being that of discovering in nature the mathematical harmonies that regulate its structure and evolution. For Kepler, then, there had to be a rational response to every reasonable problem regarding material objects. In particular, there had to be simple and immutable laws regulating the whole solar system.

Animated by this unflagging belief, Kepler spent his life looking for regularities in the structure of the solar system and the motion of the planets, utilizing the very precise observational data that Tycho Brahe had bequeathed him. We may say at once that if, on the one hand, he booked a striking success — the three laws that have made his name famous — on the other, he led science to a defeat, which was highly significant just because it was inevitable. Kepler's successes pertain to planetary motion and are valid, within well-known limits but to a high degree of precision, for the planets that orbit around our sun as well as for the comets, for double stars, and so on. Kepler's three laws represent a fundamental step forward in our knowledge of nature, and, by exploiting them to the full, Newton succeeded in unifying celestial and terrestrial physics and obtained the laws of universal gravitation. The success of Kepler's laws is in a certain sense astonishing: the planetary orbits are indeed ellipses, to a high degree of precision, and what amazes one is the fact that one of the curves Apollonius obtained by intersecting a right circular cone with planes of various inclination could reproduce the trajectory of any one planet so well. No simple closed curve known to man other than the ellipse enjoys this property. Similarly, for any planet it is true, within an extremely narrow range of error, that the areal velocity is constant and the square of the period of revolution divided by the cube of the semi-major axis of the ellipse is constant and the same for all planets.

Kepler must have thought that these laws fully confirmed his initial hypothesis and that such instances of regularities in nature could therefore only derive from a rational plan on whose basis the material world had been constructed. It thus appeared natural, and was almost taken for granted by him, that other equally simple regularities would lie hidden in other aspects of the solar system, for example in its geometric *structure*. Kepler therefore devoted himself tenaciously to the search for a geometrical or mathematical law that would reproduce the observed distances of the planets from the sun. And after a great number of failed attempts he thought he had resolved the problem, when he proposed nested boxes of regular solids as a rule for the distance.

His idea was as follows: A spherical shell contains Mercury's orbit. This orbit is circumscribed by an *octahedron*, which, in turn, is circumscribed by another spherical shell containing the orbit of Venus. Circumscribed around this is an *icosahedron*, immediately outside which lies another spherical shell containing the Earth's orbit. Circumscribed around this latter is a *dodecahedron*. After this we encounter in succession: a spherical shell containing Mars's orbit, a *tetrahedron*, a spherical shell containing Jupiter's orbit, a *cube*, and, finally, a spherical shell containing Saturn's orbit.

No simpler model was able to reproduce the distances of the planets from the sun. Today, we know of course that not even this model works, both because additional planets have been discovered, while the number of regular solids remains at five (excluding the sphere), and because the most accurate measurements we have show that Kepler's model fails already for the inner planets.

Kepler's mathematical explanation both for planetary motion and for the distances of the planets from the sun for him met two perfectly symmetrical needs: if the universe is constructed according to a rational plan, one should be able to find a simple rational answer to the problems both of motion and of structure. But we now know that, while there exists a rational approach to the problem of motion (Kepler's laws), a rational treatment of structure is missing. Accordingly, cosmologists nowadays put the structure of the solar system to one side as governed by accidental events (distribution of matter in the primeval nebula, interaction with other stars passing by chance near our system in the course of its evolution, and so on). Indeed we are now convinced that, where the history of a system plays an important role, a simple mathematical description is impossible. Even if this conception raises more problems than it solves, modern science has reached the conclusion that there are aspects of material reality that do not yield to a mathematical description in simple terms. Thus Kepler's work has left us with a twofold message: In some instances mathematics grasps the essence of reality in an amazing way, while in others it does not come anywhere near it.

Other thinkers have realized the limits on the rationality of mathematics itself. Suffice it here to quote the examples of Lagrange and Frege. Toward the end of the eighteenth century, Lagrange wrote on the subject of mathematics:

> This science is, in truth, full of contradictions, and if, notwithstanding this, it has led to great results, this is dependent only on the infinite mercy of God, who has so disposed it that its errors balance one with another. (Quoted by Waismann, 1936.)

Frege was no less harsh when he wrote at the turn of the present century:

> It is quite scandalous that science is still in the dark about the essence of number. That we still do not possess any universally accepted definition of number would perhaps be admissable, if we found ourselves in agreement on the substance. But instead, even on this — whether number is an effective group of entities, or whether it is merely a sign that a man's hand traces on the blackboard, as if it formed something in the mind, so that psychology could study its origin, or whether it is instead a purely logical construction; whether it constitutes a creation of ours and thus might disappear, or whether on the contrary it is something eternal — even on all this science has up to now not made any decision. Arithmetic does not know whether its terms deal with mere signs traced with a piece of chalk, or whether instead they deal with abstract objects . . . . Does all this not constitute a real scandal? (Quoted by Waismann, 1936.)

It is one of the lessons of contemporary scientific culture that great forward strides in the field of knowledge can be made even without the use of mathematical tools. The most striking demonstration of such an advance is the Darwinian evolution of the species. Setting off from solely macroscopic observations, Darwin succeeded in constructing the tree of life on earth, i.e., an accurate description of how evolution took place. It is enough to leaf through *The Origin of Species* to realize that mathematics played no role in its discovery. And yet its conclusions, based on comparative anatomy, have been confirmed by recent research carried out on the molecular level. As is known, the amino acids (of which there are twenty different ones) can form linear chains, each of them composed of about 150 amino acids; these are the proteins. If one stops to ask how many proteins are possible, one ends up with the biggest meaningful number in science,

$$N \simeq 20^{150} \simeq 10^{195},$$

which is much, much bigger than the number of atoms in the universe. In consequence, only a number infinitely smaller than the number of proteins that are *a priori* possible can express the number of proteins that actually exist on earth. This is tantamount to saying that the existing proteins strongly reflect the particular historical process that has generated them, beginning with the amino acids in concrete existence at the moment of their formation. For example, cytochrome c is one of the proteins that allow living beings to utilize atmospheric oxygen. There are zones of this molecule which at times can undergo advantageous changes (mutations) concomitant with other anatomic, physiological, or behavioral changes; some of these changes have been conserved by natural selection. When one gets to examine all the available sequences of cytochrome c extracted from even very different living beings, one can make a study of molecular affinities and, in the last analysis, derive an evolutionary tree. What is important to underline is that this evolutionary tree follows qualitatively what Darwin and other scientists had already obtained from paleontology and comparative anatomy. This quite striking agreement cannot be coincidence! There are, furthermore, numerous other proteins (like the hemoglobins) that yield information of the same type and which allow us to draw analogous conclusions. One may thus conclude that Darwin constructed in outline what is in effect *the true* evolutionary process on earth, and that therefore a thoroughgoing knowledge of reality is possible, at least in certain cases, even if one does not use the tool of mathematics. This conclusion would be reinforced if one considered other sciences such as geophysics and astronomy, which still use mathematics only to a very limited and partial degree even at those times when they register important advances in knowledge.

Another point which should be discussed concerns the actual role that mathematics plays in the sciences where it *is* important, such as theoretical physics. If we consider this discipline without prejudices, we are bound to come to some surprising conclusions, as will be argued next.

Theoretical physics has always made use of mathematics as an instrument of rationalization and of knowledge. This approach has led to great successes, beginning with Kepler's laws and continuing right down to the recent successful attempt to unify weak and electromagnetic interactions. The quest for the greatest possible mathematical rigor is a constant feature in theoretical research and any theory whose mathematical precision is not immediately obvious is quite rightly looked upon with suspicion. It does not, however, follow from all this that theoretical physics is the unchallenged standard bearer of supreme mathematical rigor, not only due to the fact that, as we have already seen, it is not even certain that one can speak of absolute mathematical rigor (antinomies in set theory, Gödel's theorem, uncertainty even in the definition of the most elementary entities, etc.), but also, and above all, because, during those moments of the highest physico-mathematical creativity, one often witnesses the intrusion into the formal scheme of things of conclusions that cannot be arrived at by purely deductive reasoning.

One example is possibly worth more than a host of general arguments. Dirac's prediction of antiparticles in 1928 is very illuminating and, at the same time, typical of a general situation. If one analyzes the process by which Dirac went about constructing his equation, i.e., the equation that furnished the basis for the prediction of antiparticles, it becomes apparent that the 'discovery' of Dirac's equation was definitely not the outcome of a straightforward deductive line of reasoning but rather the culmination of a complicated procedure in which Dirac was guided around obstacles by criteria of simplicity and reasonableness. This conclusion is confirmed by the fact that much more recently Dirac has proposed different alternatives to his famous equation. Even more typical and striking was, however, *the process* by which the consequences of his equation *were deduced*, since Dirac realized that one direct and inevitable consequence was . . . the end of the world, in a time of less than a millionth of a second! For, Dirac's equation predicted the existence of states in which the energy was negative, states for which no evidence has ever been found, even indirectly, in our macroscopic world.

Dirac realized that the hypothesis of a real existence of negative energy states for atoms and their components implied an unacceptable consequence. The electrons of any atom would have to 'fall' spontaneously into the negative-energy 'sea', each of them emitting at the same time an enormous amount of electromagnetic radiation (regularly endowed with positive energy). If such a process really happened, then numerical calculations reveal that it should last less than a millionth of a second! The collapse of every existing atom and the enormous accompanying flux of energy would certainly destroy matter in all its states of aggregation, in particular every form of life. Luckily, nothing of this kind ever happens in the real world, and thus the corresponding inference from Dirac's equation definitely must be rejected. To remedy this unexpected situation, Dirac was forced to modify in some way the conceptual scheme that he had created. This he did by means of a new hypothesis which assumed that all the negative-energy states were 'already occupied', in the sense of *Pauli's exclusion principle*, so that any two spin-$\frac{1}{2}$ particles of the same type, two electrons, for example, could not find themselves in states characterized by the same quantum numbers.

This new hypothesis saved matter from near-instantaneous collapse, since normal atomic electrons could not fall into a negative-energy sea, seeing that all the possible states were already occupied by other electrons.

At the price of a new hypothesis, introduced rather summarily and for reasons of expedience into the theoretical scheme, Dirac thus fashioned a new conceptual situation in which the world that we inhabit became licensed to survive indefinitely. The new hypothesis was thus an *ad hoc* one, in other words of the type that physicists should never make if they followed the advice of certain philosophers of science. All the same, the introduction of this new hypothesis had so deeply modified the theory as to imply new and surprising conclusions, which this time were not necessarily absurd.

In fact, if it is true that an electron can no longer fall into the negative-energy sea, since it is stopped from doing so by Pauli's exclusion principle, it is equally true that instead the opposite is possible, i.e., an electron from Dirac's sea that is given the necessary energy by whatever means (for example by the absorption of a gamma ray) can make a transition from the sea into an unoccupied positive-energy state. However, when this happens, the sea is no longer totally filled but contains, so to speak, a hole, that is, a nonoccupied negative-energy state. Dirac was able to show that this hole would behave, in every way that one can observe, like a particle of positive energy, which exhibits a mass equal to that of the electron but carries an electric charge of the opposite sign.

This was an amazing conclusion, because it predicted theoretically that, for every spin-$\frac{1}{2}$ particle like the electron (Dirac's prediction holds also for protons, neutrons, etc.), there must exist a new, so-called *antiparticle* having the same characteristics but the opposite electrical charge (and opposite values of other quantum numbers, such as baryon number and lepton number).

When this prediction was made, there was no proof that antiparticles of this kind existed; moreover, no thinker had ever conceived of such an idea; whereas in the case of the atom, as is well known, a similar notion had already been conceived by the ancient Greeks. Thus, here in a compelling way, the new theoretical scheme, amplified by the additional hypothesis of a negative-energy sea, establishes the necessity for the existence of antiparticles, a necessity that is in no other way evident from the intuitive point of view. Stated briefly, we may still today ask ourselves the question: Why do antiparticles exist? Our only reply is that they exist because Dirac's equation predicts that they do. At best, we are able to formulate a reply of the same tenor in the more general language of quantum field theory, but it is not a reply which implies any physical necessity on which the need for antiparticles can be based.

Only some years after 1928 did experiment demonstrate the existence of the positron, i.e., the electron's antiparticle. Still later came the discoveries of the antiproton, antineutron, and all the other antiparticles.

It seems obvious that this amazing capacity of Dirac's theory to predict important new facts derives, in all probability, from the union of two types of reasoning on which the theory is based. The introduction of a dialectic element — what we have termed Dirac's new hypothesis, which consists in fact in a radical denial of one of the premises from which sprung the reasoning that had predicted the end of the world — in some way activated the potential of theoretical reasoning to penetrate material reality and understand one of its important secrets. Without the introduction of a particle-filled vacuum, the theory not only would have failed to predict the antiparticles but, as we have said, would have implied the extremely rapid collapse of all matter. We seem here to be witnessing the linking together of two types of logic: formal logic, based on the principle of noncontradiction, and the contradictory logic of ordinary dialectics. It is probably the simultaneous use of these instruments that gave Dirac's reasoning such penetrative powers. For, it does not seem credible that the hypothesis of a sea of negative energies is valid in nature and that antiparticles are really holes in this sea, both because the sea itself would have to be of infinite depth and because it would be required to consist *simultaneously* of electrons, protons, neutrons, and all other spin-$\frac{1}{2}$ particles. The resulting framework is a strange and artificial one and does not appear capable of representing the reality of the subatomic world! Antiparticles do however in reality exist, and Dirac's theoretical scheme, burdened though it is with probable distortions in respect to the real world, has undeniably predicted them correctly.

7.3. The Role of History of Physics

The opinion was expressed by Erwin Schrödinger that the history of science is of exceptional importance for our understanding of the real nature of scientific theories. He also wrote that physics was running 'the great risk of being separated from its historical background'. This fear was expressed some thirty years ago, but today, unfortunately, we may add that it has been transformed into harsh reality. Physics research and teaching symmetrically ignore the historical developments and contrasts that gave birth to today's ideas. Commenting on this situation, Kuhn has said in speaking of the cultural formation of physicists:

> ... it is a narrow and rigid education, probably more so than any other except perhaps in orthodox theology. (Kuhn, 1962, p. 166.)

In all probability, the difficulties encountered by physics students in understanding quantum mechanics are closely linked, among other things, to a spontaneous philosophical stance of a realist nature, in other words linked to the idea that the objects dealt with by science, including atoms, in some way exist independently of man. If, among students, this attitude is often detached from a clear recognition of its nature as one involving a choice, the same world view was on the other hand explicitly at the root of a long and as yet unresolved dispute between the followers of the Copenhagen—Göttingen paradigm and those who, instead, followed radically different ideas, such as Einstein, de Broglie, Schrödinger, Planck, and Ehrenfest.

From a general historical viewpoint, one can say that the contrasts between various research programmes have certainly not been limited to quantum physics, but that the struggle between different lines of research has, instead, been an important factor in the development of many new scientific conceptions. In speaking of these clashes within science, I am not referring to those involving personalities, such as the argument between Newton and Leibnitz concerning who was the first to discover the differential calculus. (This debate made Newton say that science was "such a quarrel-some lady that one would rather deal with the law than with her".) Of greater importance were, instead, the giant conflicts of ideas such those which divided:

(i) Neptunists and Vulcanists among nineteenth-century geologists;
(ii) Wegener (who upheld the continental drift theory) and his opponents;
(iii) Freud and Jung about the meaning of psychoanalysis;
(iv) Darwin and his scientific detractors over the evolution of species;
(v) Boltzmann and the energeticists regarding atomism.

These stark contrasts were born from different prefigurations of the nascent paradigm. On the other hand, science and culture do not progress only by the incremental accumulation of new facts, but also through giant catastrophes that mark the collapse of wornout scientific theories (or old superstitions) and the birth of new theories that can interpret and explain wider ranges of phenomena. There is no automatic guarantee that any new theory will be better *in every sense* than the one it replaces, and history is not lacking in examples illustrating this fact, such as:

(i) Aristarchus of Samos and Seleucus of Babylonia placed the sun at the centre of the universe, but Ptolemy's ideas gained the upper hand and the heliocentric system was forgotten for nearly 1500 years.
(ii) For quite a number of centuries after the fall of the Roman Empire, Europe went back to the idea that the earth was flat, while in the eight centuries that separated the ancient Greek thinkers from the full glory of the Roman Empire, practically the whole of Greco-Roman culture understood and accepted that the earth was spherical.
(iii) In ancient Greece, there flourished schools of atomists, but in practice it was not until the beginning of our century that the majority of scientists accepted the existence of atoms.

The new theory is, however, always better than the old one in the sense that it can explain a greater number of empirical facts. We know little about the heliocentric theory of Aristarchus of Samos, but it is very likely that the Ptolemaic geocentric theory was more highly developed and more accurate in its predictions regarding planetary motion, eclipses, and so on.

From what has been said, the inescapable and crucial role of the history of science ought to emerge, no longer just as a cultural curiosity but as an instrument of critical understanding for any researcher in his or her own discipline.

For a deep understanding of quantum physics it is in fact necessary to consult the original articles from the first quarter of this century, a study of which provides a stimulus that is hard to anticipate for those who know only publications dealing with contemporary physics. From these articles, there clearly emerges the framework within which quantum physics originated. The different scientific approaches of Einstein and Planck, on the one hand, and

Bohr and Born, on the other, which were discerned with crystal clarity after 1923, quickly led to the consolidation of two groups of theoretical physicists, each group having a degree of internal homogeneity but being clearly distanced from the other about the goals and necessary contents of an acceptable scientific theory. 'Victory', as is well known, went to the Copenhagen and Göttingen groups, who managed to create a sufficiently consistent scientific theory which assimilated the principal contributions to quantum theory made by the physicists of the opposition group (particularly Einstein, Schrödinger, and de Broglie), at the same time that it changed their philosophical content.

This rediscovery of the head-on cultural clash between two groups of physicists of the highest caliber poses one immediate problem: that of establishing what gave rise to their different approaches to the same question, given that these differences were so strong that Einstein found it impossible to fully accept quantum mechanics, right up to the end of his life, and that it caused Heisenberg to say that the realist interpretation given by Schrödinger to his waves seemed 'disgusting'. In truth it is in events external to the strictly scientific world where the philosophical prejudices of research workers had their genesis.

By linking science and philosophy, we may attempt to fully rediscover the historical dimensions of scientific thought and to reach an understanding of the nature of the travails that gave rise to various scientific conceptions.

In this way, science itself will appear as an intertwining of historical and of objective elements, a perspective that contradicts both those attempts that would interpret science as a pure construction of the human intellect, without any objective content, and the widespread opinion that everything scientific is valid and cannot in any way be criticized unless its falsity has been demonstrated by means of scientific experiment.

For the sake of a critical understanding, the analysis that Louis de Broglie gave of nineteenth-century physics is of great interest (de Broglie, 1976, p. 10).

De Broglie distinguished two quite distinct conceptual trends, each based on what in practice were conflicting answers to the question "What is physics?" On one side he placed Fresnel's wave theory of light, Maxwell's electromagnetic theory, and Fourier's theory of heat propagation; these were theories that attempted to represent the phenomena being studied *as causal propagations in space and time*, utilizing to this end the mathematical theory of differential equations. On the other side de Broglie put classical thermodynamics, whose sole base consisted of two fundamental principles derived in a strictly empirical way (the law of energy conservation and that of entropy increase). This theory was capable of explaining a plethora of empirical observations without, however, being able to provide any causal description in space and time of the phenomena under study.

At the beginning of the present century, the second line of thought, defined mainly by Ostwald and Pierre Duhem, might have seemed to be defeated, thanks to the research carried out by Boltzmann and Gibbs, who, through the atomic hypothesis, were able to give a causal description in space and time of thermodynamic phenomena as well. Concurrently also Lorentz was expanding his research into electromagnetism in the same direction, while Planck and Einstein, too, were trying to formulate an as yet embryonic quantum theory along causal lines. This phase was however not to last long, because subsequent developments, in particular the final formulation of quantum mechanics in 1927, turned the situation completely upside down and led physics to a pessimistic assessment about the possibility of describing atomic phenomena in an intuitive fashion.

This outcome of course did not signal that the question was settled once and for all. Indeed, if we extrapolate the critical tendencies of contemporary epistemology (Popper, Feyerabend, Kuhn, Lakatos, etc.) in conjunction with new lines of research in physics that find their explicit reference points in the causal ideas of Einstein and de Broglie, then new and radical upheavals would seem to be even likely in future.

The application of a historical and philosophical critique to modern science carries with it a risk of leading us into an irrationalist territory unless it is coupled to a firm appreciation of the momentous conquests in the field of knowledge that have been made by science in general and by quantum physics in particular. It cannot, for example, be merely chance that the concept of the electron, discovered at the end of the last century, has found a place in very different theories, such as the classical Lorentz theory, the first quantum theory, special relativity, the quantum mechanics of 1927, quantum field theory, and general relativity. We are evidently dealing here with a concept whose validity transcends that of any particular scientific theory, and this is one aspect of the irreversibility of the idea of an electron, i.e., of the fact that electrons really do exist, independently of man and of his observations. The existence of irreversible cognitive elements in science was recognized for example by Max Planck when he wrote:

> The fixed unity of the world picture is, however, as I have endeavoured to show, the fixed goal which true science approaches through all its changes. In physics we may correctly state that the present picture has certain properties, and, although it varies its hue according to the individuality of the worker, no revolution in Nature or Man can obliterate these properties. This constancy, independent of all human and intellectual individuality, is what we call reality. (Planck, 1960, p. 24.)

In spite of the evident existence of irreversible cognitive elements, there is room in modern physics for quite dissimilar entities. As will be argued in the following section, when one passes from the strictly technical content (where large agreement exists) to the more conceptual one, one sees that physics resembles a labyrinth of philosophies, of diverging paths, of difficulties, and of contrasts. We have to develop an understanding of physics that is adequate for the times, along with an overall critique of the labyrinth into which we have been driven:

On the one hand, there exists today the aptitude necessary for confronting the complexity of reality, rejecting simplistic visions that do nothing but confirm our habitual representation of the world; what we need today is the most detailed map of the labyrinth possible. On the other hand, there is the fascination of the labyrinth as such, of losing oneself in it, of representing the absence of any way out as the real condition of humankind. In separating off the one attitude from the other, we wish to focus our critical attention, while still bearing in mind that we cannot always draw a clear dividing line . . . ." (Calvino, 1980, p. 96.)

These words of Italo Calvino, about literature and society, lend themselves very well to the posing of the problem of modern physics. Inspiration may be drawn from Calvino for the infusion of life into a scientific critique that *challenges the labyrinth*, as distinct from and contraposed to any line of *surrender to the labyrinth*.

If our Ariadnian thread cannot be found in a general *intensive* reading on the foundations of physics, since by so doing we come across the swamp and palafittes of which Popper has spoken, we can probably find it: (1) in the white-hot points of the crisis, the problem of wave—particle duality and the EPR paradox, which have been discussed in previous chapters of this book, or (2) in an *extensive* reading on science in its totality, a reading that cannot but generate great optimism, given that we have already managed to understand things like the existence of atoms, the equivalence of mass and energy, the evolutionary process of life on earth, and the nature of the sun and the stars, to name but a very few, although we learnt to read and write only some five thousand years or so ago — one thousandth the span of human history and one millionth the age of the planet earth.

7.4. Fragmentation of Modern Physics

Our century has been characterized from the ouset by a virtual explosion of science, not only in the sense of its rapid development in all directions, but also in the sense that increasingly wider gaps were created between one discipline and another and even among different strains of the same discipline.

The year 1902 marked the 'official' opening of the crisis in the foundations of mathematics, due to Russell's discovery of the antinomy named after him. The Königsberg Congress (1930) would testify to the depth of the disagreement between logicians, formalists, and intuitionists, a disagreement unreconciled even today.

In 1905 Einstein formulated his theory of special relativity and discovered theoretically the equivalence of mass and energy, which was to be confirmed later by Rutherford's observation (1919) of the first artificial nuclear disintegration.

The establishment of relativity was accompanied by controversy within physics as well as outside. The animosity characterizing the debate was inversely proportional to its scientific relevance (someone went so far as to write that only a Jewish physicist could have proposed a theory from which the concept of force was banned!). More elegant and civilized, yet fraught with much more serious consequences, was the rejection of the wave—particle duality formulated by Einstein in 1905. This rejection was general: One might say that no important physicist, with the exception of de Broglie, accepted Einstein's formulation. The army of his opponents, however, became eventually fragmented into a host of parties at odds with one another on the most fundamental issue of quanta, i.e., on the dual nature of atomic systems. This crisis, which remained mercifully concealed during forty-odd years dominated by positivist neutralism, has arisen once more today in a serious form.

The dramatic contradictoriness plaguing the developments of modern physics becomes apparent in many other ways:

1. From the elimination of the ether, the medium postulated by physics in the last century as supporting electromagnetic phenomena, a world of appearances and relative truths seemed to emerge; at the same time, however, energy shaped up as an indestructible entity and became the basis of all things — matter, motion, heat, light, and fields.

2. The concreteness of light relied on the property of energy all the same. The necessity to localize energy in narrow spacetime regions (photons or light quanta), a necessity imposed by the photoelectric effect, the Compton effect, and the Compton—Simon experiment, along with the impossibility to abandon wave-like aspects, however prompted Einstein to introduce empty waves (which he ironically called phantom fields) propagating in three-dimensional space but devoid of energy as a fundamental attribute.

3. Skeptical empiricism, which to some extent led Einstein to Mach and ultimately toward special relativity, seemed to resolve the problem of reality by actually dissecting it into a set of distinct opinions held by different observers, all equally valid even if all diverging. At the same time, the natural tendency to accept as *real* the totality of events simultaneous with the present instant of a given observer and to extend the concept of reality to the 'real' of all possible observers, led relativists to envisage a *four-dimensional* universe established once and for all. The time evolution of any system was therefore determined, down to the smallest details, by a rigid necessity, since for any observer equally valid as myself, my future is already unequivocally determined, is indeed already real. This Parmenidean kind of vision was blamed on Einstein by Popper in 1950 (Popper, 1976, p. 129).

4. The triumph of atomism, resulting from the theoretical investigations of Boltzmann and Einstein and the experimental research carried out by Perrin and Rutherford, was soon followed by what was essentially a devaluation of its ontological significance, brought about by the subjective formulation of quantum mechanics of the Copenhagen and Göttingen schools. The principle of complementarity, along with a certain interpretation of the 'uncertainty relations', postulated the impossibility of describing elementary phenomena

both causally and in spacetime. For many scientists, atoms had thus become only a useful conception, not unlike the network of parallels and meridians on the earth surface.

5. The *local* formulation of interactions, which was established by Einstein in special and in general relativity, implied, among other things, the elimination of action at a distance, and thus allowed one to dispense with the most mysterious aspects of classical physics and to fully assume spacetime as a nonsecondary participant in physical phenomena. However, quantum mechanics in certain instances requires this local character to disappear. The nonseparability of quantal systems (EPR paradox) appears to demonstrate the uselessness of space, practically its elimination at a microscopic level.

These contradictory developments of modern physics emphasize the importance of the relationship between physics and philosophy. Platonicism, existentialism, empiricism, idealism, pragmatism, and materialism all have a complex relationship with modern physics. At times a given philosophy can rule over a particular 'interpretation' of a theory; in other instances a definite philosophical bias might have motivated scientists responsible for the birth of a new paradigm; on still other occasions, finally, a certain philosophy can emerge from the very formal apparatus of theoretical physics (an astonishing example being the EPR paradox).

This many-faceted fractionation of physics was taking place during a period in which other sciences also achieved important advancements, but in a less traumatic way. In astrophysics, for instance, the discovery of the method of cepheids (1920) allowed first the measurement of the diameter of our galaxy, then the localization of our solar system within it, and finally the identification of a number of other galaxies and the formulation of the *big bang* theory.

In earth sciences, Wegener in 1912 formulated the theory of the continental drift, which gained the support of geologists and paleontologists but was opposed by many geophysicists. In physiology, his 1908 experiments of impregnation with gold and silver allowed Golgi to develop the theory of neurons and thus to formulate a new hypothesis on the structure and functioning of the nervous system.

Nevertheless, the central role of physics and its more advanced development have left a deeper mark on our epoch, and this has caused the coming into being of an epistemology and a pervasive philosophy of hopeless resignation, of throwing up one's arms before the labyrinth, of fleeing back into the ancient world of magic and mysticism. As a leader in this trend to live with the labyrinth, one may consider Feyerabend. In his *Against Method* (1975) one finds a desperate acceptance of an anarchist's view of science, one in which even the most negative aspects are made into rules:

> Knowledge . . . is not a series of self-consistent theories converging toward an ideal concept, is not an ideal approach, is not a step-by-step approach to truth. It is more like an *ocean*, an every growing one, of *mutually incompatible alternatives* (*perhaps even incommensurable*):

every single theory, every fable, every myth belonging to this collection compels the other to become more articulated, and all contribute, through such a competition process, to the development of our conscience. Nothing is ever decided, no concept can ever be left out from a general exposition. (Feyerabend, 1975, p. 27.)

Feyerabend does not acknowledge a privileged role for science and suggests an educational system in which not only science is taught but also a variety of types of religion, of mysticism, and of magic, so that all individuals are left free to follow whatever ideology they please (Chalmers, 1976).

More and more attempts at effecting a synthesis between physics and mysticism are being carried out within physics itself. Thus Nobel laureate Brian Josephson makes reference to the Vedic tradition of India and accepts the idea that the spirit exists before matter and is actually capable of generating it. He thinks that in this way we can reconcile the subjectivity of our individual lives with the objectivity of nature (Josephson, 1979).

Fritjof Capra is another author who borrows ideas from the religious philosophies of the Far East. In his *Tao of Physics*, a book which has been called 'immensely popular' in American universities (Harrison, 1979), Capra voices his belief that the self-consistence principle, which is at the basis of the *bootstrap* model of elementary particles, and the interdependence of all phenomena, a notion which strongly underlies the philosophies of Oriental mystics, are nothing but two different expression of the same idea (Capra, 1975).

The French physicist O. Costa de Beauregard has proposed a theory of quantal phenomena that exhibits perfect symmetry between past and future (1978). His wave functions can be used to make both statistical predictions and statistical *postdictions*; the latter are, according to him, strictly connected with parapsychological phenomena of precognition and psychokinesis. To him, the two symmetric faces of consciousness, *viz*., perception and volition, ought to both play a role in quantum measurement processes. The well-known Indian physicist E. C. G. Sudarshan (1978) also sees a connection between the metaphysics of Veda and the Platonic view of many theoretical physicists according to which the world is nothing but the concrete realization of an absolute system of thinking.

Nobel laureate Eugene Wigner is convinced that the study of quantum phenomena has produced the definitive realization that the content of our consciousness is a reality which cannot be narrowed down to material phenomena (1961).

The list of investigations and opinions of this type could go on with the mention of other names: the Dane R. D. Mattuck (1977), the Frenchman G. Klein (1980), the Britisher Paul Davies (1984), the Americans A. A. Cochran (1971) and L. Bass (1975), the Austrian H. Pietschmann (1980), and so on and on.

Although our disagreement with this type of idealistic philosophy is total, we shall show next that at least some of the previous proposals and 'conclusions' have real roots in the usual interpretation of quantum theory, in terms of knowledge and of acts of observation, *if one insists on the completeness of quantum theory.*

An elementary property of quantum mechanical wave functions is the following. If one measures the observable $\mathscr{A}$ on a quantum system α having the wave function $\Psi(x)$, and if $\{\Psi_i(x)\}$ is a set of wave functions, each describing $\mathscr{A}$ as having a fixed value [thus $\Psi_\ell(x)$ may describe $\mathscr{A}$ as having the value a_ℓ with certainty], then it is normally possible to write the state of the system as a superposition

$$\Psi(x) = \sum_i c_i \Psi_i(x).$$

The measurement of $\mathscr{A}$ on α gives the observer a knowledge that is *certain*, in the sense that a repetition of the measurement would reproduce the same value of $\mathscr{A}$ if the system had no further interactions. Therefore, if the measurement has given $\mathscr{A} = a_\ell$, then the wave function of α after this measurement will be $\Psi_\ell(x)$. There is thus a sharp jump of the wave function, during the first measurement from $\Psi(x)$ to $\Psi_\ell(x)$, the so-called *reduction of the wave packet.*

One often finds statements implying that the reduction of the wave packet is due to a change in the observer's *knowledge* about the properties of the system α generated by the measurement. In order to clarify as much as possible the full extent of the measurement problem it is therefore convenient to split up the relationship between human observer and physical object into three parts:

(A) the knowledge that the observer has (or thinks he has) of the investigated object;
(B) the wave function Ψ that according to quantum mechanics describes the object;
(C) the real structure and physical evolution of the object.

The most optimistic viewpoint one can adopt is to assume the existence of a one-to-one correspondence both between parts (A) and (B) (such that two different degrees of knowledge about the object will correspond to two different functions Ψ and *vice versa*) and between parts (B) and (C) (such that two different Ψ will correspond to two similar physical processes but with at least some objectively different peculiarities). In this manner, given Ψ, the resulting knowledge of the object on the part of the observer would be perfect.

In reality, it is very difficult to think of the description of the object given by Ψ as an absolutely complete description, and it is therefore more reasonable to assume only that two different Ψ correspond to two different physical situations, without the contrary being necessarily true. In a similar way, one may give up the idea that two separate Ψ correspond necessarily to two different degrees of knowledge of the system, since the mathematical structure of Ψ could turn out to be richer than is strictly necessary to represent our knowledge. However, it is certainly necessary to maintain that two different degrees of knowledge are represented by two separate Ψ if quantum theory is to be complete.

In conclusion, the two hypotheses:

(I_1) two different degrees of knowledge of the object on the part of the observer correspond to two different Ψ functions;

(I_2) two different Ψ correspond to two objectively different physical objects;

are the widest ones within which one can state the validity of the quantum mechanical formalism.

In this approach, the point of view according to which a change of the observer's knowledge generates a reduction of the wave packet leads one to the conclusion that, as a consequence of (I_2), changes of human knowledge can modify the physical structure of the system under investigation.

In this manner it becomes clear that the observer does not gain awareness because his interaction with a physical system generates some alteration in his state of consciousness; rather the opposite is true, because consciousness imprints on the external reality new features that it has in some way decided to generate.

It is therefore still possible to speak of one's 'knowledge' of an object but only in terms of an explicitly idealistic description that is based on some kind of action of the human mind on matter, a description which is also clearly in tune with some aspects of parapsychology.

To avoid these conclusions, one could attempt to further weaken the hypotheses (I_1) and (I_2). If one had to give up (I_1), the conclusion could be reached *ipso facto* that quantum theory is recognized as incomplete, because there would no longer be a full correspondence between our knowledge of a system and its theoretical description. Therefore, (I_1) must be maintained as valid if one wants to claim the completeness of quantum mechanics.

The only possibility left is to give up (I_2). In this case, parapsychological effects are excluded, since two different Ψ, of the type previously considered, may correspond to one and the same real system. But in this way, because of (I_1), Ψ describes only the mental state of the observer, and its evolution describes the evolution of ideas. Therefore, the state of human consciousness would develop in a strictly causal way when no 'observation' is made.

Any 'observation', on the other hand, would change human consciousness in a sudden and casual way, whence the reduction of Ψ would follow. Of course, also the 'result of an experiment' (which according to the quantum mechanical formalism corresponds to the final value of Ψ) would be a purely intellectual creation, and one could not learn anything about the 'real world' from measurements. As a consequence, the 'real world' would become a sort of ghost-behind-a-wall which cannot in any way be known, and physics would be reduced to a study of the spiritual activity of man.

This idealistic interpretation of quantum measurement is as old as quantum mechanics itself. Great physicists like Pauli, Heisenberg, and Jordan thought that the formulation of quantum mechanics had created some room for spiritualism and parapsychology, and they wrote so repeatedly; as an example, Heisenberg wrote:

> Keeping in mind the intrinsic stability of the concepts of natural language in the process of scientific development, one sees that — after the experience of modern physics — our attitude toward concepts like mind or the human soul or life or God will be different from that of the nineteenth century, because these concepts belong to the natural language and have therefore immediate connection with reality. It is true that we will also realize that these concepts are not well defined in the scientific sense and that their application may lead to various contradictions, for the time being we may have to take the concepts, unanalyzed as they are; but still we know that they touch reality. (Heisenberg, 1959, p. 200.)

These idealistic 'consequences' of quantum mechanics were also quite evident to the opposers of the final interpretation of the theory. Thus Schrödinger wrote:

> For it must have given to de Broglie the same shock and disappointment as it gave to me when we learnt that a sort of transcendental, almost psychical interpretation of the wave phenomenon had been put forward, which was very soon hailed by the majority of leading theorists as the only one reconcilable with experiments, and which has now become the orthodox creed . . . . (George, 1953, p. 16.)

Similarly, Einstein commented:

> I close the exposition . . . concerning the interpretation of quantum theory with a reproduction of a brief conversation which I had with an important theoretical physicist. He: "I am inclined to believe in telepathy." I: "This has probably more to do with physics than with psychology." He: "Yes." (Schilpp, 1970, p. 683.)

It is interesting to note that, if one took seriously the idealistic point of view, the paradoxes of quantum theory would cease to exist. For example, in the case of the EPR paradox, the generation of a mixture with nonvanishing angular momentum, even if one starts from the 'singlet' state, could be attributed to the action of the experimenter's consciousness, which imprints on the (α, β) pair the necessary new physical properties.

The idea that the reduction of the wave packet is due to the interaction of the physical apparatus with the psyche of the observer has been put to experimental test by Hall, Kim, McElroy, and Shimony (Hall *et al.*, 1977). The negative result of their experiment conveyed a rather obvious conclusion: No psychic action that could influence the obtained results was at work during the measurements. The idea investigated by Shimony and his collaborators presupposes a dualistic ontology, according to which both physical and spiritual entities are contained in nature, where these not only coexist but also interact. Their apparatus consisted of two scalers registering emissions from a radioactive source made of Na^{22}. An observer located near one of the scalers attempted to convey a message to an observer located near the other one by choosing to look at his scaler or to refrain from doing so. The results indicated that no message was conveyed, as most physicists would have anticipated.

7.5. Niels Bohr and Philosophy

To link the Danish philosopher Søren Kierkegaard to modern physics might seem to many to strain belief when one considers that Kierkegaard championed a 'qualitative dialectic' of his own and injected important mystical components into his mode of thinking. Moreover, for him the approximate knowledge of objective reality on which science is based could not be true knowledge; it was in actual fact equivalent to a pure relativsim of opinions. He considered scientific 'progress' as a continual advance toward nothing and never accepted in any way whatsoever the existence of an objective reality external to man and to his ideas; subjectivity was thus in all cases raised to the status of a decisive factor.

Nevertheless, Max Jammer has written:

> There can be no doubt that the Danish precursor of modern existentialism and neo-orthodox theology, Søren Kierkegaard, through his influence on Bohr, affected also the course of modern physics to some extent. (Jammer, 1966, p. 172.)

The reasons for which this thesis still seems very convincing today — despite the fact that Jammer limited himself to simply stating it, without adducing any adequate documentation in support of his argument, and despite certain recent attempts to depict Bohr as a scientist who was essentially independent of his cultural background — will be the subject of a brief discussion in this section. It is necessary, in order to understand the relationship between Kierkegaard and Bohr, to speak of another Danish thinker, Harald Høffding, who objectively played the role of mediator between the former's philosophical ideas and the latter's conception of physics (Svartholm, 1966).

Høffding was thirty years younger than Kierkegaard but for all his life remained deeply influenced by the latter's ideas. He spoke of this influence in many of his books and articles, going as far as defining it as 'bewitching'. Thus the reliance of the ideas of the junior Danish thinker on those of his senior compatriot is quite evident. Indeed Høffding's philosophical testament itself, dated 1925, emphasizes the fundamental importance of Kierkegaard's thinking for the genesis of his own ideas; and as early as 1892 Høffding had already published a philosophical biography of his source of inspiration, entitled *Søren Kierkegaard as Philosopher.*

As an author and as an academic at the University of Copenhagen, Høffding — who was gifted with a rich personality — exercised a strong influence on many Danish intellectuals for a number of decades. In particular, he was Niels Bohr's philosophy lecturer. In those days, Danish university students had to begin their studies with a solid and broadbased curriculum that included philosophy, psychology, logic, and history of philosophy, in whichever faculty they were registered. Naturally, students registered for a physics degree were no exception to the rule, and thus began Niels Bohr's exposure to the ideas of Høffding. But this was not the only channel by which this came about.

In his 1928 autobiography *Erindringer* (Recollections) Høffding recalls that about two decades earlier there was formed in Copenhagen a small group of prominent Danish intellectuals who met privately after academic sessions in order to have a better opportunity for discussing cultural, scientific, and philosophical problems of common interest. This small group included the physiologist Christian Bohr, the father of Niels, and the physicist Christian Christiansen who was later to become the supervisor for Bohr's degree dissertation. These two scholars were the most eminent exponents of their respective sciences in Denmark, and in addition they were men of the most refined classical culture. With Høffding, they formed a very well-balanced trio whose dominant figure was the philosopher. Somewhat later on, a fourth person, the linguist V. Thomsen, joined them. It appears that Niels Bohr together with his brother Harald in their youth had the opportunity to listen in on the discussions of this intellectual group, since these were sometimes held in the Bohr household, where in any case Høffding was a frequent guest. Høffding's influence on Bohr was therefore exerted in a number of ways: directly, on Niels Bohr as a young physics student and as an attentive listener-in on the cultural discussions and, indirectly, through his father, whom Niels Bohr throughout his life deeply respected as a man of culture. Indirectly again, the influence was exercised through Bohr's supervisor, the physicist Christiansen. And so one could go on, remembering also the role of Bohr's brother in generating enthusiasm for Høffding's ideas.

We can thus understand why, in Niels Bohr's house at Carlsberg, there were the portraits of the five persons most dear to him: his father, mother, grandfather, brother and — for the record — his mentor and inspirer, Harald Høffding. Ruth Moore has written on this subject:

> If there did exist any doubt about Høffding's influence on the life of Bohr, this would certainly be dispelled by the place given to his portrait. (Moore, 1966, p. 432.)

One could say that Bohr lived in an environment that was very strongly saturated with the ideas of Danish existentialism, which was particularly engaged in polemics against rationalism and the philosophical *grandeur* of nearby Germany.

The importance of the Høffding–Bohr connection for Bohr's scientific work comes out clearly in several ways. The Høffding–Meyerson correspondence, for example, shows in detail how the Danish philosopher followed Bohr's scientific publications with admiration and constant interest. In 1930, in one of his last published articles ('Observations on the current situation in the theory of knowledge') Høffding "could not only assert the importance for the theory of knowledge of that contribution of Bohr's that is bound up with the term of 'complementarity', but also bring out modes of thought of his own philosophy of forty years before that were clearly connected (to this)". Two years earlier, in a letter written to Emile Meyerson, Høffding had written:

> Some weeks ago, I celebrated my 85th birthday . . . . Amongst the eulogies that the press published to mark the day, an article by Niels Bohr was particularly gratifying. Bohr stated that he had found in my books ideas that had helped scientists in the 'understanding' of their work and that they had thus been of real help. This is a matter of great satisfaction to me . . . . (Høffding and Meyerson, 1939, p. 149.)

In a subsequent letter, again to Meyerson, Høffding speaks of the visits that Bohr had paid him and adds:

> I had very interesting discussions with Niels Bohr, especially on the irrationality introduced into physics by the quantum theory. (Høffding and Meyerson, 1939, p. 156.)

Various hypotheses have been advanced about the exact nature of the influence of Kierkegaard and Høffding on Bohr and, therefore, on quantum physics. Some of these hypotheses undoubtedly represent at least partial truths. Thus one may note similarities between the 'theory of the leap', philosophical in origin as it is, and the hypothesis of an atom's discontinuous jump from one energy level to another; or one may establish parallels between the role of the subject in existentialist philosophy and the uneliminable role of the observer in the quantum mechanical theory of measurement.

It is however elsewhere that we must seek the main dependence of Bohr's creation as a physicist on his cultural background. To be more specific, this dependence comes out fully in the *complementarity principle*. One of Kierkegaard's fundamental theses was a polemic against both rationalism and the Hegelian dialectic. For Hegel, reality was none other than the development of the absolute spirit, and it was just in this context that one could understand the rationality of the natural world. Now, Kierkegaard explicitly denied that the rational dialectic was applicable to the natural world (or to the spiritual one, which was of much greater concern to him). Hegel's contradictory opposites coud be reconciled in a synthesis, while, for Kierkegaard, contradictions were irreconcilable and constituted alternatives that mutually and definitively excluded one another. Thus, in any person's life, the conflict between, on the one hand, the desires, aspirations, and the needs of the subject (the *thesis*) and, on the other, the concrete situation in which that person is constrained to move, one that is often very different from, if not entirely opposed to, the one desired (the *antithesis*), quite simply provokes a state of permanent anguish that can never be resolved, in the Hegelian fashion, in a synthesis of a different and higher quality. The impossibility of overcoming the conflict between thesis and antithesis — with its consequent existential pessimism — was one of the cardinal features of existentialist philosophy. This very impossibility was generalized in succession to all situations, even to the relation between man and nature, as for example in humanity's attempt to understand material reality by means of the natural sciences. Here, too, the attempts to effect a synthesis between human conjectures and the real properties of the concrete world are destined to fail, and all that is commonly called 'scientific knowledge' is, in actual fact, a pure invention, even one determined by individual conditioning and psychology, which in no way corresponds to what exists in reality.

At this point, the relation between existentialism and Bohr's idea of complementarity ought to be clear. As we have seen in Chapter 3, this principle assumes the existence of opposed and contradictory elements that cannot be encompassed in a synthesis: such, fundamentally, are the description of atomic phenomena in space and time versus the description of the same phenomena as evolving according to the law of causality. Let us also recall that, for Bohr, the law of causality had a very precise and, in some ways, unusual meaning. A causal process is one that takes place according to strict rules, ones which do not allow the process to evolve in any different way and force it to proceed in certain precise directions. These strict rules are identified in succession as the laws of conservation of energy and of momentum.

It must at this point be remembered that, according to Bohr, energy is linked to time and momentum to space by the well-known Heisenberg relations, which relate the precisions with which these different physical quantities can be determined.

Since Planck's constant has a well-defined, nonzero numerical value, it will always be impossible to know (to measure) simultaneously with absolute precision either one of the pairs:

(energy-time), (momentum-space).

From this it follows, according to Bohr, that physicists who study atomic phenomena can decide to describe them in space and time, carrying out this program with the appropriate measurement instruments, i.e., measuring spatial and time coordinates with absolute precision. However, by so doing, they of necessity lose any possibility of verifying the law of causality, because that would require the simultaneous measurement of energy and momentum with absolute precision, which Heisenberg relations prevent them from doing.

Alternatively, in a different context and with different measuring instruments, physicists can opt to verify the exact validity of the law of causality by measuring energy and momentum with absolute precision and checking that they are indeed conserved. However, by so doing they of necessity lose any possibility of describing the phenomena as evolving in space and time, again because of the Heisenberg relations.

On account of the foregoing, physicists must always oscillate between causality and spacetime, rather than being able to use both of them. The element of irrationality is introduced by the finite value of the quantum of action *h*. Also in this case, then, one is in the presence of contradictory opposites that are mutually irreconcilable and thus not utilizable in a synthesis that would provide a complete description. The parallel with Kierkegaard's philosophy is striking and, because of what has been said above, certainly not an accidental one.

Max Jammer has devoted a very interesting section to the influence of philosophy on the genesis of quantum mechanics. He asserts that:

> . . . philosophical considerations in their effect upon the physicist's mind act more like an undercurrent beneath the surface than like a patent well-defined guiding line. It is the nature of science to obliterate its philosophical preconceptions, but it is the duty of the historian and philosopher of science to recover them under the superstructure of the scientific edifice. For this task biographies, correspondence, and auto-biographical remarks supply more information in general than scientific publications themselves. (Jammer, 1966, p. 166.)

As we have shown in previous chapters, the paradoxes of quantum mechanics pose the problem of the actual existence of physical objects in space and time independently on the presence of humans and their observations. But this is one of the fundamental problems of classical philosophy which has generated a branching of philosophy into tendencies described as idealistic, materialistic, dualistic, and so on.

In the following we shall discuss the ideas of an important contemporary philosopher of science, Karl Popper, who has devoted considerable attention to the problems of quantum mechanics and has drawn from his studies the conclusion that realism is still an acceptable philosophy, in spite of all that has been written about its 'incompatibility' with the predictions of quantum theory.

In his autobiography, Karl Popper recalls hearing a lecture by Einstein in Vienna in 1919 and being much struck by his admission that, if the experiments on the weight of light had had a negative outcome, the theory of general relativity would have had to be rejected. Popper felt that this was the true scientific attitude and that a new theory should be subjected to every possible experimental check and accepted only if it could survive all examinations of this type. This conviction lies at the root of the deep interest that Popper has always shown for physics. From the thirties onwards he has dissented most strongly from Bohr and Heisenberg, while always remaining favorably disposed toward Einstein. Whilst fully recognizing the great importance of the successes of quantum theory, Popper maintains that physics is today in a state of crisis for two reasons, both of which stem from the Copenhagen way of thinking: the intrusion of subjectivism into physics and the assumption that quantum theory has allowed us to reach a final and complete truth. Popper affirms that he is a realist in the following graphic words:

I am a realist in a very simple sense: I conjecture that I shall soon die, as we all shall sooner or later .... And I expect the world to go on after my death, and also your deaths. This is a simple way of saying that I am a *metaphysical* realist. For it is, at least for us, impossible empirically to test our expectation that the world will go on after we are dead. (Popper, 1985a, p. 3.)

This statement of the metaphysical nature of realism holds for Popper in general and not only in the example he considers. It is worthwhile observing though that not all forms of realism are metaphysical. We have demonstrated that the reality and separability hypotheses, if taken together, constitute a falsifiable (and therefore certainly not a metaphysical) form of realism.

In another sense, however, Popper's realism is tied to falsification. Scientific theories are for him our conjectures and our inventions, which are so audacious that they may conflict with reality. When this happens, we *know* that a reality exists, which is, in fact, that certain thing which can tell us that our ideas are false.

Popper opposes the idea that physical theories are none other than convenient methods of calculations and predictions of experimental results. He traces the cause of this idea back to the 'Copenhagen doctrine', according to which quantum theory is intrinsically incomprehensible, due to the fact that we can understand only classical concepts such as those of wave or particle. He comments on this: "I think this is a mistaken and even a vicious doctrine" (Popper, 1982, p. 42). Regarding von Neumann's theorem on the nonexistence of hidden parameters, Popper declares he does not believe it to be valid. He goes on to propose that the expression 'hidden parameters' be abandoned, since basically all reality is *a priori* 'hidden', and the aim of science is to discover it. Popper does not even like the complementary concept, which he views as a mere ideological element that has no place in a scientific theory. He goes on to maintain that the wave—particle relationship has not been sufficiently explored, but that there are many reasons for believing in de Broglie's ideas. The wave, then, in certain cases could exist without a particle, while there could never be a particle without an associated wave. It cannot however be said that Popper has always defended this position consistently. In 1982 he wrote:

There is thus no symmetry or duality between particles and waves: the waves describe dispositional *properties* of the particles. As Alfred Landé puts it: "If this ... is the famous *duality*, then swans must always be ... dualistic birds, since on the one hand they *are* individuals or particles; on the other hand, they *have* the quality of wavy necks; so that one will for ever remain uncertain whether swans are really particles or waves. This duality is a contrast of incontrastibles." I fully agree, except that I would have preferred a more obviously dispositional property than wavy necks; for example, the swans' disposition to make, under certain circumstances, wavy movements with their necks. (Popper, 1982, p. 127.)

According to this latter point of view, then, wave properties are *propensities*, dispositions of particles to behave in a certain manner (the undulatory one!) and nothing more. This position is not really defensible: It is enough to think of the experiments of Jánossy—Naray and Clauser, discussed in Chapter 3, to realize that a photon's propensity to divide into two, along the two paths (transmitted and reflected) that exit from the semitransparent mirror, should manifest itself only if the photon has a future encounter with an interference system, and never when it meets a counter!

Popper also shows great interest in the EPR paradox, to which he, together with his pupil Angelidis, has devoted a great deal of energy over the last few years, and it may be said that he comes out decisively in favor of the famous EPR argument, about which he has this to say:

Although this clear and simple argument is decisive, it has not been accepted as such. I believe that the reason for its rejection is simply Bohr's authority, not counter-argument. For, in contrast to Einstein's simple and clear ideas, Bohr's counter-argument is very hard to understand. Einstein never understood it (as he told me), and I have not met any physicist who could present it in anything like a clear and simple form. I do think that I have managed to understand the gist of it, but I am not at all certain. (Popper, 1982, p. 149.)

Popper's recent work on the EPR paradox attempts to demonstrate that the paradox does not exist and that quantum theory is perfectly compatible with realism and separability. However, if what we have argued in Chapter 5 is true, such a research program is *a priori* destined to fail.

Popper's greatness does not and cannot lie in a series of proposals about the nature and problems of contemporary physics that are 'all correct', but rather in an overall framework of ideas of exceptional interest that have filtered through science and epistemology in their progression (and regression) during the course of this century. One has to heed Popper's ideas if one wants to understand anything essential about modern science. To quote but one example, his conceiving of scientific development as resulting from the continual interplay of conjectures and refutations, just by considering every conjecture as an almost completely free variable, an arbitrary invention of whoever makes it (but one which must none the less be susceptible to the severest check by experiment), opens up an important — for us essential — vista for the relation between physics and philosophy. Specifically, if any conjecture is a free *creation* of whoever makes it, one may expect that the prejudices of its creator become part of the creation itself. How and why Bohr introduced Danish existentialism into quantum theory thus becomes clear!

7.6. Quantum Physics and Biological Sciences

Today it is very popular to think that physics is a purely technical activity practised by physicists in their laboratories, that it proceeds according to well-defined rules, both experimentally and theoretically, and that it is basically neutral with respect to cultural tendencies, philosophy, social problems, and so on. Actually, the history of science is replete with examples to the contrary, among which we shall briefly discuss one dealing with the nature of life to argue that a theory developed according to some particular philosophical line naturally tends to export its 'truth' to other disciplines.

The fundamental work of Charles Darwin, *Origin of Species*, first published in 1859, had profound cultural implications and consequences. The statement that man, along with the other living creatures, has been produced by an evolutionary process strongly contradicted all those anthropocentric views that from the times of Aristotle had dominated the human culture.

Among the people who did not accept Darwin's ideas was Christian Bohr, the father of Niels Bohr and a professor of physiology at the University of Copenhagen. His opinions are summarized by Rosenfeld thus:

> Christian Bohr, in the wake of the reaction against mechanistic materialism at the beginning of this century, had put up a vigorous advocacy of the teleological point of view in the study of physiology: without the previous knowledge of the function of an organ, he argued, there is no hope of unravelling its structure or the physiological processes of which it is the seat. (Rozental, 1967, p. 132.)

There is a striking similarity between Christian Bohr's teleological views and the opinions expressed, decades later, by Niels Bohr about the consequences of quantum mechanics for biological sciences. In fact, Niels Bohr applied the concept of complementarity to biology and stressed that we should accept both "the physico-chemical side of vital processes, governed by the kind of causality we are accustomed to herald as the truly scientific one, and the properly functional aspect of these processes, dominated by teleological or finalistic causality" (*Ibid.*).

Bohr argued that, in order to have a complete knowledge of the physics and chemistry of a living organism, one should make such detailed experiments that the organism would be killed. From this he deduced a relation of complementarity (rigid contradiction) between life and scientific description: both aspects are important and should be considered, but they are mutually contradictory and this contradiction is in principle not eliminable. Therefore Bohr concluded:

> Owing to this essential feature of complementarity, *the concept of purpose, which is foreign to mechanical analysis, finds a certain field of application in biology*. Indeed, in this sense *teleological* argumentation may be regarded as a legitimate feature of physiological description

which takes due regard of the characteristics of life in a way analogous to the recognition of the quantum of action in the correspondence argument of atomic physics. (Bohr, 1958, p. 10.)

In this way, the teleological conceptions of Bohr's father were thought to be justified.

If in Bohr the attack against the theory of natural evolution is indirect (Darwin is never mentioned), a more direct reference is made by Heisenberg (1959). In the chapter on '*The Relation of Quantum Theory to Other Parts of Natural Science*', Heisenberg discusses Darwin's theory of evolution and concedes that there can be no doubt that this theory contains a great deal of truth. Discussing, however, the claim of many biologists that the marriage of the concepts of history and evolution with those of physics and chemistry will account for all biological phenomena, he writes:

One of the arguments frequently used in favour of this theory emphasizes that wherever the laws of physics and chemistry have been checked in living organisms they have been found to be correct; there seems definitely to be no place at which some '*vital force*' different from the forces in physics could enter. On the other hand, it is just this argument that has lost much of its weight through quantum theory. (Heisenberg, 1959, p. 103.)

In subsequent pages Heisenberg sympathizes with those biologists who cannot see how quantum theory plus history could ever account for concepts like perception, function of an organ, and affection and concludes:

If this view is corrected, the combination of Darwin's theory with physics and chemistry would not be sufficient to explain organic life, but still it would be true that living organisms can to a large extent be considered as physicochemical systems — as machines, as Descartes and Laplace have put it — and would, if treated as such, also react as such. One could at the same time assume, as Bohr has suggested, that our knowledge of a cell being alive may be complementary to the complete knowledge of its molecular structure. (Heisenberg, 1959, p. 104.)

Also for Heisenberg then life becomes a mystery which no physical law can fully explain.

Similarly, Pauli approvingly referred to the 'interesting applications' that Bohr made of his gnoseological viewpoint of complementarity to the fields of biology and psychology.

Born joined Bohr and Heisenberg in showing how important the new quantum conceptions could be in making room within other sciences for idealistic conceptions. He wrote:

If even in inanimate nature the physicist comes up against absolute limits, at which strict causal connection ceases and must be replaced by statistics, we shall be prepared, in the realm of living things, and emphatically so in the processes connected with consciousness and will, to meet insurmountable barriers, where mechanistic explanation, the goal of the older natural philosophy, becomes entirely meaningless. (Born, 1962, p. 32.)

Finally, Jordan devoted a large part of his book *Das Bild der modernen*

Physik to the discussion of biological problems and stressed that the acausal nature of quantum mechanics allowed one to consider life and consciousness as something primary which could not be accounted for by scientific laws From this conclusion, Jordan went on to justify teleology and religion.

Coming back to Bohr, it is interesting to notice that he regarded as likely the existence in biological phenomena of something analogous to the quantum of action in atomic physics and which could be used to justify the introduction of an element of irrationality into biology as well. He wrote:

> . . . we should doubtless kill an animal if we tried to carry out the investigation of its organs so far that we could tell the part played by the single atoms in vital functions. In every experiment on living organisms there must remain some uncertainty as regard the physical conditions to which they are subjected, and the idea suggests itself that the minimal freedom we must allow the organism will be just large enough to permit it, so to say, to hide its ultimate secrets from us. On this view, the very existence of life must in biology be considered as an elementary fact, just as in atomic physics the existence of the quantum of action has to be taken as a basic fact that cannot be derived from ordinary mechanical physics. (Bohr, 1958, p. 9.)

The same concept was underlined quite some years later by Bohr:

> . . . the existence of life itself should be considered, both as regards its definition and observation, as a basic postulate of biology, not susceptible of further analysis, in the same way as the existence of the quantum of action, together with the ultimate atomicity of matter, forms the elementary basis of atomic physics. (Bohr, 1958, p. 21.)

Very interesting is also the following statement:

> . . . we may refer especially to the veritable schism between science and religion which accompanied the development of the mechanical conception of nature at the time of the European Renaissance. On the one hand, many phenomena, hitherto regarded as manifestations of divine providence, appeared as consequences of general immutable laws of nature . . . . In emphasizing the necessity in unambiguous communication of paying proper attention to the placing of the object—subject separation, modern development of science has, however, created a new basis for the use of such words as knowledge and belief. (Bohr, 1958, p. 80.)

We are dealing here, as is obvious, with a new use of the idea of complementarity. Knowledge and faith, while being mutually irreducible, are both accepted at different moments in people's lives, in the same way as causality and spacetime are irreducible one to the other, but valid at different moments in our description of the physical world. Bohr's point of view, as one might expect, did not remain limited to a number of stances adopted orally and in writing; instead it penetrated deeply into modern biology thanks to the work of a young collaborator of his, the physicist Max Delbrück (Kay, 1985).

The son of a well-known historian and the grandson of a theologian, Delbrück arrived in Copenhagen in 1931, when the golden years of the development of quantum theory were already a thing of the past. A reading of the text of Bohr's 1932 lecture 'Light and Life' led to a profound change in Delbrück's life. He became won over to the idea that in biological phenomena there exists an ultimate truth which in principle is recognizable but, in analogy with the quantum of action *h*, is not capable of being fully known and which hides within it the secret of life. From that moment on, Bohr's philosophy of life became the deepest motivation for Delbrück's research and his 'source of meaning, harmony and continuity'. He hoped to push the analysis of biological phenomena so far that it would run up against a paradox analogous to the one that classical physics encountered when applied to atomic phenomena. Animated by this conviction, he created a research group at the California Institute of Technology, from which there later emerged James Watson who, together with the Briton Francis Crick, was to discover DNA in 1953. This was a discovery of exceptional importance, to which Delbrück himself had contributed indirectly, but one which marked the necessary establishment of a point of view contrary to the one which the complementarity reasoning had sought to affirm. As Lily E. Kay has noted discerningly in this respect:

> . . . since Delbrück's quixotic efforts in genetics . . . were primarily aimed at discovering a biological analogy of the indeterminancy relation, the evaluation of Delbrück's life work by the narrowly epistemic criteria traditionally used in the history of science would have to be rather unfavourable, for in his epistemological mission Delbrück did not succeed. (Kay, 1985.)

The initial hopes of Delbrück's research centered on bacteriophages, but as early as 1942 the electron microscope allowed him to recognize their structure (a central body and a quite separate tail), in conflict with the idea that they could be the *irreducible* units of life.

While Delbrück was developing molecular biology, Bohr was unceasingly producing new applications of his complementarity principle to different aspects of living organisms. Bohr's proposals in this direction are contained in speeches held at: the International Congress on Light Therapy (1932); the International Physics and Biology Congress Commemorating Galvani (1937); the International Congress of Anthropological and Ethnological Sciences (1938); the Symposium Marking the Second Centenary of Columbia University (1955); the Conference Organized by the Royal Danish Academy of Sciences (1955); and the Conference of the Danish Medical Sciences (1957).

An application of complementarity to human mental activity allowed Bohr to justify *free will*:

> In fact, according to the generalized interpretation of the psyco-physical parallelism, the *freedom of the will* is to be considered as a feature of conscious life which corresponds to functions of the organism that not only evade a causal mechanical description but resist even a physical analysis carried to the extent required for an unambiguous application of the statistical laws of atomic mechanics. Without entering into metaphysical speculations, I may perhaps add that an analysis of the very concept of explanation would naturally begin and end with a renunciation as to explaining our own conscious activity. (Bohr, 1958, p. 11.)

Bohr also introduced a complementarity relationship between 'thought' and 'feeling':

We all know the old saying that if we try to analyze our own emotions, we hardly possess them any longer, and in that sense we recognize between physical experiences, for the description of which words such as 'thoughts' and 'feelings' are adequately used, a complementarity relationship similar to that between the experiences regarding the behaviour of atoms . . . . (Bohr, 1958, p. 27.)

Bohr went on to find other applications of complementarity, to the 'instinct' — 'reason' relationship and to that between European civilization and the culture of primitive societies.

7.7. Forms of Physical Realism

The universe of science appears nowadays to imitate the fractures and decompositions of modern art, literature, and society: It has become a multifarious macrocosm for which any unifying assessment appears absolutely inadequate.

If the main tendency is toward partitioning, one also finds opposite trends, within which the urgency for a unitary vision of science and of the material world is constantly rediscovered and reassessed. In particular, attempts to critically reconsider the foundations of modern physics (relativity, quantum mechanics, and probability theory) have been under way ever since 1952 and appear presently to undergo a rapid growth.

These attempts often stem from an antitragic vision of the world, devoid of religious connotations as well as of anthropomorphic and anthropocentric illusions. They simply state the existence of a four-dimensional labyrinth of processes upon which the temporal labyrinth of human history, in particular of the history of physics, is superimposed. From here they attempt to rise to a level from where the labyrinth is seen to dissolve into a pattern of interconnections, rationality, and harmony.

In 1932 von Neumann proposed his famous theorem stating the mathematical incompatibility of the formal apparatus of quantum mechanics with a causal completion of the same theory. This theorem is something like a first commandment of the Copenhagen—Göttingen philosophy ('Thou shalt have no philosophy but me'.), which attempted to dispose of causality once and for all. The great physicists of Copenhagen and Göttingen took an energetic stand in favor of von Neumann's theorem, as one could easily document by quoting passages from books and articles in which Bohr, Born, Pauli, Heisenberg, and Jordan expressed their full support for this theorem.

Einstein, adamant in his refusal to accept the reigning interpretation of quantum physics, was intellectually isolated at Princeton to such an extent that he once stated: "Here at Princeton they consider me an old fool" (quoted by Born, 1969, p. 161).

Nevertheless, it was Einstein who nudged physics into taking its most important step in the direction of causality with his famous 1935 paper, written in collaboration with Podolsky and Rosen. This work did not tackle directly the problem posed by von Neumann (which was probably still ignored by Einstein) but gave a counter solution, i.e., a solution favorable to the causal completion of quantum mechanics, thus to the identical same problem, *starting* however *from different hypotheses.*

Einstein was historically (if not logically) defeated by Bohr, who inspired such allegiance to his principle of complementarity among physicists of the day that they were content, for its sake, to leave Einstein's proposal unacknowledged. Thus there prevailed what later turned out to be the wrong solution of the problem. The realization that von Neumann was in error first dawned on a few specialists at the beginning of the fifties, when Bohm *actually constructed* a causal completion of quantum mechanics (he accomplished exactly what according to von Neumann's theorem could in principle not be done) and when de Broglie went back to his original critical approach toward the Copenhagen—Göttingen ideas. The root of von Neumann's error, namely the insufficient generality of the hypotheses underlying his theorem, was unraveled by Bell (1966).

Thus nearly forty years were necessary to dispose of von Neumann's theorem. The profound distortions that had in the meantime developed in theoretical physics should be examined through the lens of an accurate historical analysis, which would no doubt reveal that a fundamental role was played by the strictly ideological and philosophical ideas of the Copenhagen and Göttingen schools. In any event, today we can confidently declare that the road toward a causal completion of quantum mechanics lies open before us and that various groups of theoretical physicists are already working in the proper direction with interesting results (e.g., see van der Merwe *et al.*, 1988).

Let us now come to the spacetime description. It is sufficient to open any book on quantum mechanics to find, within the first few pages, the hypothesis (which is presented as an 'inevitable conclusion') that atomic and subatomic reality is not describable as evolving in space and time. A few authors will patiently explain that the categories of space and time derive from our experience with macroscopic objects and that there is thus no guarantee that the same concepts would also hold for the micro-level of reality, so very far removed from our direct knowledge. Now, up to this point, such an assertion could still be right, even though we see the spacetime description working perfectly well in a macrocosm twenty-six orders of magnitude larger than that of our daily experience, while it should presumably cease to work properly in a microcosm 'only' ten orders of magnitude removed from ours. But while this last objection is not compelling, what *is* conclusive is the fact that a detailed analysis of major experiments on the dual nature of quantal systems has revealed an almost unlimited scope for the fashioning of a *unitary* description of the phenomena involved, one in which every subatomic system (photon, electron, neutron, etc.) can be treated as an entity that exhibits not only localization, all the energy and momentum of the system being confined to a tiny region of space (as they are for a corpuscle), but also an extended phenomenon (the wave), devoid of any energy or inomentum, traveling in close association with the 'corpuscle'. With the aid of this concept we can explain:

(1) the double-slit experiment;
(2) the experiments of Jánossy and Naray;
(3) the experiment of Clauser as well as that of Mandel and Dagenais;
(4) the experiments of neutron interferometry;
(5) the experiment of Pfleegor and Mandel;
(6) the experiment of Blake and Scarl;
(7) fluorescence in the vicinity of mirrors;
(8) superfluorescence;
(9) x-ray interferometry;
(10) experiments on the Bohm—Aharonov effect;
(11) experiments on electron interference.

It is also extremely interesting that this proposal for the description of micro-phenomena has consequences that can actually be studied concretely at an empirical level. From this feature stem the full-fledged scientific credentials of our proposal, even in Popper's sense of the word, since it can be falsified. Now, a negative outcome of any future experiment appears unlikely, since, if it occurred, one would have to consider as accidental the agreement of a single, simple spacetime model with a large collection of experimental facts obtained under very different experimental conditions. *The existence of this model exposes the falsehood of the alleged human incapability to interpret microscopic phenomena in space and time.* In a theoretical and conceptual sense, one can thus already speak of an existing falsification of the Copenhagen—Göttingen paradigm.

Among the positive aspects of our proposal, one must cite the possible discovery of a new aspect of objective reality, the *empty wave*, and the obvious positive potential of this idea, whose full development requires a series of theoretical and experimental investigations.

We have concluded then that a causal interpretation of atomic phenomena is possible, thus negating von Neumann's theorem. It has also been inferred that a spacetime description of the same phenomena can be given. Moreover, nothing prevents us from unifying these two conclusions to furnish a single causal description in spacetime, thereby finally defeating all the prohibitions that the Copenhagen and Göttingen schools have placed in our way.

Let us consider the statement: "The world exists even if no human being is observing it in any way. In particular, the world existed even before man appeared on earth".

However natural and obvious this statement might sound, it cannot be *proved* as valid to whomever might not wish to believe it. For, establishing the truth of any proposition involves setting it off from other propositions that are accepted as true and deducing the validity of the proposition in question by means of a rigorous argument. It is obvious that this mode of reasoning requires starting points, i.e., axioms, which are accepted as self-evident. Since the total number of propositions that can be formulated by the human mind, however large, is certainly finite, it is clear that not *all* of them can be proved; of necessity, some will have to be postulated as valid in order to avoid circular arguments. The proposition 'objective reality exists' is, in point of fact, one of those assertions that one must assume or reject *a priori.* At most, 'objective reality' can be reduced to strictly analogous concepts, such as 'real external world' or 'being' or some other notion: no one has ever validated them rigorously by taking as his starting point something that does not already assume the existence of objective reality in some way.

In short, *from a strictly logical point of view*, the existence of objective reality is a postulate, and whoever decides to construct a logical system may include or exclude it from among his postulates.

This does not mean that the present author is indifferent to the choice between believing and not believing in objective reality. There are other reasons, which cannot be reduced to a short, rigorous argument, but which are none the less convincing, for accepting objective reality. I shall limit myself to recalling just one of these. A scientific theory constructed on an antirealist basis, just because it comes into conflict with common sense, has a tendency to be rigorously excluded from the cultural heritage of nearly every human society, except among a small group of 'experts', which thereby sets itself apart as a cultural elite.

We have stated that the world exists even if no human being is observing it. At this level of specificity, however, realism is such a general philosophy as to be nearly useless. In fact, assuming that there is not a better definition of 'objective reality' is like postulating the existence of the universe, which may be important conceptually to note but is of little use to a researcher who has most often to deal with minuscule *parts* of the universe (solar system, the earth, everyday objects, cells, atoms, etc.). What is of greater interest is the assumption that *objects* exist, one independently of another. Experience teaches that it makes sense to speak of the separate existence of two objects, but that the distinction between them cannot be an absolute one.

Indeed, all that exists is subject to transformation, and every single object, even when it *appears* to us as immutably equal to itself, is in actual fact changing, albeit slowly with respect to our time scale. However, for short time intervals, it can sometimes be *of use* to neglect the evolution of the system being studied: all sciences are, after all, based on idealizations of this type. Arithmetic, for instance, is based on the freezing of a process in its present mode. Thus we talk of 'one stone' and convey thereby that the stone's property of being single and compact is a constant one, remaining unaltered in time; in this way, we take our 'one stone' and place it outside time.

When saying 'one stone' we also imply that the stone's property of being one and constant remains the same at different points in space, i.e., that its spatial position is unimportant as regards its being 'one stone'; in this manner, we also place our stone outside space. A good understanding of what is involved may be gained by referring, for example, to *two* stones. The fact that there are now two of them does not depend on their relative positions, which may be varied without alterating the number. This conclusion, which would not seem open to debate or criticism to an ancient philosopher, now seems less obvious in the light of modern knowledge. For, we know in the first place that every object is a field source (at least a source of a gravitational field); secondly, every field contains energy (which is tied to the potential energy of a second body placed in the field of the first); and, finally, the very matter of which each object is composed can be seen as a very condensed form of energy, according to Einstein's formula $E = mc^2$.

The difference between an object and its field is not after all so fundamental, if both may be seen as modes of being of energy. The difference is quantitative, in the sense that the halo (the field) surrounding a body is much more rarefied, from the point of view of energy, than is the body itself. If we now again consider the two objects, taking into account their 'halo', we realize that the mutual immersion of each in the field of the other brings about deformations in the overall field that vary with distance. In a certain sense, the 'two objects' are 'one sole object', and arithmetic appears as a schematization which, right from the start, neglects mutual interactions. This approach represents of course an excellent approximation for small numbers, but it does not indefinitely continue to do so. As the number n of the objects increases, the number of interactions increases according to the law $n(n-1)/2$, i.e., like n^2; and if n is very large (let's say 10^{30}) then the total interaction becomes so significant that the objects lose their isolation and their enumerability. In this manner, there is created a barrier between arithmetic, which adds objects indefinitely, and material reality.

The lesson of modern science is therefore that the *absolute* independence of the behavior of two objects on the distance between them is an illusion, since in reality there always remain some form of interaction. The scope of this conclusion is however specified and limited by further knowledge regarding the nature of interactions. There exist in nature at least four types of interaction (gravitational, electromagnetic, nuclear, weak), which *all* have the property that they decrease with distance. The gravitational and electromagnetic interactions give rise to potential energies that decrease in inverse ratio to the distance. The nuclear and weak decrease even more rapidly with distance (an exponential decrease). A fifth possible interaction (the superweak) behaves like the weak in its dependence on distance.

All we know *about force fields* leads us to the conclusion that two microscopic objects (two photons, an e^+e^- pair, etc.) at a macroscopic distance from each other are physically separate, in the sense that the dependence of their behavior on the mutual interaction is so weak that it does not modify the first several significant figures of any measurable quantity. This idea ensures that one can speak substantively of objects and phenomena as singularly real and not merely of some generic and undifferentiated 'objective reality' of the universe.

It was demonstrated in 1965 by Bell that there exists a measurable quantity Δ that must satisfy the inequality $\Delta \leqslant 2$ if the principle of separable reality holds, whereas Δ is predicted by quantum mechanics to have the value 2.83. This disparity has led to the conclusion that quantum mechanics is incompatible with the principle of separable reality, which is the natural basis of modern scientific realism. But this conclusion stems from a theoretical formalization that was based on needs deriving neither from experiment nor from some form of historical debate, because before 1965 no one had understood that there was any violation of the principle of separable reality. All this does not *of necessity* imply that the conclusion in question is mistaken, since on other occasions modern physics has been able to use mathematics as a rationalizing instrument (one might think of the prediction of antiparticles). Given however, the exceptional strangeness of an inference that contradicts the essential core of what modern scientific realism might be, and given the fact that on other occasions, too, modern physics has used its formal apparatus badly (here mention of von Neumann's theorem is essential), one might suspect that the alleged violation of separability does not exist in nature. Fortunately the question may be decided through observation: it is sufficient to measure Δ to know whether it is 2.83 or whether the Bell inequality $\Delta \leqslant 2$ holds, as required by the principle of separable reality.

Now, various experiments have already been carried out, all of them confirming, if taken literally, the result predicted by quantum mechanics. There are, however, valid reasons that strongly caution against adopting an uncritical belief in the results of these experiments. Indeed, in light of the critical considerations that were advanced in Chapter 6, the problem of the validity of Bell's inequality still seems open to a variety of solutions. To my mind, it seems essential to defend the principle of separable reality, which has all the marks of providing the firmest basis for any modern scientific realism. Naturally, physics at its frontiers may in principle arrive at a demonstration of the nonvalidity of this principle. And in a certain sense it may be objected that 'frontier physics' has already done just that. But this presumed falsification depends on the historically determined, logically arbitrary ideological contents of the quantum paradigm and not on unchallengeable empirical evidence. It is obvious that a clarification must take place in this latter arena.

There are a number of comments on the principle of separable reality that it is now important to add:

(α) The validity of the principle of separable reality (PSR) is forcefully indicated by our knowledge of physical interactions (but not by the formal apparatus, by the kinematics of quantum mechanics, we might almost say). It is of great importance to note that the PSR constitutes a *nonmetaphysical* basis for realism. Thus the objections of the positivists, according to which realism is essentially a metaphysical philosophy, insofar as it cannot identify an experiment that could falsify it, are seen to be invalid. Thanks to the PSR, we can today say: Show us unequivocably that Bell's inequality ($\Delta \leqslant 2$) does *not* hold in nature, or that any other of the infinite number of inequalities that stem from the PSR fails, and our realism will have been demonstrated to be false.

(β) The PSR implies the validity of $\Delta \leqslant 2$, while quantum mechanics violates this inequality. This means that the formal and axiomatic apparatus of quantum mechanics (together with the rules that give empirical meaning to the symbols) is incompatible at the experimental level with the PSR. Such a shattering philosophical negativity *associated with the mathematical structure of a theory, totally independent of its 'interpretation'*, has up to now probably never yet been seen.

(γ) To the question "What is space?", the PSR gives a clear answer. Space manifests itself in physical phenomena by keeping them separate (i.e., more physically independent one of another), the more the larger the distance that separates them. This very natural function of space would no longer hold if the PSR were not valid in nature, which is a little like saying that space itself would disappear. The outcome would be a weird world in which two macroscopic objects (for which classical physics guarantees the validity of the PSR) are separated by a large spatial interval, while two atoms, one belonging to the first and the other to the second object, could be separated by distance zero!

(δ) The hypothesis that Einstein, Podolsky, and Rosen placed at the foundation of their reasoning was, in essence, the PSR. They, however, deduced only half its possible consequences, limiting themselves to examining the case of a complete quantum mechanics. Einstein, Podolsky and Rosen concluded that quantum mechanics could not be complete. There exists however a second line of reasoning which assumes the incompleteness of quantum mechanics. The *same* absurd consequences are deduced in this case too, thereby showing that the completeness (or incompleteness) of quantum mechanics constitutes a false issue and that there remains an incompatibility (expressed by $\Delta \leqslant 2$) between the PSR and quantum mechanics.

The perspective of the unlimited validity of the principle of separable reality seems fascinating for at least two reasons.

Firstly, the PSR has not been seriously challenged by any science other than physics. It seems to hold, without any constraint on its validity, in astrophysics as in geophysics, in genetics as in psychology. The only proposals that the validity of the PSR be limited come from the subject of parapsychology, which seems to become more discredited the deeper one goes into the analysis of this discipline's activity. Secondly, the PSR is eminently harmonizable with a causal view of space and time which recent research on wave—particle duality has shown to be possible. It would therefore be possible to construct on the basis of this principle a unified approach to atomic and subatomic phenomena.

If in future a change takes place in quantum theory in order to get rid of nonlocality, this will probably not be a minor change. As we saw in a previous chapter, state vectors of the second type are responsible for nonlocal effects. Their elimination implies a drastic modification of the superposition principle, that is, of the linear nature of quantum laws. This, in turn, would however very probably imply an automatic resolution of the measurement problem (the reduction of the wave packet, which is the passage from a superposition to a mixture of states, would no longer be necessary); and also the nonlocal effects for single systems, discussed in our second chapter (de Broglie's paradox), would immediately disappear as soon as one accepts the lack of completeness of quantum theory.

It is an interesting historical fact that the problem of separability played no role in that rich and tumultuous historical process which led in 1927 to the formulation of quantum mechanics. It now appears that the most important philosophical implication of the new theory, which is its lack of separability, was born as a byproduct of other researches and was fully appreciated only eight years after the final formulation of the theory. It is no surprise that Albert Einstein refused to accept quantum mechanics as a fully satisfactory theory of atomic phenomena *precisely because of this problem*. In a letter to Max Born dated December 2, 1947, Einstein admits that the theory contains a considerable amount of truth, but writes:

> However I cannot seriously believe in it because the theory is incompatible with the principle that physics is to represent reality in time and space, without spookish long-distance effects. (quoted by Born, 1969, p. 162.)

References

Books

Bell, J. S.: 1987, *Speakable and Unspeakable in Quantum Mechanics*, Cambridge University Press, Cambridge.

Bohr, N.: 1958, *Atomic Physics and Human Knowledge*, Wiley, New York.

Bohr, N.: 1961, *Atomic Theory and the Description of Nature*, Cambridge University Press, Cambridge.

Born, M.: 1969, *Physics in my Generation*, Springer, New York.

Born, M.: 1962, *Atomic Physics*, Blackie and Son, London and Glasgow.

Calvino, I.: 1980, *Una Pietra Sopra*, Einaudi, Torino.

Capra, F.: 1975, *The Tao of Physics*, Shambhala, Berkeley.

Chalmers, A. F.: 1976, *What is this Thing called Science?*, University of Queensland Press, Brisbane.

Davies, P.: 1984, *God and the New Physics*, Penguin Books, Harmondsworth.

de Broglie, L.: 1976, *Recherches d'un demi-siècle*, Albin Michel, Paris.

de Broglie, L.: 1982, *Les incertitudes d'Heisenberg et l'interprétation probabiliste de la mécanique ondulatoire*, Gauthier-Villars, Paris.

Eddington, A.: 1928, *The Nature of the Physical World*, Macmillan, New York.

Feyerabend, P. K.: 1975, *Against Method: Outline of an Anarchist Theory of Knowledge*, NLB, London.

Fine, A.: 1986, *The Shaky Game: Einstein Realism and the Quantum Theory*, University of Chicago Press, Chicago.

Forman, P.: 1971, 'Weimar Culture, Causality and Quantum Theory, 1918—1927: Adaptation by German Physicists and Mathematicians to a Hostile Intellectual Environment', in R. McCormmach (ed.), *Historical Studies in the Physical Sciences*, Vol. 3, University of Pennsylvania Press, Philadelphia.

France-Culture: 1980, *Science et conscience*, Stock, Paris.

Frank, P.: 1957, *Philosophy of Science*, Prentice-Hall, Englewood Cliffs.

George, A. (ed.): 1953, *Louis de Broglie, physicien et penseur*, Albin Michel, Paris.

Good, I. J. (ed.): 1962, *The Scientist Speculates*, Heinemann, London.

Heisenberg, W.: 1959, *Physics and Philosophy*, Harper and Row, New York.

Høffding, H., and Meyerson, E.: 1939, *Correspondance*, Munksgaard, Copenhagen.

Hora, H., and Shepanski, J. R. (eds): 1975, *Directions in Physics*, Wiley, Sidney.

Jammer, M.: 1966, *The Conceptual Development of Quantum Mechanics*, McGraw-Hill, New York.

Kline, M.: 1980, *Mathematics: The Loss of Certainty*, Oxford University Press, Oxford.

Kuhn, T. S.: 1962, *The Structure of Scientific Revolutions*, University of Chicago Press, Chicago.

Madden, E. H. (ed.): 1960, *The Structure of Scientific Thought*, Routledge & Kegan Paul, London.

Moore, R.: 1966, *Niels Bohr*, Knopf, New York.

Pietschmann, H.: 1980, *Das Ende des naturwissenschaftlichen Zeitalters*, Zsolnay, Wien.

Planck, M.: 1949, *Scientific Autobiography and Other Papers*, The Philosophical Library, New York.

Planck, M.: 1960, *A Survey of Physical Theory*, Dover, New York.

Popper, K.: 1976, *Unended Quest, An Intellectual Autobiography*, Fontana/Collins, Glasgow.

Popper, K.: 1982, *Quantum Theory and the Schism in Physics*, Hutchinson, London.

Redhead, M.: 1987, *Incompleteness, Nonlocality, and Realism*, Clarendon Press, Oxford.

Rozental, S. (ed.): 1967, *Niels Bohr*, North-Holland, Amsterdam.

Schilpp, P. A. (ed.): 1970, *Albert Einstein: Philosopher-Scientist* (3rd edn), Open Court, La Salle.

Scott, W. T.: 1967, *Erwin Schrödinger*, University of Massachusetts Press, Amherst.

Selleri, F.: 1987, *Paradossi e realtà*, Laterza, Roma-Bari.

Squires, E.: 1986, *The Mystery of the Quantum World*, Adam Hilger, Bristol.

van der Merwe, A., Selleri, F., and Tarozzi, G. (eds): 1988, *Microphysical Reality and Quantum Formalism*, Kluwer Academic Publishers, Dordrecht.

Waismann, F.: 1936, *Einführung in das mathematische Denken*, Gerold, Wien.

Articles

Bass, L.: 1975, *Found. Phys.* **5**, 159.

Bell, J. S.: 1966, *Rev. Mod. Phys.* **38**, 447.

Centrone, M.: 1978, *Ann. Fac. Lett. Filos. Univ. Bari* **21**, 267.

Cochran, A. A.: 1971, *Found. Phys.* **1**, 235.

Costa de Beauregard, O.: 1978, *Phys. Lett.* **67A**, 171.
Dirac, P. A. M.: 1975, 'The Development of Quantum Mechanics' in H. Hora *et al.* (eds); see book list.
Hall, J., C. Kim, B. McElroy and A. Shimony: 1977, *Found. Phys.* **7**, 759.
Harrison D.: 1979, *Am. J. Phys.* **47**, 779.
Kay, L. E.: 1985, 'The Secret of Life: Niels Bohr's Influence on the Biology Program of Max Delbrück', *Rivista di storia d. scienza* **2**, 487.
Klein, G.: 1980, *La mesure en mécanique quantique*, Centre de Recherches Nucléaires de Strasbourg.
Josephson, B.: 1980, '*L'expérience de la conscience et sa place en physique*', in *Science et Conscience*; see book list.
Mattuck, R. D.: 1977, in *Psychoenergetic Systems* **2**, 188.
Popper, K.: 1985a, 'Realism in Quantum Mechanics', in G. Tarozzi and A. van der Merwe (eds); see book list.
Popper, K.: 1985b, 'Evolutionary Epistemology', *ibid.*
Reichenbach, H.: 1960, *Are there atoms*?, in E. H. Madden (ed.); see book list.
Sudarshan, E. C. G.: 1978, 'Evolution of Mind: Projection to the future', Indian Institute of Science, Bangalore, preprint.
Svartholm, N.: 1966, 'Sören Kierkegaard und die moderne Physik', University of Göteborg, preprint.
Wigner, E. P.: 1961, 'Remarks on the Mind—Body Question', in I. J. Good (ed.); see book list.

Index

Adam, A., 178
Aerts, D., 261
Aharonov, Y., 178, 261
Alefeld, B., 178
Allen, L., 32, 177
Anaxagoras, 27
Andrade e Silva, J., 158, 178–9
Angelidis, T., 351
Apollonius, 328
Arecchi, F.T., 178
Aristarchus of Samos, 26, 335
Aristotelian 'potentia', 107
Aristotle, 237–8, 352
Aspect, A., 178–9, 228, 234, 261
Aspect–Dalibard–Roger experiment, 233
Aspect–Grangier–Roger experiment, 230, 232–3
Auger, P., 32

Badurek, G., 178
Ballentine, L.E., 71, 178
Baracca, A., 210, 261
Bartell, L.S., 122, 172, 178, 261
Barut, A.O., 261
Bass, L., 341, 364
Bassini, M., 179
Bauer, E., 113, 121
Bauer, G., 5
Bayer, D., 180
Bayth, W., 179
Belinfante, F.J., 71
Bell, J.S., 49, 51, 71, 217, 242, 257, 319, 357, 363
Bell's inequality, ix, 206, 208–10, 214–16, 219–20, 227, 238, 240, 244, 248–9, 252, 260–1, 267, 276, 278–9, 290–1, 296–7, 299, 308–9, 314, 322, 361–2
Beltrametti, E.G., 122
Ben-Menahem, Y., 32
Bergia, S., 261
Bergson, E., 9
Bethe, H.A., 8
Beyer, H.J., 263
Bialynicki-Birula, I., 136, 178
Biraben, F., 178–9
Birkhoff, G., 63, 71, 108
Bitsakis, E., 261
Blackett, P.M.S., 12
Blake, G.D., 178
Blake–Scarl experiment, 130–1, 358
Bloch, I., 319
Bloch vector, 310
Blokhintzev, D.I., 71
Bocchieri, P., 178
Boersch, H., 178
Bohm, D., 36, 51, 69–71, 118, 122, 124, 127, 178–9, 200, 240–2, 261–2, 357
Bohm–Aharonov effect, 167–71, 255–7, 259–60, 358
Bohr, C., 15, 346, 352–3
Bohr, H., 346
Bohr, N., vii, 2, 8, 12, 15–16, 20–5, 31, 35, 75, 95–8, 106, 108, 114, 121–2, 135, 137, 178, 187–9, 191, 236, 261–2, 266, 319, 321–2, 325, 336, 345–7, 349, 351–2, 354–7, 363
Bohr magneton, 307
Bohr's complementarity, ix, 3, 11–13, 16, 21, 70, 97–105, 107, 119, 265, 267, 348, 353
Bohr's correspondence principle, 123
Bohr's virtual wave, 138
Bohr-Fock debate, 103, 118
Bohr-Kramers-Slater theory, 98–9, 102, 109, 130
Bohr-Sommerfeld quantization, 90

Boltzmann, L., 5, 7, 9, 13–14, 18, 23, 236, 335, 337, 339
Bonse, U., 77, 121–2, 135, 178
Bopp, F., 175, 178
Born, M., vii, 2, 7, 8, 11–13, 15, 17, 19–23, 25, 29–32, 35, 71, 94, 96, 100, 102, 121–2, 137, 184, 188, 321, 336, 353, 353, 356, 363
Born's probabilistic interpretation, 42, 80, 93, 95–7, 269
Bose, S.N., 2
Bothe, W., 122
Bothe–Geiger experiment, 99, 109
Bragg angle, 78, 233
Bragg's formula, 90
Brahe, T., 328
Brewster angle, 231
Brownian motion, 10, 34, 246, 285
Bruno, G., 27
Bub, J., 71
Buonomano, V., 251, 262
Burba, D.A., 319
Busch, P., 319

CHSH additional assumption, 217–21, 311
Cagnac, B., 144, 178–9
Calvino, I., 338, 363
Canals-Frau, D., 262
Cannata, F., 261
Cantelaube, Y., 178
Capasso, V., 71, 210, 262
Capra, F., 341, 363
Carazza, B., 191, 262
Carnap, R., 328
Caser, S., 311, 319
Cassinelli, G., 122
Cauchy problem, 246
Centrone, M., 328
Chalmers, A.F., 364
Chambers, R.G., 178
Chebotayev, D.P., 144, 180
Chishaev, A.V., 180
Christiansen, C., 346
Chubarov, M.S., 178, 319
Clauser, J.F., 76, 122, 178, 242, 259, 262, 269, 280, 308, 319
Clauser experiment, 231, 234, 351, 358
Clauser–Horne additional assumption, 221–5, 311
Clauser–Horne factorizability condition, 243, 279, 281–2, 284–5, 289
Clausius, R., 5
Clifton, R.K., 319
Cochran, A.A., 113, 122, 341
Cohen-Tannoudji, C., 178
Compton, A.H., 122
Compton, K.T., 2, 12, 84–5, 97
Compton effect, 101, 339
Compton scattering, 99
Compton–Simon experiment, 99, 109, 339
Condon, E.U., 12
Corchero, E.S., 262
Costa de Beauregard, O., 116, 178, 236–8, 262, 341, 365
Courant, R., 22, 31
Crick, F., 355
Croca, J.R., 158–61, 178
Croce, B., 325
Cufaro-Petroni, N., 196, 246–7, 262

D'Alembertian equation, 167
D'Espagnat, B., 71, 178, 261, 263, 291, 318–20
Dagenais, M., 76, 122, 179
Dalibard, J., 178
Dalla Chiara, M.L., 262
Darwin, C., 330–1, 335, 352–3
Datta, A., 319
Davidson, F., 2, 179
Davies, E.B., 319
Davies, P., 341, 364
Davisson, C.J., 90, 122
de Broglie, L., vii–ix, 2, 4, 15, 17–19, 22, 24–5, 31, 33, 51, 69–72, 85, 92–3, 100, 103, 116, 121–2, 126–7, 130, 135, 137, 159–62, 165, 178–9, 188, 262, 323, 334, 336–7, 339, 344, 350, 357, 364
de Broglie, M., 18
de Broglie waves, 90, 95, 176, 257
de Broglie's apodization experiment, 92
de Broglie's box paradox, viii, 36–9, 77, 111–12, 238, 260, 322, 363
de Broglie's duality, 86–92, 125, 162
de Broglie's formula, 77, 80, 90, 136, 169
De Caro, L., 319
de Muynck, W.M., 262
Debye, P., 2, 8, 15
Delbrueck, M., 12, 354–5
Democritus, 27
Descartes, R., 242, 353
Destouches-Fevrier, P., 69, 72
Dewdney, C., 122, 127–8, 161, 179
Dicke, R.H., 141, 179
Diderot, D., 27
Diner, S., 177

Dirac, P.A.M., 2, 12, 15, 17, 20, 23–5, 31, 73, 109, 121, 248, 262, 323, 331, 334, 365
Dirac's aether, 246
Dirac's equation, 327, 332–3
Doppler effect, 91
Drexhage, K.H., 133, 143
Drummond, P.D., 319
Duane's selection rule, 174–5
Duhem, P., 17, 337
Duncan, A.J., 262–3, 319
Dyson, F.J., 2

EPR elements of reality, 189, 201, 266, 322
EPR experiment, 217, 271, 308–9
EPR paradox, viii, ix, 39, 116, 118, 186, 302–7, 322, 338, 340, 344, 362
EPR reality criterion, 181–3, 200, 202, 205, 212, 235–6, 264–5, 268, 270
Eberhard, P.H., 204, 262
Eddington, A., 364
Ehrenfest, P., vii, ix, 2, 8–10, 25, 31, 41, 323, 334
Einstein, A., vii–ix, 2, 4, 8–18, 22, 24–5, 29–31, 33–4, 36, 39, 51, 93, 97, 99–103, 112, 116, 122, 130, 135, 160, 187–8, 190–1, 203, 254, 262, 319, 322–3, 325–7, 334–40, 344, 349, 351, 356–7, 360, 363
Einstein locality, 206–8, 210, 213, 214, 216, 227, 244, 249, 251–2, 256–7, 264–5, 267–9, 285, 291–2, 294–5, 297–9, 301–2, 309
Einstein's empty wave, 81–6, 98, 109, 137–8
Einstein's relations, 89
Einstein-de Broglie duality, 102, 123–4, 129
Elbaz, C., 179
Ellis, J., 262
Elsasser, W., 12
Engelke, C.E., 179
Engelke, C.W., 179
Epstein, P.S., 8, 190
Euclid's fifth postulate, 327
Ewald, P.P., 8
Exner, F., 30

Falciglia, F., 262, 319
Faraday cage, 168
Fargue, D., 177
Fayerabend, P.K., 364
Fermat's principle, 90
Fermi, E., 2, 12
Ferrero, F., 311, 319
Ferrero Melgar, M., 32
Feyerabend, P.K., 337, 340–1
Feynman, R.P., 2, 31, 248, 262
Fine, A., 262
Finkelstein, J., 319
Fock, V.A., 121–2
Fock's relativity, 102–7
Ford, K.W., 30
Forman, P.L., 28–32, 71, 177, 364
Fortunato, D., 71, 195, 258, 262
Fourier's theory of heat, 17, 336
Fowler, H.A., 179
Fowler, R.H., 23
Franck, J., 2, 12, 15
Franck, P., 364
Franck–Hertz experiment, 12
Fraunhofer interference, 172
Freedman, S.J., 262
Freedman's inequality, 220, 231, 308
Freedman's strong inequality, 234–5, 309
Freedman–Clauser experiment, 219, 230
Frege, G., 330
Fresnel's theory of light, 17, 336
Freud, S., 335
Friberg, S., 143, 148, 179
Fry, E.S., 262
Fry–Thompson experiment, 232
Furry, W.H., 190, 193–4, 262

Galilei, G., 26, 236, 328
Ganesh Vaidyanathan, A., 179
Garg, A., 214, 262, 297, 319
Garg–Mermin model, 301–2
Garuccio, A., 141, 144, 161, 178–9, 235, 259, 262, 319
Garuccio's superinequality, 297
Garuccio–Rapisarda additional assumption, 225, 311
Garuccio–Rapisarda experiment, 224, 226, 228, 233
Garuccio–Rapisarda–Vigier experiment, 142
Gaussian function, 203
Gaussian trasmission filter, 172
Geiger, H., 122
Geiger counter, 95
George, A., 31, 71, 121, 364
George, E.A., 179
Gerlach, W., 72, 83, 122
Germer, L.H., 90, 122
Ghirardi, G.C., 259, 262, 319
Gibbs, J.W., 18, 23, 337
Giovanelli, R., 144, 179

Glauber, R.J., 177–8
Gleason, A.M., 69, 72
Goedel's theorem, 327, 331
Goeppert-Mayer, M., 12
Golgi, C., 340
Good, I.J., 121, 364
Gordon, P., 179
Goudsmit, S., 41, 72
Gozzini, A., 133–5, 179
Grangier, P., 179, 234, 261
Greenberger, D.M., 179, 319
Grib, A.A., 262
Grynberg, G., 144, 178–9
Gudder, S.P., 69, 72
Gueret, Ph., 179

Haken, H., 148, 177
Hall, J., 344, 365
Hall, M.J.W., 319
Hamilton, W.A., 179
Hamilton's equations, 123
Hamilton–Jacobi equation, 126, 241
Hamiltonian operator, 136
Hanbury-Brown, R., 179
Hanbury-Brown–Twiss effect, 309
Hardies, M.L., 179
Harrison, D., 365
Hart, M., 135, 178
Hartkaemper, A., 318
Hasenhoerl, F., 13
Hassan experiment, 234
Hawking, S.W., 262
Hegel, G.W.F., 100, 347
Heideman, A., 179
Heilbron, J.L., 32, 177
Heisenberg, W., vii, viii, 2, 8, 12, 15, 17, 20–3, 25, 27, 30, 32, 35, 69–71, 94–6, 99, 100, 102, 107–10, 123, 137, 243, 321–2, 326, 336, 344, 353, 356, 364
Heisenberg's relations, 100, 101, 104–5, 121, 182, 280, 348–9
Heisenberg's uncertainty principle, ix, 3, 91
Heitler, W., 8, 137
Helmholtz, H.L., 5
Herbert, N., 179
Hertz, H., 102
Hertz's equation for dipole radiation, 133
Heyting, A., 328
Heywood, P., 204, 262
Hilbert, D., 6, 12, 22, 31
Hilbert space, 52, 210, 260
Hiley, B.J., 32, 127–8, 161, 179, 240, 261–2
Hilfer, E.S., 179
Hoffding, H., 15, 100, 345–7, 364
Holt, R.A., 262
Holt-Pipkin experiment, 231
Home, D., 319
Hong, C.K., 179, 319
Hooker, C.A., 204, 262
Hora, H., 31, 364
Horlick-Jones, T., 210, 262
Horne, M.A., 242, 262, 269, 280, 319
Houtermans, G.F., 12
Hulet, R.G., 179
Hurwitz, A., 6

Iaci, G., 262
Inglis, D.R., 319

James, W., 15
Jammer, M., 31–2, 71, 82, 93, 121, 190–1, 261, 263, 345, 349, 364
Janossy, L., 74–6, 122, 178
Janossy–Naray experiment, 86, 106, 113, 116, 137, 351, 358
Jauch, J.M., 72, 256, 263
Jauch–Piron's theorem, 61–9
Jaynes, E.T., 140, 176–7
Jeans, J.H., 15
Joensson, C., 127, 179
Jordan, P., 2, 12, 17, 22–3, 25, 27, 30–2, 35, 109, 236, 321, 344, 353–4, 356
Josephson, B., 341, 365
Jung, C.G., 335

Kaiser, H., 179
Kant, I., 6
Kasday, L., 256
Kay, L.E., 355, 365
Kaye, R.D., 179
Kemble, E.C., 190, 262
Kemperman inequality, 299
Kepler, J., 26, 328, 330
Kepler's laws, 329, 331
Kierkegaard, S., 15, 100, 345–7, 349
Kim, C., 344, 365
Kirchhoff, G., 5, 322
Klein, A.G., 179
Klein, F., 6, 12
Klein, G., 341, 365
Klein, M.J., 32
Klein, O., 2, 15, 20
Kleiner, A., 10
Kleinpoppen, H., 263
Kleppner, D., 179
Knight, P.L., 132, 179

Kochen, S., 69, 72
Kopernicus, N., 26
Kostro, L., 179, 318
Kramers, H.A., 2, 15, 20–1, 95, 97–8, 108, 122, 178
Kraus, K., 262, 319
Kuhn, H., 133, 179
Kuhn, T.S., 1, 31–2, 177, 334, 337, 364

Ladenburg, R., 2, 82
Lagrange, G., 330
Lahti, P., 261–3
Lakatos, I., 337
Lamb, W.E., Jr., 176, 178
Landau, L.D., 15
Landé, A., 8, 32, 121, 174–5, 177, 350
Langevin, P., 2, 9
Laplace, P.S., 268, 353
Laplacian determinism, 104, 285
Larmor, J., 15
Laue scattering, 78, 95, 159, 175
Lee, T.D., 303
Leibnitz, G.W., 334
Leighton, R.B., 31
Lenard, Ph., 82–3
Lenz, W., 8
Lepore, V.L., 319
Ley, M., 179
Liddy, D.E., 319
Lie algebra, 245
Lieber, M., 180
Liebowitz, B., 179
Lo, T.K., 319
Lochak, G., 124, 177–8
Loinger, A., 178
London, F., 113, 121
Lorentz, H.A., 2, 9, 18, 337
Lorentz condition, 167
Loschmidt, J., 237
Loudon, R., 148, 178–9
Lyman signal, 234

Mach, E., 9, 19, 236, 322, 339
Madden, E.H., 364
Mandel, L., 76, 122, 143, 148, 177–9, 319
Mandel–Dagenais experiment, 358
Margenau, H., 190, 214, 263
Marlow, A.R., 178
Marshall, T.W., 179, 263, 319
Marx, K., 9
Matteucci, G., 179
Mattuck, R.D., 341, 365
Maupertuis' variational principle, 90
Maxwell, J.C., 14, 102, 236
Maxwell distribution of velocities, 84
Maxwellian electrodynamics, 17, 124, 167, 326–7, 336
Mayer, E., 83, 122
McElroy, B., 344, 365
McLean, R.G., 260, 263
Mehlhop, W., 257, 263
Mermin, N.D., 214, 262–3, 299, 319
Meyerson, E., 347, 364
Meystre, P., 261
Michelson, A.A, 74
Millikan, R.A., 83, 121–2
Milonni, P.W., 132, 179–80
Minkowski, R., 7, 12
Minkowski space, 239
Misra, B., 69, 72
Mittelstaedt, P., 261–3, 319
Moellenstedt, G., 179
Moessbauer effect, 135–6
Monzoni, V., 261
Moore, R., 31–2, 346, 364
Mueckenheim, W., 247–8, 263
Mugur-Schaechter, M., 71
Mycielski, J., 136, 178

Namiki, M., 179
Naray, Zs., 74–6, 122
Ne'eman, Y., 245, 263
Nernst, W., 2
Neumann, H., 318
Newton, I., 328, 334
Newtonian mechanics, 327
Nietzsche, F., 9
Nikolayev, E.P., 178, 319
Nordheim, L.W., 12
Noyes, P., 173, 179

Ochs, W., 72
Oliver, B.J., 319
Opat, G.I., 179
Oppenheimer, R.J., 12
Ostwald, W., 14, 17, 236, 337
Otake, Y., 179
Ou, Z.Y., 319
Overhauser, A.W., 179

Pappalardo, L., 262
Parmenidean conception, 339
Pascazio, S., 253–4, 263, 311, 319
Paty, M., 244, 263
Pauli, W., 2, 8, 12, 15, 17, 19, 20, 25, 31–2, 35, 40–1, 72, 94,

137, 321–2, 344, 353, 356
Pauli matrices, 42, 68, 199, 206
Pauli's exclusion principle, 332
Pearle, P.M., 291, 319
Peierls, R., 118, 121
Perrie, W., 263
Perrie *et al.* experiment, 234
Perrin, J., 339
Pfleegor, R.L., 122, 179
Pfleegor–Mandel experiment, 90–1, 141–2, 157, 159, 162, 358
Philippidis, C., 127–8, 161, 171, 179
Piccioni, O., 257, 263
Pietschmann, H., 341, 364
Pipkin, F.M., 262–3
Piron, C., 72
Planck, M., vii, ix, 2, 4–8, 13–16, 18, 25, 29–33, 39, 83, 93, 95, 100, 174, 322–3, 334–5, 337, 364
Planck's constant, 70, 93, 120, 125, 188, 348
Planck's law, 83, 87, 96
Podolsky, B., 122, 262, 319
Poincare' group, 246
Poisson distribution law, 144
Poissonian Statistics, 310
Polder, D., 180
Popper, K.R., 263, 268–9, 318, 323–4, 337–9, 349–50, 358, 364–5
Pound, R.V., 135, 180
Pozzi, G., 179
Prigogine, Y., 240
Prosser, R.D., 128–9, 180
Przibram, K., 121
Ptolemaic cosmology, 26
Ptolemaic geocentric theory, 335
Ptolemy, 26–7

Rabinowitsch, 12
Rapisarda, V., 141, 179, 262
Rastall, P., 245, 263
Rauch, H., 77, 121–2, 178, 180
Raychaudhuri, A., 319
Rayski, J., 116, 238–9, 263, 319
Rebka, G.A., 135, 180
Redhead, M.L.G., 204, 262, 319
Reichenbach, H., 321, 365
Reid, M.D., 309, 319
Reignier, J., 319
Reynaud, S., 178–9
Riemannian 4-space, 245
Rietdijk, C.W., 116, 239, 263
Rimini, A., 262
Roehrlich, F., 267, 320
Roger, G., 178
Rosen, N., 122, 262–3, 319
Rosenfeld, L., 20, 100, 102–3, 121, 352
Ross, M., 320
Rothe, K.W., 178
Roy, S.M., 320
Roy–Singh inequality, 290
Rozental, S., 31–2, 121, 364
Ruark, A.E., 190, 263
Russell, B., 338
Russell's paradox, 327
Rutherford, E., 15, 338–9
Rydberg atoms, 310

Sachs, M., 263
Salomon, J., 69
Sanchez-Gomez, J.L., 310, 320
Sands, M., 31
Santos, E., 72, 263, 309, 319–20
Sanz, A.L., 310, 320
Sapogin, L.G., 175–6, 180
Sargent, M., iii, 176, 178
Scalera, G.C., 252–4, 262–3
Scarl, D.B., 178, 180
Scheer, J., 180
Schepanski, J.R., 31, 137, 141
Schilpp, P.A., 31–2, 261, 364
Schmidt, M., 180
Schommers, W., 180
Schopenauer, A., 9
Schottky, W., 30
Schroedinger, E., vii–ix, 2, 4, 7, 8, 13–16, 22, 24–5, 28, 30–2, 39, 100, 122, 188, 190, 204, 263, 322–3, 334, 336, 344
Schroedinger's cat paradox, viii, 238
Schroedinger's equation, 125, 136, 161, 168, 170–1, 191, 240, 259, 326
Schroedinger's interpretation, 12, 92–6, 269
Schuurmans, M.F.H., 180
Schwarz, G.M., 263
Schwarz inequality, 58
Scott, W.T., 32, 364
Screaton, G.R., 261
Scully, M.O., 148, 176, 178
Seidel, L., 5
Seleucus of Babylonia, 335
Selleri, F., 71–2, 122, 141, 161, 177–80, 235, 261–3, 311, 318–20, 364
Selleri–Zeilinger model, 312, 315–6
Shaknov, I., 263
Shepanski, J.R., 364
Shimony, A., 136, 180, 214, 262–3, 319,

344, 365
Shishaev, A.V., 144
Siegwart, D., 320
Simon, A.W., 84, 122
Singh, V., 320
Siragusa, G., 178
Six, J., 320
Skalsey, M., 320
Slater, J.C., 21, 95, 97, 122, 178
Smith, S.R., 180
Solomon, J., 72
Sommerfeld, A., 2, 4, 6–8, 15, 17, 19, 20, 23–5, 31–2, 40–1, 174
Specker, E.P., 69, 72
Spencer, W.P., 179
Spengler, O., 29, 32, 325
Spinoza, B., 247
Squires, E., 320
Stapp, H.P., 116, 242–4, 263, 319
Stern, O., 72
Stern–Gerlach experiment, 40, 306
Stirling experiment, 308
Stroud, C.R., Jr., 319
Sudarshan, E.C.G., 341, 365
Suppes, P., 179, 319–20
Svartholm, N., 365
Szczepanski, A., 180

Tagliagambe, S., 103, 121
Tambakis, N., 261
Tarozzi, G., 144, 178, 180, 261–2, 319–20
Tarozzi experiment, 156–7
Tersoff, J., 180
Thompson, R.C., 262
Thomsen, V., 346
Thomson, G.P., 90, 122
Thomson, J.J., 15, 82
Timson, D.R.E., 263
Toernqvist, N., 302, 320
Tonomura, A., 171, 180
Toraldo di Francia, G., 262
Treimer, W., 77, 122
Twiss, T.Q., 179

Uhlenbeck, G.E., 41, 72
Umberger, D.K., 180

van Stekelenborg, J.T., 262
van Vleck, J.H., 2
van der Merwe, A., x, 178, 261–3, 319, 357
van der Waerden, B.L., 121, 178
Varga, P., 178
Vasilenko, L.S., 144, 180
Vigier, J.-P., 141, 178–9, 246, 263
Voigt, W., 6
Voltaire, 27
von Hippel, 12
von Jolly, P., 5
von Lindemann, F., 6
von Meyenn, H., 31
von Mises, R., 30
von Neumann, J., 2, 12, 23, 36, 39, 63, 71, 108, 112–13, 121, 193, 319, 328
von Neumann's theorem, ix, 35–6, 44–61, 67, 69, 118–19, 186, 267, 276, 350, 356–8, 361
von Weizsaecker, C.F., 108
Vrehen, Q.H.F., 178, 180

Waismann, F., 330, 364
Walls, D.F., 309, 319
Walther, H., 178
Wan, K.K., 260, 263
Watson, J., 355
Weber, T., 262
Wegener, A.L., 335, 340
Weidlich, W., 148, 177
Weisskopf, V., 12
Wentzel, G., 8
Werner, S.A., 179–80
Wheeler, J.A., 114, 116, 118, 122, 178, 238
Wheeler's delayed choice, viii, 115, 117
Wick, G.C., 72
Wien, W., 15, 30
Wightman, A.S., 72, 204, 263
Wigner, E.P., 12, 72, 112, 214, 263, 280, 282, 289, 312, 320, 341, 365
Wigner–Araki–Yanase theorem, 265
Wolf, E., 177–8
Wolfe, H.C., 190, 263
Wootters, W.K., 122, 180, 320
Wu, C.S., 263
Wu–Shaknov experiment, 255–6

Yanase, M.M., 263, 320
Yang, C.N., 303
Yasin, A., 319
Yuen, H.P., 180, 320

Zanotti, M., 320
Zeilinger, A., 136, 162, 180, 234, 263, 311, 320
Zermelo, E., 237
Zurek, W.H., 122, 180, 320
Zweifel, P.F., 113, 122